Master Electrician's Review

Master Electrician's Review

Richard E. Loyd

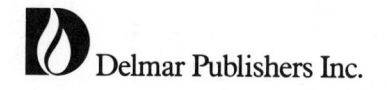

Delmar Publishers Inc.

I(T)P™

NOTICE TO THE READER

Publisher does not warrant or guarantee any of the products described herein or perform any independent analysis in connection with any of the product information contained herein. Publisher does not assume, and expressly disclaims, any obligation to obtain and include information other than that provided to it by the manufacturer.

The reader is expressly warned to consider and adopt all safety precautions that might be indicated by the activities described herein and to avoid all potential hazards. By following the instructions contained herein, the reader willingly assumes all risks in connection with such instructions.

The publisher makes no representations or warranties of any kind, including but not limited to, the warranties of fitness for particular purpose or merchantability, nor are any such representations implied with respect to the material set forth herein, and the publisher takes no responsibility with respect to such material. The publisher shall not be liable for any special, consequential or exemplary damages resulting, in whole or in part, from the readers' use of, or reliance upon, this material.

Cover design by design M design W

Delmar Staff:
Senior Executive Editor: Mark W. Huth
Assistant Editor: Nancy Belser
Project Editor: Elena M. Mauceri
Production Coordinator: Dianne Jensis
Production Editor: Brian Yacur
Art/Design Coordinator: Cheri Plasse

For information, address Delmar Publishers Inc.
3 Columbia Circle, Box 15-015
Albany, New York 12212-5015

COPYRIGHT © 1994
BY DELMAR PUBLISHERS INC.

The trademark ITP is used under license.

All rights reserved. No part of this work covered by the copyright hereon may be reproduced or used in any form, or by any means—graphic, electronic, or mechanical, including photocopying, recording, taping, or information storage and retrieval systems—without written permission of the publisher.

Printed in the United States of America
Published simultaneously in Canada
by Nelson Canada,
a division of The Thomson Corporation

10 9 8 7 6 5 4 3 2 1   XXX   00 99 98 97 96 95 94

ISBN: 0-8273-5852-0

New and Revised Titles for 1994

Test Equipment/ AVO Multi-Amp
ISBN 0-8273-4923-8

Technicians Guide to Programmable Controllers, 3E/ Cox
ISBN 0-8273-6238-2

Industrial Electricity, 5E/ Nadon
ISBN 0-8273-6074-6

Journeyman Electrician's Review/ Loyd
ISBN 0-8273-5725-7

To request examination copies, call or write to:

Delmar Publishers Inc.
3 Columbia Circle
P.O. Box 15015
Albany, NY 12212-5015

Phone: 1-800-347-7707 • 1-518-464-3500 • Fax: 1-518-464-0301

Dedication

I would like to dedicate this book to my wife, Nancy, as she has worked tirelessly by my side assisting with the research and development of the material for this book. Without her, this book would not be possible. I also would like to dedicate this book to Mary Standley, who has been my grammarian and typist. Without Mary I could never have completed this book on time. I would like to also thank all of the Code experts with whom I travel throughout the United States. Their knowledge shared with me over the years has made this book possible, and especially Richard L. Lloyd (deceased), the #1 circuit rider and roadrunner of all times. Richard has been my mentor and my inspiration for the past forty years in the industry. He always unselfishly shared his knowledge with me. All of the historical data that he had through the many years in serving on each of the *NEC®* Code panels and as chair of the correlating committee, his world of knowledge has been a great asset to me. I appreciated speaking by his side with him and traveling with him over the past seven years.

Table of Contents

Foreword.................................. ix
Preface xi
About the Author xii

CHAPTER ONE
   Examinations and National Testing
      Organizations 1
   Developing an Item (Question) Bank 1
   Preparing for an Examination 2
   Sample Examination—75 questions 4

CHAPTER TWO
   National Fire Protection Association
      Standards......................... 20
   National Electrical Code® (*NEC*®)......... 20
   Local Codes and Requirements............ 21
   Approved Testing Laboratories............ 21
   Question Review 21

CHAPTER THREE
   Basic Electrical Mathematics Review....... 28
   Calculator Math 28
   Working With Fractions.................. 28
   Working With Decimals 31
   Working With Square Root 31
   Powers 31
   Ohm's Law............................ 32
   Question Review 35

CHAPTER FOUR
   Introduction to the *National Electrical Code*®
      (*NEC*®) 42
   Know Your Code Book 42
   Question Review 44

CHAPTER FIVE
   General Wiring Requirements............. 51
   Question Review 58

CHAPTER SIX
   Branch Circuits and Feeders 64
   Question Review 73

CHAPTER SEVEN
   Services 600 Volts or Less................ 80
   Services Over 600 Volts.................. 86
   Question Review 86

CHAPTER EIGHT
   Overcurrent Protection................... 92
   Question Review 94

CHAPTER NINE
   Grounding 101
   Question Review 104

CHAPTER TEN
   Wiring Methods 111
   Question Review 122

CHAPTER ELEVEN
   General Use Utilization Equipment......... 129
   Question Review 131

CHAPTER TWELVE
   Special Equipment and Occupancies........ 137
   Question Review 137

CHAPTER THIRTEEN
   Communication Installations.............. 143
   Question Review 143

CHAPTER FOURTEEN
   Electrical and *NEC*® Question Review 150
   Practice Exam 1—25 questions............ 151
   Practice Exam 2—25 questions............ 157
   Practice Exam 3—25 questions............ 162
   Practice Exam 4—25 questions............ 167
   Practice Exam 5—25 questions............ 172
   Practice Exam 6—25 questions............ 177
   Practice Exam 7—25 questions............ 183
   Practice Exam 8—25 questions............ 188
   Practice Exam 9—25 questions............ 193
   Practice Exam 10—25 questions........... 200
   Practice Exam 11—25 questions........... 205
   Practice Exam 12—25 questions........... 213
   Practice Examination—50 questions........ 218
   Final Examination—75 questions 227

Appendix 1: Symbols 245
Appendix 2: Basic Electrical Formulas........ 248
Appendix 3: 1993 Highlighted Code Changes ... 251
Appendix 4: Answer Key 280

Index 303

Foreword

In ancient times electricity was believed to be an act of the gods. Not even Greek and Roman civilizations could understand this element called electricity. It was not until about 1600 A.D. that any scientific theory was recorded, when after 17 years of research William Gilbert wrote a book on the subject titled "De Magnete." It was almost another 150 years before major gains were made in understanding electricity and that it might some day be controlled by humans. At this time Benjamin Franklin, who is called the grandfather of electricity or electrical science, and his close friend, Joseph Priestley, began to gather the works of the many worldwide scientists who had been working independently over many years. Benjamin Franklin traveled to Europe to gather this information, and with that trip our industry began to surge forward for the first time. Franklin's famous kite experiment occurred in 1752. Coupling the results of that with other scientific data gathered, he decided he could sell installations of lightning protection to every building owner in Philadelphia. He was very successful in doing just that! We credit Franklin for the terms conductor and nonconductor replacing the words electric and nonelectric.

With the accumulation of the knowledge gained from these many scientists and inventors of the past, many other great minds emerged and lended their names to even more discoveries. Their names are familiar as other terms related to electricity. Alessandro Volta, James Watt, Andre Marie Ampere, George Ohm, and Heinrich Hertz are names of just a few great minds that enabled our industry to come together.

The next major breakthrough came about 100 years ago. Thomas Edison, the holder of hundreds of patents promoting DC voltage, and Nikola Tesla, the inventor of the three-phase motor and the promoter of AC voltage, began an inimical competition in the late 1800s. Edison was methodically making progress with DC current along the Eastern Seaboard and Tesla had exhausted his finances for his experimental project with AC current in Colorado, when Chicago requested bids to light the great "Columbian Exposition of 1892." To light the world's fair in Chicago

Electricity is here! *(Courtesy of Underwriters Laboratories Inc.)*

1893 Great Columbian Exposition at the Chicago World's Fair. The Palace of Electricity astonished crowds; it also astonished electricians by repeatedly setting fire to itself. *(Courtesy of Underwriters Laboratories Inc.)*

would require building generators and almost a quarter million lights. Edison teamed with General Electric and submitted a bid using DC, but George Westinghouse and Nikola Tesla submitted the lowest bid using AC and were awarded the contract. As spectacular as the lighting were the arcing, sparking, and fires started by the display with accidents occurring daily. Edison was embellishing the dangers of AC current by electrocuting animals while promoting DC current as being much safer. But the public knew better because many fires and accidents were also occurring with DC current.

About this time three important events occurred simultaneously. The Chicago Board of Fire Underwriters hired an electrician from Boston, William Merrill, as an electrical inspector for the exposition. Mr. Merrill saw the need for safety inspections and started "Underwriters Laboratories." A group of insurance underwriters, fire investigators, and engineers got together and wrote the first version of the *National Electrical Code®*. A group from Buffalo, New York hired Westinghouse and Tesla to build AC generators for Niagara Falls. AC current had won over DC current. The rest is history.

Preface

Although the primary purpose of this book is to provide electrical students with a concise, easily understandable study guide to the 1993 edition of the *National Electrical Code®* (*NEC®*) and the application of electrical calculations, this text is written recognizing that the electrical field has many facets and the user may have diverse interests and varying levels of experience. However, the interests of the relatively experienced electrician preparing himself or herself to enter the electrical industry as a legally competent licensed master/contractor, the young electrician, and especially the advanced electrical student apprentice were carefully considered in prepaing the text. It provides a ready source of the basic information on the *NEC®*. This text can be used for reference or as study material for students and electricians who are preparing for an examination for licensing. It is designed as an aid to those studying for the nationally recognized examinations such as "The National Assessment Institute" (NAI) and "Block and Associates Inc," but also will help anyone preparing for any electrical examination and will provide a quick, easily understood study guide for those needing to update themselves on the *National Electrical Code®* and basic mathematical formulas and calculations. The text is brief and concise for easy application for classroom study or home study.

ACKNOWLEDGMENTS

The author would like to give special thanks to:

The National Assessment Institute
Crouse-Hinds
Alflex
Ray Mullin, Bussmann Manufacturing, Division of Cooper Industries
Patricia H. Horton, Allied Tube and Conduit Company
Charles Forsberg, Carlon, A Lampson and Sessions Company
Bill Slater, Raco Incorporated, Subsidiary of Hubbell Inc.
Jim Pauley, Square D Company
Mike Holt, Mike Holt Enterprises
Roger Sanstedt, B-Line Systems Inc.

About the Author

Richard E. Loyd is a nationally known author and consultant specializing in the *National Electrical Code®* and the model building codes. He is president of his own firm, R&N Associates Inc. located in Perryville, Arkansas. He and his wife, Nancy, travel throughout the country presenting seminars and speaking at industry conventions. He also serves as a code expert at 35 to 40 meetings per year at the International Association of Electrical Inspectors (IAEI) meetings throughout the United States. Mr. Loyd represents the NEMA (National Electrical Manufacturers Association) Section 5RN Steel Rigid Conduit and Tubing as a *NEC®* consultant. Mr. Loyd currently serves on the *NEC®* Code Making Panel-8 representing the American Iron and Steel Institute (AISI). CMP-8 is responsible for eighteen Articles in Chapter 3 and Tables 1 through 4 in Chapter 9. He is currently vice chair of the NBEE (National Board of Electrical Examiners). Mr. Loyd is actively involved doing forensic inspections and investigations on a consulting basis and serves as an expert witness in matters related to codes and safety. He is currently secretary/treasurer of the Arkansas Chapter of Electrical Inspectors Association. He is currently a contributing editor for Intertec Publications (EC&M Magazine).

He is an active member in IAEI (International Association of Electrical Inspectors Association), NFPA (National Fire Protection Association), IEEE (The Institute of Electrical & Electronics Engineers Inc.) where he serves on the Power Systems Grounding Committee (Green Book), SBCCI (Southern Building Code Congress Intl.), ICBO (International Congress of Building Officials), and BOCA (Building Officials & Code Administrators Intl.). Mr. Loyd is currently licensed as a master contractor/electrician in Arkansas and Idaho and is a NBEE certified master electrician.

Mr. Loyd served as the chief electrical inspector and administrator for Idaho and Arkansas. He has served as chair of NFPA 79 Electrical Standard for Industrial Machinery, as a member of Underwriters Laboratories Advisory Electrical Council, as chair for Educational Testing Service (ETS) multistate electrical licensing advisory board, and as a master electrician and electrical contractor. He has been accredited to instruct for licensing certification courses in Idaho, Oregon, and Wyoming, and has taught basic electricity and *National Electrical Code®* classes for (BSU) Boise State University, Idaho.

Chapter One

EXAMINATIONS AND NATIONAL TESTING ORGANIZATIONS

Examinations are an old phenomenon. They have been given to many disciplines to evaluate one's competency in their field of specialty. Until recently they were prepared by others within their chosen field, based on the knowledge of each person preparing the examination. Today many examinations are professionally developed by state agencies specializing in examination preparation and national testing organizations.

In recent years many local jurisdictions—city, county, and state—have dropped their electrical examinations in favor of the national testing agencies that prepare and administer electrical examinations for those local entities. However, there are still many local jurisdictions developing and administering electrical examinations much in the same manner as has been done since the beginning of electrical examinations. These examinations are often prepared and administered by local electrical inspectors and are often developed with question material related to areas in which the inspector often finds violations or to areas in the *National Electrical Code*® where the inspector feels are most difficult to understand. Rarely do local examinations fully evaluate an entry-level craftsman to his or her competency in this new career field. National testing organizations, such as the National Assessment Institute or Block & Associates, have carefully taken the advantages of localized testing methodology and incorporated the latest state-of-the-art technology to develop a test that is fair to the taker as well as a good evaluation of the taker's knowledge in this specific career field. These testing organizations use those experienced in the trades in pools and in sit-down committees to develop a task analysis to determine what the examination will evaluate and to then develop questions from throughout those tasks. These organizations strive to develop question criteria that is clear and not confusing, questions that have only one answer, and questions in which an entry-level candidate into the journeyman electrical field or into the master contractor field should know as minimum standards to enter those fields. These examinations are generally multiple choice. The study arrangement in this book is designed to prepare the user to successfully pass these national examinations. However, it also should provide those preparing for localized tests on these same subjects.

DEVELOPING AN ITEM (QUESTION) BANK

The development of examinations that effectively measure the minimum competency of an entry-level candidate is a process that uses many concepts. National testing firms, such as the National Assessment Institute and Block & Associates and formerly ETS (the Electrical Testing Service for multistate licensing), have developed examinations that have become a standard in the industry today. The first step in developing an examination is a task or job analysis. The task analysis must evaluate the ability, knowledge, and skill to perform the tasks related to the job in a way that will not endanger public health, safety, and welfare. They must be relevant to the actual electrical practice in the field. The first step is then to establish the criteria, the subject areas of the trade that are most important and need to be tested. An acceptable method for determining these various job-related tasks would be a survey of licensed tradespeople performing these tasks in the field on a day-to-day basis, or an assembly of competent experts in these fields meeting in a forum to list the content outline and determining the degree of importance of each task to be evaluated by the examination, a blueprint that shows which percent of the exam for each subject task. This blueprint ties the test to the job performance, and the test outline shows the importance of each subject. When done properly, the task analysis will substantiate the validity of the exam.

The next task is to hold workshops or group meetings to develop an item (question) bank for a group of questions. This task of developing the item bank is normally performed by a group of interested experts working in workshops to develop the item bank based on the *National Electrical Code*® and other electrical related texts or reference books. As the item bank is developed, it is important that each item be clear, precise, and have only one correct answer. On completion of the item bank, there should be a pool of 500–800 questions. The questions are entered into the computer, each file based on the task analysis for each job task and weighted according as to the difficulty level of each question. The electrical board developed from contract states subscribing to these national tests then determines the passing scores, the item selection, and test form assembly. When this has been completed, the examination is ready for administration. Following the test administration to the candidates, a post-test examination analysis is then conducted to determine the effectiveness of the examination. In the post-test analysis, which is conducted following every exam, the effectiveness of each item is examined. Inadequate items are returned to the workshop to be rewritten or corrected or are eliminated from future exams. Each item is evaluated from developed criteria from previous exams, and the difficulty level is verified. Postexamination analysis provides critical and important information for future workshops and the validity of the exams. This analysis provides information that cannot be gained in any other manner. Information continues to be developed throughout the life of the item bank. This information is fed into the computer and the continually proves and improves the validity of future examinations. The candidates themselves provide the most important information in the continuing development and improvement of the examination.

PREPARING FOR AN EXAMINATION

The first step in preparing for an examination is to obtain the examination information from the Authority Having Jurisdiction to which you are about to make application for an examination. National testing firms generally furnish each authority having jurisdiction a bulletin containing generic material that applies nationwide, with specific unique material related only to the jurisdiction in which you are about to take an examination. This bulletin contains information that must be read carefully about the eligibility and procedure for registering for the test. This includes your verification of experience and education requirements and usually requires an application fee. It also contains the deadlines that are important for submitting the application and fees, and the examination dates related. For instance, you may be required to have your application and fees submitted up to 60 days before the examination date. The fees vary from state to state. They generally require the application fee be made to the jurisdiction and the exam administration fee to be made directly to the national testing firm. However, this procedure varies and must be read carefully. Once your application has been approved, your name will be placed on the roster for the next administration of the exam in your area. You will be mailed a registration form that will serve as an admission letter. Do not lose this because it will be required when you arrive at the test center or test location to take the exam. If you fail to bring this letter, you will not be permitted to take the exam. Also, it is very important that you read this bulletin to determine what materials are required that you bring and what materials are allowed, usually the *National Electrical Code®* or handbook in the edition in which the test is developed. Other electrical reference books may be permitted. You are usually required to bring your own pencils and a silent, hand-held, nonprogrammable, battery-powered calculator. Some jurisdictions also require a hands-on practical test. If it is required in the area in which you are making application, the outline of that work should also appear in the bulletin. Read it carefully because the practical test can mean the difference between passing and failing the examination. One such jurisdiction required that the candidate bring conduit, a conduit bender, hacksaw, reaming tool, tape measure, and continuity or ohm-meter. Also contained in the candidate bulletin will be the requirements for candidates with physical disabilities. These disabilities must be submitted with documentation at the time the application is made. Special considerations are generally permitted to accommodate religious or disability needs.

Finally, in the candidate bulletin is a content outline, the number of questions in each content area, the study reference material that the test has been developed from, such as the *National Electrical Code®* book; *Alternating Current Fundamentals*, Duff & Herman, 1986, Delmar Publishers Inc.; *Direct Current Fundamentals*, Loper & Tedsen, 1986, Delmar Publishers Inc.; the *American Electrician's Handbook*; and a sample of questions. It is suggested that you cover the answers and take this sample test in much the same manner as you would take the real examination. It gives you an idea of what the test will be like and what the format of the test is if you were not previously familiar with multiple-choice examinations. This is an important part of preparing for the electrical examination. Take the candidate bulletin seriously.

The next step in preparing for the examination is to study this book. The review questions at the end of each chapter are taken from throughout the *National Electri-*

cal Code® and not necessarily only from the content of each chapter. The *National Electrical Code®* is a reference document. As a reference document, it is necessary that you obtain a good proficiency in using the Table of Contents and the Index to be able to quickly locate the article and section of the Code. For each review question that you answer, you should not only research and find the correct answer, but you should also reference the *NEC®* section in which the answer was found. As you go through all the review material, take the sample tests throughout the document. This will enable you to pass the test the first time you take it. Anything less may prove disappointing. You should be aware that if you fail the examination, many jurisdictions require a waiting period of six months to one year before you can retake the exam.

The next section of this book contains a sample examination similar to the examination that you will take in the jurisdiction that you are studying for and similar to the examination given by many national testing organizations. This test should be taken in a quiet room. You should time yourself for four hours. It should be taken all at one time. To do otherwise will not give you a true evaluation of your present knowledge and of the areas in which you are weakest where you need to concentrate your efforts and study time to gain the competency needed to pass your examination.

You may have purchased this book for various reasons. Some individuals study merely to improve their understanding in their chosen field, but most do so because they are required to do so to pass their performance evaluation on their job or to pass an evaluation in a state or local examination. The specific reason is unimportant. The important thing is that you are about to undertake the study of electricity that, if done properly, will improve your competency in this field. The better you prepare, the better you will do on any evaluation or examination. Preparedness has more than one effect. First of all, your IQ can fluctuate 30 to 40 points between any given date. The best way to improve an individual's performance is to reduce the anxiety level. This book and other Delmar Publishing Inc. books related to this subject are designed to improve an individual's understanding on how to use the *National Electrical Code®* as a reference document, how to use the basic electrical formulas as applicable, and to duplicate as nearly as possible the test format used by many national testing organizations and local jurisdictions. Whether you are preparing for an examination or just to improve your skill, good study habits will give you the maximum efficient use of your time. To improve your study habits, consistent study time is important. You should allow yourself the same amount of time each day or each week to study this course. Consistent study time improves your retention.

Test Day

The time has finally arrived for you to take your examination. Be sure you allow plenty of time to arrive on time, but do not arrive too early. Discussing the test with other examinees may raise your anxiety level. Nervousness is contagious! Be sure you have your receipt for your application and your entrance ticket; be sure you have all materials required by the jurisdiction or testing agency, including study books and a hand-held calculator.

When a test is multiple choice, you generally have four alternative answers. The odds are four to one that you can get the right answer. If you can eliminate two of the alternatives, your chances are now increased to 50/50. This is extremely beneficial when referencing the Code, as you can quickly eliminate half of the possibilities leaving you only two to research.

Read the directions carefully. Many mistakes are made merely because of misunderstanding. If you have any questions, discuss them with the proctor before the test time starts. Any discussions after the test starts will take away from the time you have to answer the questions. Allow yourself ample time for each question. Answer the question first before you check any alternatives. This way you can evaluate your answer against any alternatives that you may have found in the *NEC®*. Don't spend too much time on any one question. If you find the question extremely difficult and you are unsure of the answer or cannot find it in the *NEC®*, skip it and only go back to it if you have time at the end of the test. Your first choice is usually best. Read the questions very carefully. Make sure you know key words that might change the meaning of the question. Note any negatives, such as "Which of the following are *not* . . . ," ". . . shall *not* be permitted . . .". It may be beneficial to underline key words as you read the question. This has a tendency to channel your thoughts in the right direction. Dress properly. The test facility should be approximately 70°F. Too many clothes or uncomfortable clothes have a tendency to make you drowsy and makes concentration difficult. After you have answered all the questions, recheck your work. Make sure you have not made clerical mistakes. Go over the extremely difficult questions. Questions on which you know you have guessed you may want to mark in some way so if there is time at the end of the test, you can do additional research. Good luck!

4 Master Electrician's Review

SAMPLE EXAMINATION

Examination Instructions:

For a positive evaluation of your knowledge and preparation awareness, you must

1. locate yourself in a quiet atmosphere (room by yourself)
2. have with you at least two sharp No. 2 pencils, the 1993 *NEC®*, and a hand-held calculator
3. time yourself (three hours), with no interruptions
4. after the test is complete, grade yourself honestly and concentrate your studies on the Sections of the *NEC®* in which you missed the questions.

Caution: Do not just look up the correct answers, as the questions in this examination are only an exercise and not actual test questions. Therefore, it is important that you be able to quickly find answers from throughout the *NEC®*.

Directions: Each question is followed by four suggested answers A, B, C, and D. In each case select the **one** that best answers the question.

1. Liquidtight flexible metal conduit can be used in which of the following locations?
   A. in areas that are both exposed and concealed
   B. in areas that are subject to physical damage
   C. in connection areas for gasoline dispensing pumps
   D. in areas where ambient temperature is to be greater than 200°C

   Answer:_____   Reference:_____

2. The **maximum** number of quarter bends allowed in one run of nonmetallic rigid conduit is
   A. 2
   B. 4
   C. 6
   D. 8

   Answer:_____   Reference:_____

3. The **maximum** size of flexible metallic tubing that can be used in any construction or installation is
   A. 3/8 inch
   B. 1/2 inch
   C. 3/4 inch
   D. 1 inch

   Answer:_____   Reference:_____

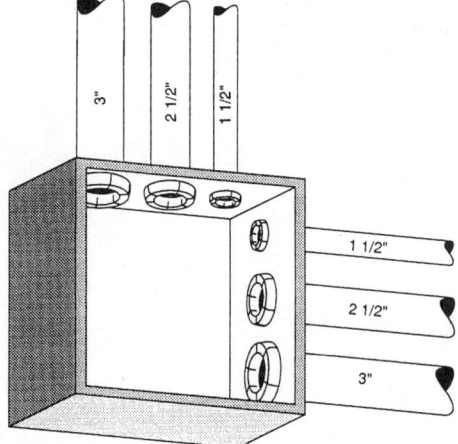

4. The size of the pull box in the diagram above should **not** be less than
   A. 12 inches × 14 inches
   B. 16 inches × 18 inches
   C. 22 inches × 22 inches
   D. 24 inches × 24 inches

   Answer: _____   Reference: _____

5. A metal raceway system installed in the washing area of a car wash must be spaced at **least** how far from the walls?
   A. ⅛ inch
   B. ¼ inch
   C. ¾ inch
   D. 1 inch

   Answer: _____   Reference: _____

6. Unless otherwise indicated, busways should be supported at intervals **not** to exceed how many feet?
   A. 3 feet
   B. 5 feet
   C. 7 feet
   D. 10 feet

   Answer: _____   Reference: _____

7. Which of the following percentages of conduit fill should be used when four type THW conductors are being installed?
   A. 55 percent
   B. 53 percent
   C. 40 percent
   D. 31 percent

   Answer: _____   Reference: _____

8. Flexible appliance cords of 10-amp capacity are considered adequately protected if the circuit overcurrent device is set at a **maximum** value of how many amperes?
   A. 15
   B. 20
   C. 30
   D. 40

   Answer: _____   Reference: _____

9. All of the following sizes of solid aluminum conductors must be made of an aluminum alloy **except**
   A. No. 8 AWG
   B. No. 10 AWG
   C. No. 12 AWG
   D. No. 14 AWG

   Answer: _____   Reference: _____

10. An insulated bushing is required to be used on a raceway entering a cabinet if the ungrounded conductors entering the raceway are at **least**
    A. No. 10 AWG
    B. No. 8 AWG
    C. No. 6 AWG
    D. No. 4 AWG

    Answer: _____   Reference: _____

11. The allowable distance between supports of nonmetallic-sheathed cable installed in an on-site constructed one-family dwelling is a **maximum** of how many feet?
    A. 2½ feet
    B. 3 feet
    C. 4 feet
    D. 4½ feet

    Answer: _____   Reference: _____

12. It is permissible to use No. 4 THW CU AWG for service in dwelling units to a **maximum** load of
    A. 100 amps
    B. 95 amps
    C. 70 amps
    D. 65 amps

    Answer: _____   Reference: _____

13. Where conductors of less than 600 volts emerge from the ground, they must be protected by enclosures or raceways that extend from below grade to a point how many feet above the finish grade?
    A. 5 feet
    B. 6 feet
    C. 7 feet
    D. 8 feet

    Answer: _____   Reference: _____

14. Which of the following statements is (are) true about the installation of six 3-wire cables in a 6-foot wireway with 18% fill?
    I. The cables must be laced together to form a single bundle.
    II. The cables must be derated to provide a maximum allowable load current of a value less than the standard ampere-rating of the cables.
    A. I only
    B. II only
    C. Both I and II
    D. Neither I nor II

    Answer: _____   Reference: _____

15. The service rating for a residence that uses No. 2 TW copper service entrance conductors is
    A. 95 amps
    B. 110 amps
    C. 115 amps
    D. 125 amps

    Answer: _____   Reference: _____

16. Surge arrester grounding conductors that run in metal enclosures should be
    A. bare
    B. insulated
    C. bonded on one end of the enclosure only
    D. bonded at both ends of the enclosure

    Answer: _____   Reference: _____

17. The number of overcurrent devices in a single cabinet of a lighting and appliance panelboard should **not** exceed
    A. 30
    B. 36
    C. 42
    D. 48

    Answer: _____   Reference: _____

8   Master Electrician's Review

18. When it is impractical to locate a service head directly above the point of attachment, the **maximum** allowable placement distance from the point of attachment is how many feet?
    A. 1 foot
    B. 2 feet
    C. 3 feet
    D. 4 feet

    Answer: _____   Reference: _____

19. An underground service to a small controlled water heater is to be installed. The single branch circuit used for this service requires that copper conductors be at **least**
    A. No. 12 AWG
    B. No. 10 AWG
    C. No. 8 AWG
    D. No. 6 AWG

    Answer: _____   Reference: _____

20. Service conductors can have less than 3 feet of clearance from which of the following:
    A. doors
    B. tops of windows
    C. fire escapes
    D. porches

    Answer: _____   Reference: _____

21. Service conductors that pass over rooftops should have a vertical clearance of not less than how many feet?
    A. 3 feet
    B. 7 feet
    C. 8 feet
    D. 10 feet

    Answer: _____   Reference: _____

22. The interruptive rating marking on all circuit breakers must be at **least**
    A. 1,000 amps
    B. 2,000 amps
    C. 2,500 amps
    D. 5,000 amps

    Answer: _____   Reference: _____

23. Illumination is mandatory for service equipment or panelboards in a dwelling unit if the service to the unit exceeds how many amps?
    A. 100
    B. 200
    C. 400
    D. required for all services regardless of amperage

    Answer: _____   Reference: _____

24. The *National Electrical Code®* requires ground-fault protection of service equipment under which of the following conditions?
    I. 1,000 amps or more, not to exceed 600 volts in a main disconnect
    II. Four-wire delta, 480/277-volt, three-phase, solidly grounded
    A. I only
    B. II only
    C. Both I and II
    D. Neither I nor II

    Answer:_____   Reference:_____

25. Which of the following is the correct connection sequence for a three-phase, **delta-connected** motor to be connected to 240-volt supply?

	L1	*L2*	*L3*
A.	To 1 and 7	To 2 and 8	To 3 and 9; 4, 5, and 6
B.	To 1, 7, and 6	To 2, 8, and 4	To 3, 9, and 5
C.	To 1, 7, and 4	To 2, 8, and 5	To 3, 9, and 6
D.	To 1	To 2	To 3; 7 and 4; 8 and 5; 9 and 6

    Answer:_____   Reference:_____

26. Impedance is present in which of the following circuits?
    A. AC only
    B. DC only
    C. Resistance only
    D. Both AC and DC

    Answer:_____   Reference:_____

27. Excluding fuses and exceptions, how many overcurrent protection devices, such as trip coils, relays or thermal cutouts, are required on a three-phase motor?
    A. 5
    B. 3
    C. 2
    D. 0

    Answer:_____   Reference:_____

28. All of the following are induction-type motors **except**
    A. wound-rotor
    B. split-phase
    C. capacitor
    D. universal

    Answer:_____   Reference:_____

10 Master Electrician's Review

Feeder

Branch circuit A

Branch circuit B

Branch circuit C

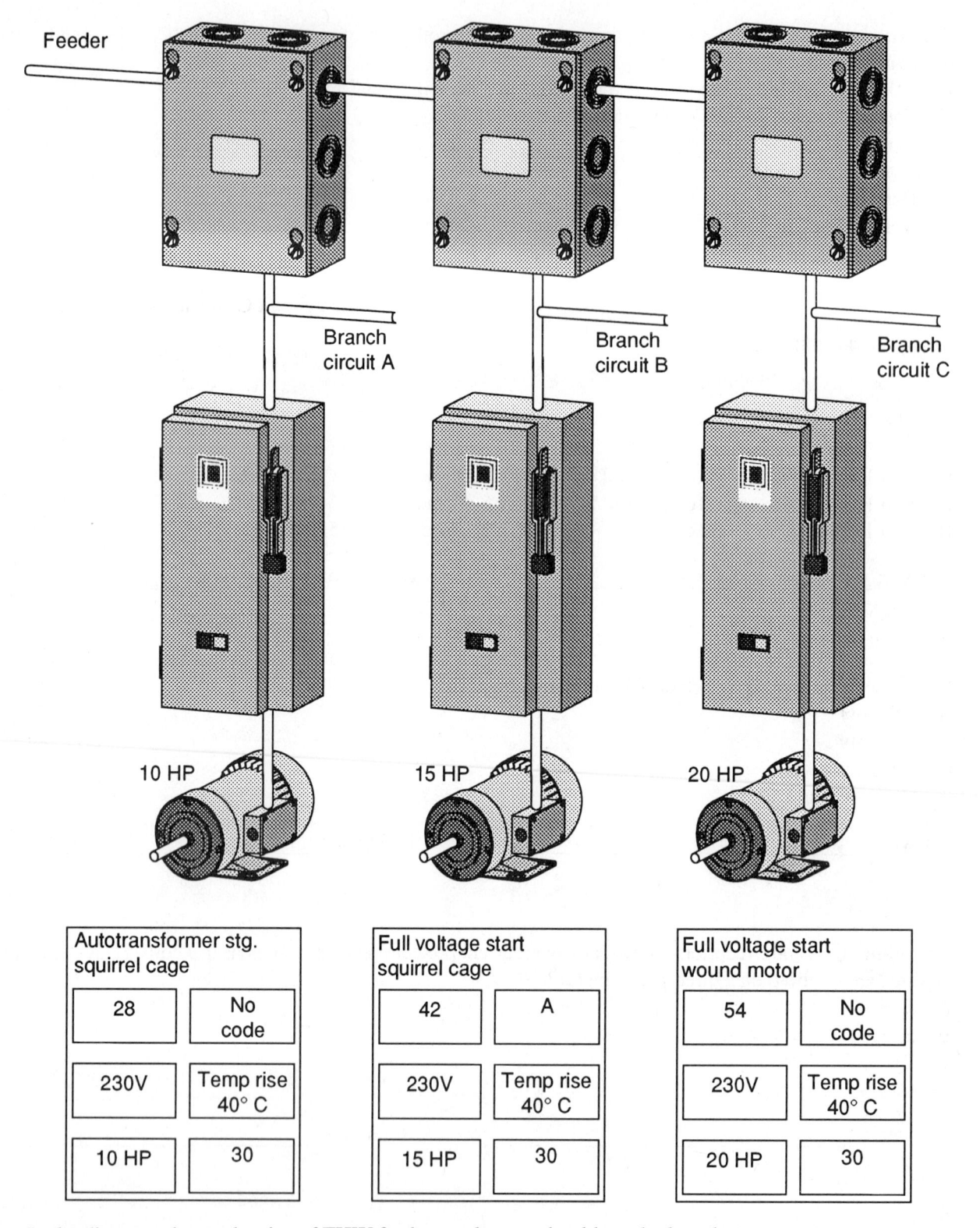

10 HP

15 HP

20 HP

Autotransformer stg. squirrel cage	
28	No code
230V	Temp rise 40° C
10 HP	30

Full voltage start squirrel cage	
42	A
230V	Temp rise 40° C
15 HP	30

Full voltage start wound motor	
54	No code
230V	Temp rise 40° C
20 HP	30

29. In the diagram above, the size of THW feeder conductors should **not** be less than
    A. No. 4
    B. No. 2
    C. No. 1
    D. No. 1/0

    Answer: _____   Reference: _____

30. In the diagram on page 10, the size of type THW conductors for branch circuit A should **not** be less than
    A. No. 10
    B. No. 8
    C. No. 6
    D. No. 4

    Answer:_____   Reference:_____

31. In the diagram on page 10, the size of type THW conductors for branch circuit C should **not** be less than
    A. No. 8
    B. No. 6
    C. No. 4
    D. No. 3

    Answer:_____   Reference:_____

32. The attachment plug and receptacle can be used as the controller for a portable motor that has a **maximum** horsepower rating of
    A. ⅓ hp
    B. ½ hp
    C. 1 hp
    D. 2 hp

    Answer:_____   Reference:_____

33. Snap switches can be grouped or ganged in outlet boxes if voltages between adjacent switches do **not** exceed how many volts?
    A. 200
    B. 300
    C. 400
    D. 500

    Answer:_____   Reference:_____

34. Each doorway that leads into a transformer vault from the building interior should have a tight-fitting door with a **minimum** fire rating of how many hours?
    A. 1
    B. 2
    C. 3
    D. 5

    Answer:_____   Reference:_____

12 Master Electrician's Review

35. The conduit run between warm and cold locations should be
    A. sealed
    B. sleeved
    C. insulated
    D. provided with drains

    Answer: _____   Reference: _____

36. Except for points of support, recessed portions of lighting fixture enclosures should be spaced at **least** how many inches from combustible material?
    A. ¼ inch
    B. ½ inch
    C. 1 inch
    D. 3 inches

    Answer: _____   Reference: _____

37. Surface-mounted fluorescent lighting fixtures that contain a ballast and are to be installed on combustible, low-density cellulose fiberboard should be spaced **not less** than how many inches from the surface of the fiberboard?
    A. ½ inch
    B. 1 inch
    C. 1½ inches
    D. 2 inches

    Answer: _____   Reference: _____

38. The minimum amount of on-site fuel for an emergency generator should be sufficient for **not less** than
    A. 2 hours at full demand
    B. 4 hours at 75% demand
    C. 6 hours at 50% demand
    D. 8 hours at 25% demand

    Answer: _____   Reference: _____

39. Which of the following statements about dimmers installed in theaters is (are) true?
    I. Dimmers installed in ungrounded conductors must have overcurrent protection not greater than 125% of the dimmer rating.
    II. The circuit supplying auto-transformer type dimmers must not exceed 150 volts between conductors.
    A. I only
    B. II only
    C. Both I and II
    D. Neither I nor II

    Answer: _____   Reference: _____

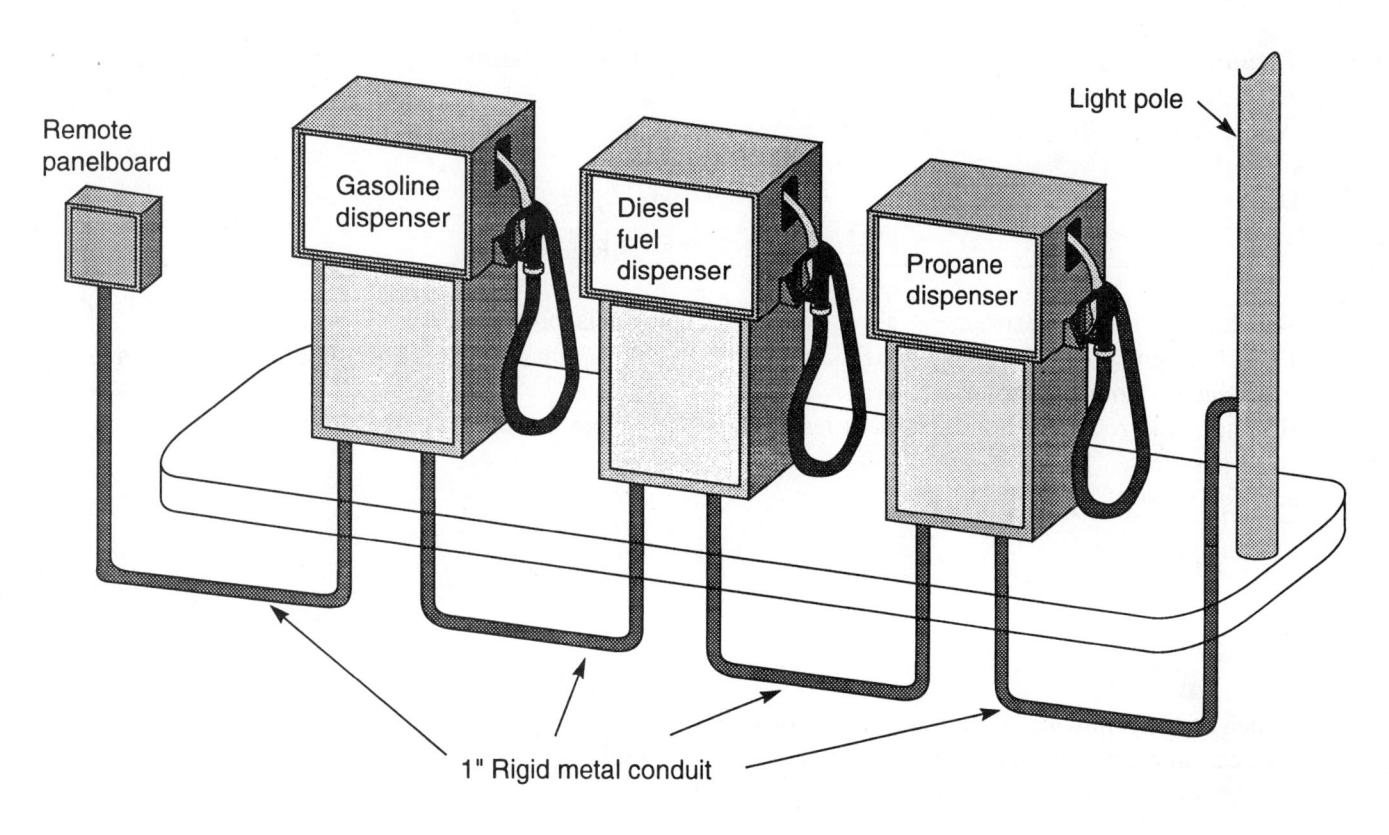

40. In the diagram above, how many conduit seals are required?
    A. 2
    B. 4
    C. 6
    D. 8

    Answer: _____   Reference: _____

41. An area must be classed as a Class II hazardous location if it contains which of the following?
    A. combustible dust
    B. flammable gases
    C. ignitable fibers
    D. ignitable vapors

    Answer: _____   Reference: _____

42. If the wiring is to be placed 18 inches above the floor, which of the following wiring methods is **prohibited** for a commercial automotive repair shop?
    A. metal-clad cable
    B. electrical metallic tubing
    C. rigid nonmetallic conduit
    D. nonmetallic sheathed cable

    Answer: _____   Reference: _____

43. The **minimum** number of receptacles in a patient-bed location of a hospital's general care area should be
    A. 1 duplex
    B. 2 duplex
    C. 3 duplex
    D. 3 single

    Answer: _____   Reference: _____

44. Which of the following statements about lightning protection is (are) true?
    I. If lightning protection is required for an irrigation machine, a driven ground rod should be connected to the machine at a stationary point.
    II. Lightning rods should be spaced at least 6 feet away from noncurrent carrying metal parts of electrical equipment.
    A. I only
    B. II only
    C. Both I and II
    D. Neither I nor II

    Answer: _____   Reference: _____

45. Which of the following will effectively ground a 230-volt, single-phase, residential air conditioning unit that replaces an older unit supplied by a two-wire, 230-volt circuit?
    I. A new circuit containing a grounding conductor
    II. An 8-foot ground rod installed at the unit
    A. I only
    B. II only
    C. Both I and II
    D. Neither I nor II

    Answer: _____   Reference: _____

46. When connected to a made-grounding electrode, grounding electrode conductors need **not** be sized larger than
    A. No. 8
    B. No. 6
    C. No. 4
    D. No. 2

    Answer: _____   Reference: _____

47. Which of the following is the most acceptable grounding electrode?
    A. a driven steel approved ground rod
    B. well casing
    C. building foundation steel
    D. an underground cold water metal piping system

    Answer: _____   Reference: _____

48. Which of the following terms can be used in place of "resistance" in the phrase "resistance is to ohms"?
    A. reactance
    B. inductance
    C. impedance
    D. capacitance

    Answer: _____   Reference: _____

49. The *National Electrical Code®* requires which of the following for the kitchen small appliance load in dwelling units?
    A. two 20-amp circuits
    B. two 15-amp circuits
    C. one 20-amp circuit
    D. one 15-amp circuit

    Answer: _____   Reference: _____

Questions 50 and 51 refer to a 120-gallon water heater with 220-volt, 4,000-watt units and an interlocking thermostat.

50. Installation of this water heater will require branch-circuit conductors of size
    A. 8
    B. 10
    C. 12
    D. 14

    Answer: _____   Reference: _____

51. The maximum branch circuit overcurrent protection for this water heater is
    A. 20 amps
    B. 25 amps
    C. 30 amps
    D. 40 amps

    Answer: _____   Reference: _____

Questions 52 and 53 refer to a 230-volt circuit that is 500 feet long, displays a resistance of 1 ohm, and supplies a load of 10 amps.

52. What is the wattage of the load in this circuit?
    A. 2,000
    B. 2,200
    C. 2,300
    D. 2,400

    Answer: _____   Reference: _____

53. How much wattage is lost in this circuit?
    A. 100
    B. 120
    C. 240
    D. 220

    Answer: _____   Reference: _____

54. Ground-fault protection of equipment is a requirement for solidly grounded wye services that are rated at more than 150 volts to ground but do not exceed 600 volts, phase-to-phase, for each service disconnecting means rated at 1,000 amps or more. The **maximum** amperage for this ground-fault protection relay should be set at
    A. 1,000 amps
    B. 1,200 amps
    C. 1,500 amps
    D. 2,000 amps

    Answer: _____ Reference: _____

Number	Size (MCM)	Type	Function
2	500	THW	Phases A and B
1	400	THW	Neutral
1	300	THW	Phase C
1	No. 3	THW	Ground Conductor

55. The raceway required for installation of the conductors above should **not** be less than
    A. 2 inches
    B. 2½ inches
    C. 3 inches
    D. 3½ inches

    Answer: _____ Reference: _____

Questions 56 and 57 refer to three, three-phase induction motors, rated respectively at 28, 16, and 12 amps.

56. The motors would require a feeder rated at how many amps?
    A. 56
    B. 63
    C. 70
    D. 112

    Answer: _____ Reference: _____

57. The motors are started at line voltage and are protected with time-delay fuses. The **maximum** feeder time-delay fuse should be rated at how many amps?
    A. 70
    B. 80
    C. 90
    D. 100

    Answer: _____ Reference: _____

58. The combined load of several 240-volt fixed space heaters on a 20-amp circuit should **not** exceed how many kilowatts?
    A. 2.4
    B. 2.6
    C. 3.8
    D. 4.8

    Answer: _____ Reference: _____

59. If three resistors with values of 5, 10, and 15 ohms, respectively, are connected in parallel, the combined resistance of the units will be
    A. 1.63 ohms
    B. 2.73 ohms
    C. 20.0 ohms
    D. 30.0 ohms

    Answer: _____   Reference: _____

60. Conductors that supply an 8-kilowatt, 230-volt, fixed electric space heater consisting of resistance elements should have a kilowatt rating of **not** less than
    A. 6.4
    B. 8
    C. 10
    D. 12

    Answer: _____   Reference: _____

61. The *National Electrical Code*® covers the installation of electrical conductors and equipment in all of the following locations **except**
    A. public and private buildings
    B. floating dwelling units
    C. buildings used by a utility for warehousing
    D. centers of transmission and distribution of electrical energy

    Answer: _____   Reference: _____

62. "As-built" drawings refer to
    A. approximate locations
    B. suggested locations
    C. exact locations
    D. general references

    Answer: _____   Reference: _____

63. The most recent editions of the *National Electrical Code*® are revised every
    A. year
    B. 2 years
    C. 3 years
    D. 4 years

    Answer: _____   Reference: _____

64. Galvanized rigid conduit is used to protect conductors in a cable tray. Where the conductors enter and leave the conduit within the cable tray,
    A. junction boxes are required even if the conduit is bushed.
    B. junction boxes are not required if the conduit is bushed.
    C. *no* bushings are required if the conduit is smaller than 1 inch.
    D. *no* bushings are required if the conductors are smaller than 38 copper.

    Answer: _____   Reference: _____

18   Master Electrician's Review

65. Surface type cabinets for electrical equipment in damp or wet locations shall be mounted so at least _____ inch air space between the wall or other supporting surface exists.
    A. 1/8
    B. 1/4
    C. 3/8
    D. 1/2

    Answer: _____   Reference: _____

66. Generally, rigid metal conduit shall be fastened in place within _____ feet of each box, outlet, cabinet or fitting.
    A. 1
    B. 2
    C. 3
    D. 4

    Answer: _____   Reference: _____

67. For six size 12 AWG conductors, the minimum trade size octagonal box that can be used is
    A. 4 inches × 1 1/4 inches
    B. 4 inches × 1 1/2 inches
    C. 4 inches × 2 1/8 inches
    D. 4 inches × 2 1/4 inches

    Answer: _____   Reference: _____

68. A motor controller that is installed with the expectation of its being submerged occasionally for short periods, shall be installed in a rated enclosure type No. _____.
    A. 3
    B. 3S
    C. 4X
    D. 6

    Answer: _____   Reference: _____

69. Where recessed high-intensity discharge fixtures are installed indoors and operated by remote ballasts,
    A. thermal protection is not required.
    B. only the fixture requires thermal protection.
    C. only the ballast requires thermal protection.
    D. both the fixture and the ballast require thermal protection.

    Answer: _____   Reference: _____

70. Any pipe or duct system foreign to the electrical installation must **not** enter a transformer vault. The _____ is **not** considered foreign to the vault.
    A. vault automatic fire protection
    B. building water main
    C. air duct passing through the vault
    D. roof drain piping

    Answer: _____   Reference: _____

71. A dry type transformer rated at 480 volts can be installed on a building column and shall **not** be required to be
    A. accessible.
    B. readily accessible.
    C. rated less than 112½ kVA.
    D. rated less than 600 volts.

    Answer: _____   Reference: _____

72. A capacitor is located indoors. It must be enclosed in a vault if it contains more than a minimum of _____ gallon(s) of flammable liquid.
    A. 1
    B. 2
    C. 3
    D. 5

    Answer: _____   Reference: _____

73. A single-phase hermetic refrigerant motor compressor has a rated load current of 24 amps and a branch circuit selection current of 30 amps. The branch circuit conductors are copper, type TW. They operate at 80° Fahrenheit, and they are the only conductors in the conduit to this compressor. The smallest possible branch circuit conductors must be at least size _____ AWG.
    A. 12
    B. 10
    C. 8
    D. 6

    Answer: _____   Reference: _____

74. A three-phase, squirrel-cage motor with full-voltage reactor starting has no code letter. The calculated fuse size to protect the branch circuit of the motor would be sufficient for the starting current of the motor. The nontime delay fuse to protect the circuit of this motor shall be sized at _____ % of full-load current.
    A. 150
    B. 200
    C. 250
    D. 300

    Answer: _____   Reference: _____

75. A dry type transformer is to be installed indoors. If rated more than _____ kilovolt-amps, the transformer must be installed in a fire-resistant transformer room.
    A. 25.0
    B. 35.0
    C. 100.5
    D. 112.5

    Answer: _____   Reference: _____

Chapter Two

NATIONAL FIRE PROTECTION ASSOCIATION STANDARDS

The National Fire Protection Association has acted as the sponsor of the *National Electrical Code®* (*NEC®*) as well as many other safety standards. The most widely used electrical code in the world is the *NEC®*. The official designation for the *National Electrical Code®* is ANSI/NFPA 70.

NATIONAL ELECTRICAL CODE® (*NEC®*)

ANSI/NFPA 70-1993 the *National Electrical Code®*, *NEC®*, was first developed about 100 years ago by interested industry and governmental authorities to provide a standard for the safe use of electricity. It is now revised and updated every three years. It was first developed to provide the regulations necessary for safe installations and to provide the practical safeguards of persons and property from the hazards arising from the use of electricity. It still provides these regulations and safeguards to the industry today. The *NEC®* is developed by 20 different code making panels composed entirely of volunteers. These volunteers come from industry, testing laboratories, inspection agencies, engineering, users, government, and the electrical utilities. Suggestions for the content come from various sources, including individuals like you. Anyone can submit a proposal or make a comment on a proposal submitted by another individual.

For more specifics on the *NEC®* process, you can contact the National Fire Protection Association, Batterymarch Park, Quincy, MA, 02269. Request their free booklet "The NFPA Standards Making System."

The *National Electrical Code®* is advisory as far as NFPA and ANSI are concerned but is offered for use in law and regulations in the interest of life and property protection. The name *National Electrical Code®* might lead one to believe that this document is developed by the federal government. This is not so; the *NEC®* only recognizes uses of products. It approves nothing. The *NEC®* has no legal standing until it has been adopted by the *authority having jurisdiction* (see *NEC®* definition Article 100), usually a governmental entity. Therefore, we must first check with the local electrical inspector to see which edition of the *NEC®* has been adopted. Compliance within and proper maintenance will result in an installation essentially free from hazard but not necessarily efficient, convenient, or adequate for good service or future expansion of electrical use. The *National Electrical Code®* is not an instruction manual for untrained persons nor is it a design specification. However, it does offer design guidelines.

NEC® Section 90-2 defines the scope of this document. It is intended to cover all electrical conductors and equipment within or on public and private buildings or other structures, including mobile homes, recreational vehicles, and floating buildings, and other premises such as yards, carnival, parking, and other lots, and industrial substations. Installations of conductors and equipment that connect to the supply of electricity, other installations of outside conductors on the premises, and installations of optical fiber cables are clearly covered by the *NEC®*. The *NEC®* is not intended to cover installations on ships, watercraft, railway rolling stock, automotive vehicles, underground mines, and surface mobile mining equipment. The *NEC®* is not intended to cover installations governed by the utilities, such as communication equipment, transmission, generation, and distribution installations on right-away. **NOTE:** For the complete list of the exemptions and coverage see 1993 *NEC®* Section 90-2. Mandatory rules are characterized by the word "Shall." Explanatory rules are in the form of fine print notes (FPN). All tables and footnotes are a part of the mandatory language. Material identified by the superscript letter "x" include text extracted from other NFPA standards and documents, such as NFPA 99 for Health Care Facilities, and NFPA 30 Flammable and Combustible Liquids Code. A complete list of all NFPA documents referenced can be found in the Appendix A. New revisions are identified by a vertical line on the margin where inserted and deleted text is identified by an enlarged black dot "Bullet."

To use the *NEC*® one must first have a thorough understanding of Article 90 "The Introduction," Article 100 "Definitions," Article 110 "Requirements for Electrical Installations," and Article 300 "Wiring Methods." The rest of the code book can be referred to on an as-needed basis.

LOCAL CODES AND REQUIREMENTS

Although most municipalities, countries, and states adopt the *National Electrical Code*®, they may not have adopted the latest edition, and in some cases the authority having jurisdiction (AHJ) may be using an older edition and not the 1993 edition. Most jurisdictions make amendments to the *NEC*® or add local requirements. These may be based on environmental conditions, fire safety concerns, or other local experience. An example of one common local amendment is that all commercial buildings be wired in metal raceways (Rigid, IMC, or EMT). The *NEC*® generally does not differentiate between wiring methods in residential, commercial, or industrial installations; however, many local jurisdictions do. Metal wiring methods, especially in fire zones, are another common amendment. Some major cities have developed their own electrical code, e.g., Los Angeles, New York, Chicago, and several metropolitan areas in Florida. In addition to the *National Electrical Code*® and all local amendments, the designer and installer must also comply with the local electrical utility rules. Most utilities have specific requirements for installing the service to the structure. There have been many unhappy designers and installers who have learned about special jurisdictional requirements after making the installation, thus incurring the costly corrections at their own expense.

If you are preparing for a state or city examination, you must check with the AHJ and see exactly what the examination content is based on. If it is a locally developed examination, the content may vary widely. If it is a nationally developed examination, it will generally be based on the latest version of the *National Electrical Code*® (1993 *NEC*®). However, some jurisdictions have supplemental examinations that cover their unique amendments, and in some jurisdictions a practical hands-on examination is given in addition to the written portion. (See Chapter One of this book for information related to some of these unique requirements.)

APPROVED TESTING LABORATORIES

Underwriters Laboratories (UL) has long been the major product testing laboratory in the electrical industry. In addition to the testing, UL is a long-time developer of product standards. By producing these standards and contracting for follow-up service after undergoing a listing procedure, a manufacturer is authorized to apply the UL label or to mark their product. In the last decade, numerous other electrical testing laboratories have arrived and are being officially recognized. Underwriters Laboratories is not the only testing laboratory evaluating electrical products. There are many testing laboratories operating today, such as Electrical Testing Laboratories (ETL), Applied Research Laboratories (APL), and Canadian Standards (CSA). Some jurisdictions evaluate and approve laboratories; others accept them based on reputation. OSHA is now evaluating and approving testing laboratories. It is the responsibility of the entity responsible for specifying the materials to verify that the product has been evaluated by a testing laboratory acceptable by the AHJ where the installation is being made. It is the installer's responsibility that the product is installed in accordance with the products listing (*NEC*® Section 110-3(b)).

CHAPTER 2 QUESTION REVIEW

1. Is liquidtight flexible nonmetallic conduit permitted to be used in circuits in excess of 600 volts?

   Answer: _____

   _____

   _____

   _____

   Reference: _____

2. Who is the authority having jurisdiction?

   Answer: _____
   _____
   _____
   _____

   Reference: _____

3. What are the four types of installations not covered by the *NEC*®?

   Answer: _____
   _____
   _____
   _____

   Reference: _____

4. How is explanatory information characterized in the *NEC*®?

   Answer: _____
   _____
   _____
   _____

   Reference: _____

5. What section of the *NEC*® requires equipment to be installed in accordance with manufacturer's instructions?

   Answer: _____
   _____
   _____
   _____

   Reference: _____

6. When does the *NEC*® become the legal document?

   Answer: _____
   _____
   _____
   _____

   Reference: _____

7. Are all installations by utilities exempt from the *NEC®*? Explain.

   Answer: _____

   _____

   _____

   _____

   Reference: _____

8. What does the vertical line in the margin of the *NEC®* indicate?

   Answer: _____

   _____

   _____

   _____

   Reference: _____

9. What is the latest amended edition of the *NEC®* used in all jurisdictions? When was the latest edition of the *NEC®* published?

   Answer: _____

   _____

   _____

   _____

   Reference: _____

10. Who set up the first meeting and when and where was it held that resulted in the *National Electrical Code®*?

    Answer: _____

    _____

    _____

    _____

    Reference: _____

11. Can Type NM cable be used for a 120/240-volt branch circuit as temporary wiring in a building under construction where the cable is supported on insulators at intervals of not more than 10 feet?

    Answer: _____

    _____

    _____

    _____

    Reference: _____

12. Which chapter of the *NEC®* is independent of all other chapters?

    Answer: _____

    _____

    _____

    _____

    Reference: _____

13. Is the *National Electrical Code®* considered a training manual?

    Answer: _____

    _____

    _____

    _____

    Reference: _____

14. Are all mining facilities exempt from the *NEC®*?

    Answer: _____

    _____

    _____

    _____

    Reference: _____

15. A comment often made by those involved in electrical design or installation states that the *NEC®* is a minimum. Where in the *NEC®* does it state that it is a minimum permitted for electrical installations?

    Answer: _____

    _____

    _____

    _____

    Reference: _____

16. As the inspector was making an inspection and looked at the size of overcurrent device and conductors supplying a motor, the inspector asked, "Is this motor circuit rated for continuous duty?" What is continuous duty? What is a continuous load?

    Answer: _____

    _____

    _____

    _____

    Reference: _____

17. When arriving into a nearby city to make an installation, the electrical inspector informed you that they had not adopted the last two editions of the *National Electrical Code®*. Which edition would that city be enforcing?

    Answer: _____

    _____

    _____

    _____

    Reference: _____

18. In wiring a small residence, the owners inform you that they wanted a doorbell mounted on the front and back of the house. What class wiring would this small 24-volt doorbell circuit be wired in and what article governs that wiring method?

    Answer: _____

    _____

    _____

    _____

    Reference: _____

19. We have been asked to bid a nursing home where patients will be in varied degrees of mobility. Some will come and go freely, cook their own meals, and do their own housekeeping; others may require meals to be prepared and minor medical attention, such as someone to be sure that they take their medicine regularly. Other patients may be ambulatory and require oxygen or doctor's care. Which article of the *National Electrical Code®* covers a nursing home facility of this type?

    Answer: _____

    _____

    _____

    _____

    Reference: _____

20. In establishing a grounding electrode on a new installation, the owner says that although an underground metal water pipe exists, the owner prefers that it not be used and that a standard 8-foot ground rod be used instead. Which section of the *National Electrical Code®* governs grounding electrodes?

    Answer: _____

    _____

    _____

    _____

    Reference: _____

26 Master Electrician's Review

21. In wiring a kitchen of a residential home, the inspector informs you that the cord on the disposal is too long. Where are the requirements for the cord and attachment plug for a home disposal?

    Answer: _____

    _____

    _____

    _____

    Reference: _____

22. In working in an industrial facility, you have been informed that some control circuitry will be installed to cut down on noise interference. A fiber optical cable and fiber optics will be used. Are fiber optics covered in the *National Electrical Code*®?

    Answer: _____

    _____

    _____

    _____

    Reference: _____

23. You recently have been asked to wire an irrigation pump on a nearby farm. On arriving, you find the farmer has a three-phase irrigation pump; however, the utility can only serve single-phase power. The farmer tells you he has heard about a phase converter that can be purchased at relative low cost that will convert single-phase power to three-phase power. Which section of the *National Electrical Code*® would cover an installation of this type?

    Answer: _____

    _____

    _____

    _____

    Reference: _____

24. In making an installation, you are asked to connect deicing equipment to the rain gutters on a house. Which section of the *NEC*® covers these requirements?

    Answer: _____

    _____

    _____

    _____

    Reference: _____

25. In making an installation at a new golf course clubhouse, several ornamental pools are to be installed in and around the golf course with electrical lighting in the pools and around the pools. Which section of the *National Electrical Code*® covers this installation? What part of that article covers this area?

    Answer: _____

    _____

    _____

    _____

    Reference: _____

Chapter Three

BASIC ELECTRICAL MATHEMATICS REVIEW

In this chapter, we will check our knowledge of the basic mathematical calculations needed to perform the simple day-to-day mathematical tasks necessary to perform our work as electrical contractors or electricians. Many tasks reviewed in this chapter will appear simple. However, if you have difficulty in solving any of these problems, then it may be necessary for you to obtain a more complete study guide on basic electrical mathematics. This chapter is not a mathematical teaching aid but only a refresher to see that your mathematical skills are sufficient to solve the problems in this text.

CALCULATOR MATH

Many of us were taught and believe that mathematical study should be done manually. But in today's world, with the introduction of hand-held calculators into the industrialized world in the past 25 years, it would be somewhat silly to depend on all hand calculations. Most testing jurisdictions permit the use of hand-held, silent, battery-powered, nonprogrammable calculators to use during the test. Therefore, it is advantageous to work the problems in this chapter using the same calculator that you will take to the examination. Many calculators are on the market, priced from advertising giveaways to complex scientific notation calculators. However, for your purposes any quality hand-held calculator will do. It should have the standard engineering functions, but it is not necessary to have one with scientific capabilities. The button arrangement should be good quality. It should be a battery powered and not a solar-powered calculator, because the room lighting might be inadequate for solar power. You should have a memory function and most modern mathematical functions, including the basic keys, input, error correction, combining operations, calculator hierarchy (calculations with a constant), roots, powers, reciprocal, factorial, percents, natural logarithm and natural antilogarithm, trigonometric functions, and error indication, accuracy and rounding, good memory usage for store, recall and memory exchange, conversion factors from English to metric, and temperature conversions. A calculator of this quality is available in most discount stores for under $20.00. Get the calculator that you plan to use and familiarize yourself with it so you are familiar with its operations before the examination.

Although programmable calculators are not permitted in most testing organizations, there are good programmable hand-held calculators on the market as shown in Figures 3–1 and 3–2. Although these are unacceptable in an examination, they can provide a convenient tool for the contractor or electrician or those studying to improve their skills in the electrical industry.

WORKING WITH FRACTIONS

Assuming that your basic addition, subtraction, multiplication, and division skills are still sharp, we will begin with a brief review of fractions. A fraction is a part of a number. For instance, if we have a pie, we have one whole pie. If we cut that pie into four pieces, then each piece is one-fourth of a pie. The four pieces together equal four fourths, or one. If we cut that whole pie into five pieces, then each piece would be one-fifth of the pie. The five pieces together would be one pie.

If we have two pieces of a pie that has been cut into four pieces, we would have two-fourths. This fraction can be expressed in two different ways, by $2/4$ or $1/2$. To reduce a fraction is to change it into another equal fraction. To do this, you divide the numerator (top number) and the denominator (bottom number) by the same number. For instance, to reduce $2/6$, divide the 2 and the 6 by 2. Then the numerator would be 2 divided by 2, or 1. The denominator would be 6 divided by 2, which equals 3. Therefore, the reduced fraction is $1/3$; $2/6 = 1/3$.

Example: $\dfrac{2}{6} = \dfrac{2/2}{6/2} = \dfrac{1}{3}$

To reduce larger fractions, such as $24/96$ to the lowest form, we can divide both the 24 and the 96 by 24 (to get

Figure 3–1 *Courtesy of Calculated Industries, Inc.*

¼) or we can divide the 24 and the 96 by 8 (which gives ³⁄₁₂) and then divide the 3 and the 12 by 3, which gives ¼, the answer.

Example: $\dfrac{24}{96} = \dfrac{24/24}{96/24} = \dfrac{1}{4}$

Reducing Mixed Numbers to Improper Fractions

In a proper fraction, the numerator is always smaller than the denominator, such as ⅓, ⅙, ¹⁄₁₂. Mixed numbers are made of two numbers, a whole number and a fraction, for example: 2½, 5½, 3⁹⁄₁₆, numbers that we use every day in our work. To change these to improper fractions, the numerator will always be larger than the denominator, such as ¹²⁄₃ or ⁹⁄₆ or ²⁴⁄₁₀. For example, to change 2½ to an improper fraction, the 2 will be a certain number of halves (2 = ⁴⁄₂), which we then add to the fraction we are given (½). Thus, 2½ = ⁴⁄₂ + ½ = ⁵⁄₂.

Example: $2\dfrac{1}{2} = \dfrac{2 \times 2 + 1}{2} = \dfrac{5}{2}$

To change a mixed number to an improper fraction, multiply the denominator of the fraction by the whole number and add the numerator of the fraction. Place this answer over the denominator to make an improper fraction.

Example: $5\dfrac{1}{2} = \dfrac{2 \times 5 + 1}{2} = \dfrac{11}{2}$

Changing Improper Fractions to Mixed Numbers

To change an improper fraction, such as ¹³⁄₃, to a mixed number, divide the numerator by the denominator (13 divided by 3). Any remainder is placed over the denominator (13 divided by 3 = 4 with a remainder of 1). The resulting whole number and proper fraction form a mixed number. Thus, ¹³⁄₃ = 4 + ⅓ = 4⅓.

Example: $\dfrac{13}{3} = 4 + \dfrac{1}{3} = 4\dfrac{1}{3}$

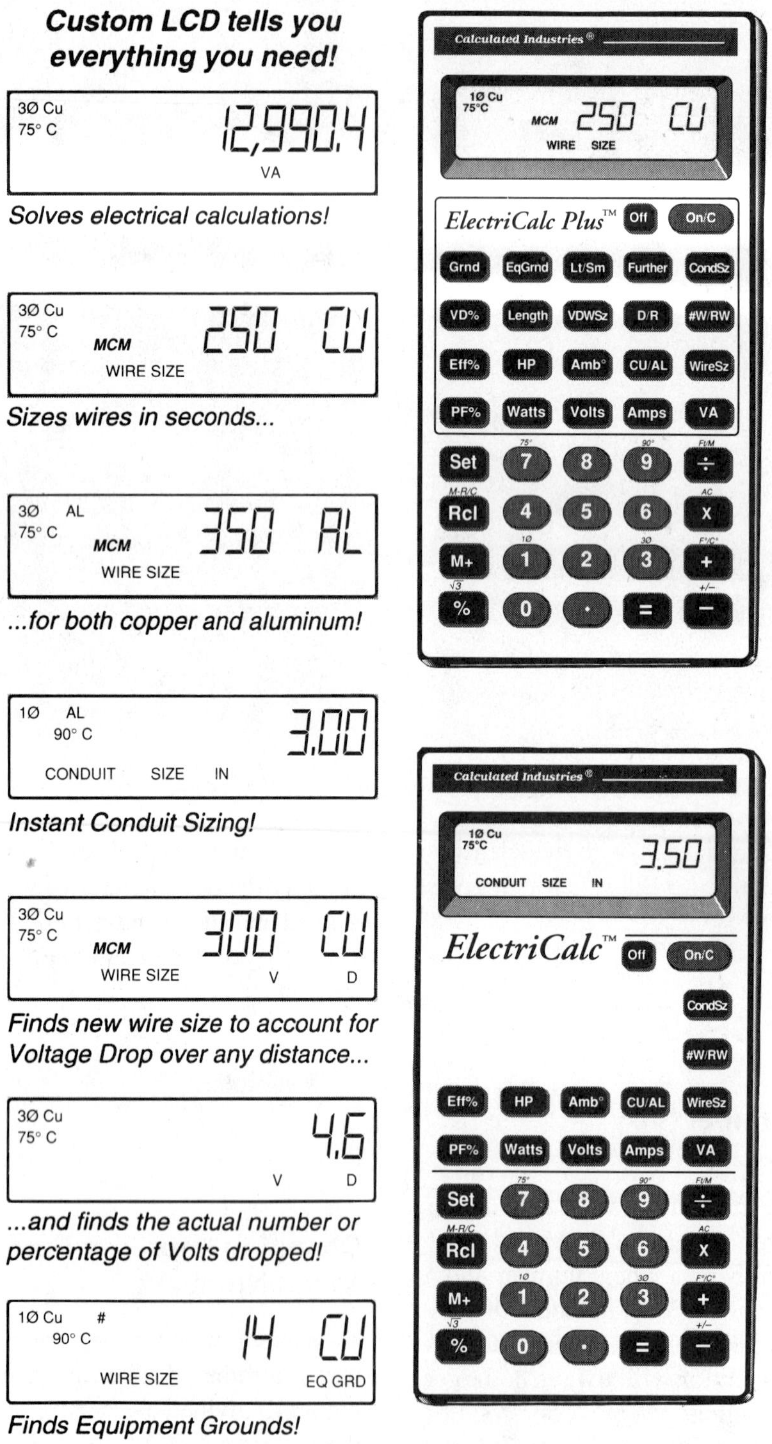

30 Master Electrician's Review

Figure 3–2 *Courtesy of Calculated Industries, Inc.*

Multiplication of Fractions

To multiply fractions, place the multiplication of the numerators over the multiplication of the denominators and reduce to lowest terms. For example: ½ × ¾. Answer: ½ × ¾ means that 1 × 3 = 3 and 2 × 4 = 8; thus, the answer is ⅜. Example: ½ × 5. Answer: ½ × 5 means that 1 × 5 = 5 and 2 × 1 = 2, giving an improper fraction of 5/2, which then has to be reduced to 2½.

Examples: $$\frac{1}{2} \times \frac{3}{4} = \frac{1 \times 3}{2 \times 4} = \frac{3}{8}$$

$$\frac{1}{2} \times 5 = \frac{1 \times 5}{2 \times 1} = \frac{5}{2} = 2\frac{1}{2}$$

To cancel numerator and denominator means to divide both numerator and denominator by the same number. As an example, when multiplying ⅜ × 4/9, notice that the 3 on top and the 9 on the bottom can both be divided by 3. Cross out the 3 and write a 1 over it. Cross out the 9 and write a 3 under it. Also, the 4 on the top and the 8 on the bottom can both be divided by 4. Cross out the 4 and write a 1 over it. Cross out the 8 and write a 2 under it. The solution then is 3 reduced to 1 × 4 reduced to 1 (1 × 1) over 8 reduced to 2 × 9 reduced to 3 (2 × 3), thus giving ⅙.

WORKING WITH DECIMALS

A decimal is a fraction in which the denominator is not written. The denominator is one. Many answers to electrical problems will be fractions, such as ¼ of an amp, ½ of an amp, of ½ of a volt, ⅓ of a volt, etc. These will be correct mathematical answers, but they will be worthless to an electrician. The electrical measuring instruments give values expressed in decimals, not in fractions. In addition, the manufacturers of components give the values of parts in terms of decimals, such as 3.4 amps, etc., on the nameplate of a motor. Suppose we worked out a problem and found the current in the circuit should be ⅛ of an amp, and using the ampmeter we test the circuit and find that .125 amps flow. Is our circuit correct? How will we know? How can we compare ⅛ to .125? The easiest way is to change the fraction ⅛ to its equivalent decimal and then compare the decimals. To change a fraction into a decimal, divide the numerator by the denominator, (1 divided by 8 = .125).

$$\frac{1}{8} = \frac{1}{8.000} = .125$$

WORKING WITH SQUARE ROOT

Roots are the opposite of powers. A square root is the opposite of the number to the second power. The sign of the square root is $\sqrt{}$. To find a square root, find a number that when multiplied by itself is the number inside the square root sign. For example, to find the value of the square root of 36, ask yourself what number times itself is 36. The answer is 6, so 6 is the square root of 36.

If you can find the square root of a number with the method that uses averages, suppose that you did not know that the square root of 144 = 12. When you divide the number by its square root, the answer is the square root. If you cannot find this answer, then guess as close as possible. A good guess for 144 would be 10, because 10 × 10 = 100, which is close to 144. Divide 144 by 10 and the answer is 14 + a remainder. Average the guess, 10, and the answer to the division problem, 14. 10 + 14 = 24, 24/2 = 12, which is the correct answer.

Follow these steps to find the square root of the larger number: Guess the answer; divide the guess into the large number; average the guess and the answer to the division problem; and check. For example, find the value of the square root of 1,024.

Step 1: Guess: in the list of square roots, 30 × 30 = 900. This is too small, but it is easy to divide by.

Step 2: Divide 1,024 by 30. The answer is 34, with a remainder.

Step 3: Find the average of 30 and 34. 30 + 34 = 64; 64/2 = 32.

Step 4: Check: multiply 32 × 32. The answer is 1,024. Thus, 32 is the square root of 1,024. When you use this method to find square roots, always guess a number that ends in 0. It is easier and faster to divide by these numbers. If the average is not the square root of the number, use the average as a new guess and try again.

POWERS

A power is a product of a number multiplied by itself one or more times. For an example, what is the value of 3^2? The exponent is 2; write 3 two times, or $3^2 = 3 \times 3$;

$3^2 = 9$. What is the value of 5^3? The exponent is 3; write 5 three times; in other words, $5^3 = 5 \times 5 \times 5 = 125$. These powers are terms often used in the electrical field and should be known.

OHM'S LAW

E = voltage; I = current (amperes); R = resistance

$$E = IR; \quad I = \frac{E}{R}; \quad R = \frac{E}{I}$$

As an example, all numbers are relative. For example, if E = 12, I = 3, and R = 4, then $3 \times 4 = 12$, $12/3 = 4$, and $12/4 = 3$. All numbers are relative. Remember, in any electrical circuit, the voltage force is the current through the conductor against its resistance. The resistance tries to stop the current from flowing. The current that flows in the circuit depends on the voltage and the resistance. The relationship among these three quantities is described by Ohm's Law. Ohm's Law applies to an entire circuit or any component part of a circuit.

Voltage Drop Calculations

We have discussed the methods for finding the resistance by using Ohm's Law. Now we can find the voltage drop for circuit loads using this same form.

$$E = I \times R$$

Voltage Drop Formula

Single phase voltage drop = amperes × length × resistance × (2)

Voltage drop % = voltage drop × 100/volts

Three-phase voltage drop—calculate as a single-phase circuit and multiply resultant by .866.

$$V_{(D)} = I \times 2L \times R$$

I = Amperes
L = Length of circuit
R = Resistance values from *NEC*® Chapter 9 Table 8
Resistance = R × [1 + a(temperature − 75)]
a = 0.00323 for copper, .000330 for aluminum
Temperature = ambient temperature

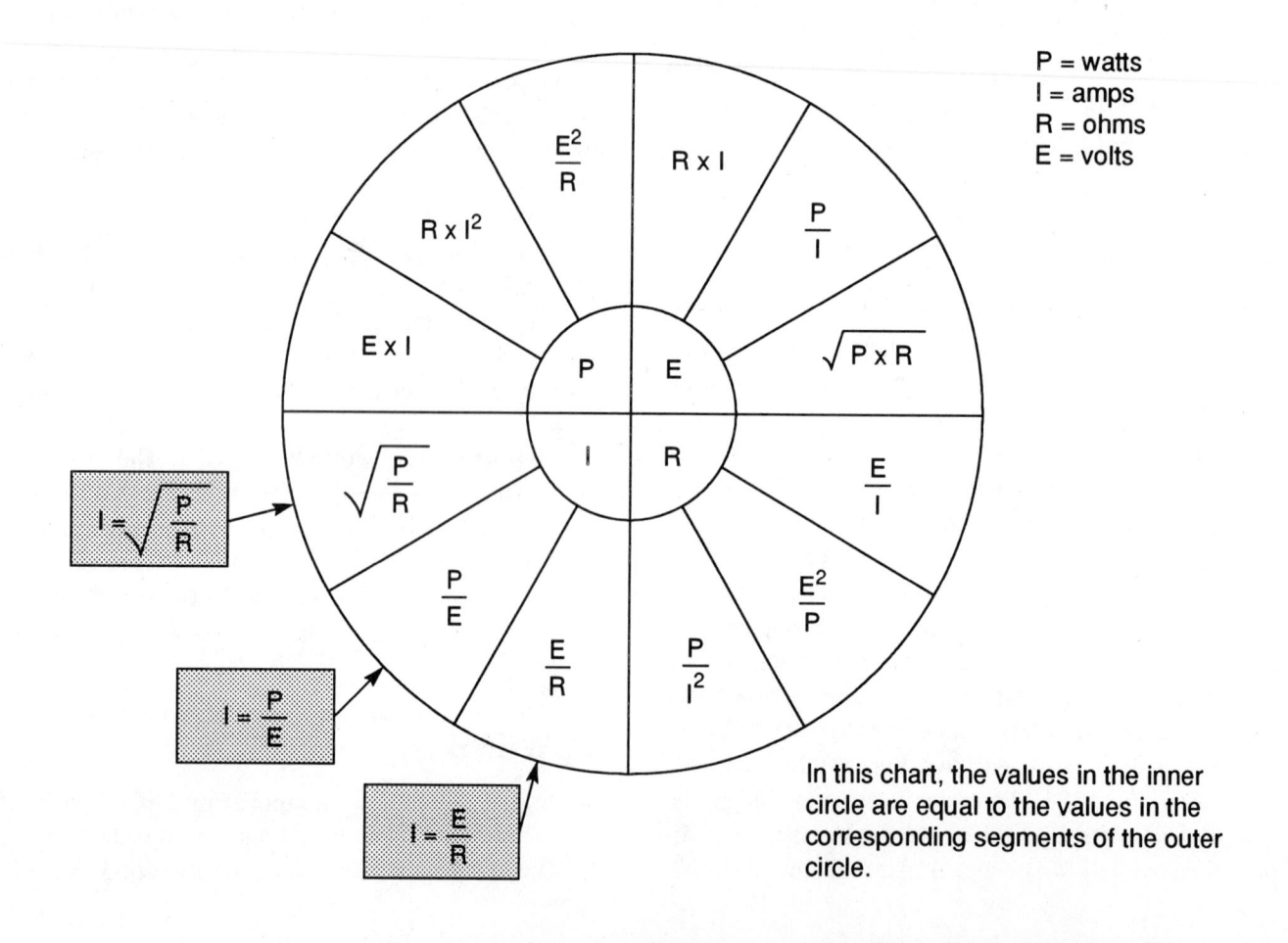

P = watts
I = amps
R = ohms
E = volts

Ohm's Law wheel showing:
Inner circle: P, E, I, R
Outer segments: $\frac{E^2}{R}$, $R \times I$, $R \times I^2$, $\frac{P}{I}$, $E \times I$, $\sqrt{P \times R}$, $\sqrt{\frac{P}{R}}$, $\frac{E}{I}$, $\frac{P}{E}$, $\frac{E^2}{P}$, $\frac{E}{R}$, $\frac{P}{I^2}$

$I = \sqrt{\frac{P}{R}}$

$I = \frac{P}{E}$

$I = \frac{E}{R}$

In this chart, the values in the inner circle are equal to the values in the corresponding segments of the outer circle.

The *NEC®* references prescribe voltage drop values or percentages for feeder conductors and branch circuit conductors; however, it should be noted these appear in a fine print note (FPN) and are only suggested and not mandatory requirements. The *NEC®* references in fine print note 2, Article 215 as not to exceed 3% of the farthest outlet of power, heating, or lighting loads or combinations of such loads, and the maximum voltage drop for both feeders and branch circuits to the farthest outlet is not to exceed 5% and will provide reasonable efficiency of operation.

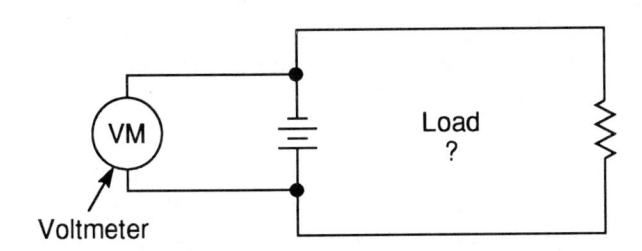

Example: A direct current circuit has 2 amperes of current flowing in it. A voltmeter reads 10 volts line to line. How much resistance is in the circuit? Answer: 5 ohms.

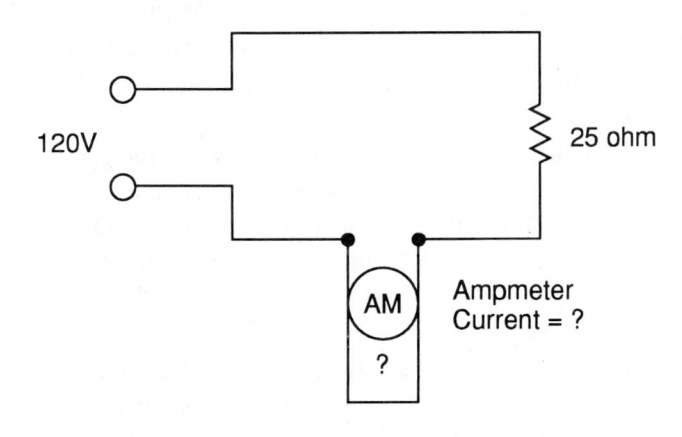

$$R = \frac{E}{I} \quad R = \frac{10}{2} \quad R = 5 \text{ ohms}$$

Example: If the voltage is 120 volts and the resistance is 25 ohms, what amount of current will flow in the circuit? Answer: 4 amperes.

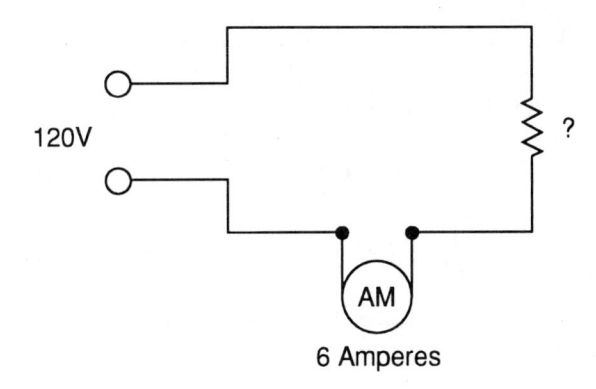

$$I = \frac{E}{R} \quad I = \frac{100}{25} \quad I = 4 \text{ amperes}$$

Example: If the potential across a circuit is 120 volts and the current is 6 amperes, what is the resistance? Answer: 20 ohms.

$$R = \frac{E}{I} \quad R = \frac{120}{6} \quad R = 20 \text{ ohms}$$

A series circuit may be defined as a circuit in which the resistive elements are connected in a continuous run (end to end). It is evident that because the circuit has no branches, the same current flows in each resistance. The total potential across the circuit equals the sum of potential drops across each resistance.

Example: $E_1 = IR_1$
$E_2 = IR_2$
$E_3 = IR_4$
$E = E_1 + E_2 + E_3$
$R = R_1 + R_2 + R_3$

The total potential of the circuit is:

$$E = IR_1 + IR_2 + IR_3 = I(R_1 + R_2 + R_3)$$

$$I = \frac{E}{R_1 + R_2 + R_3}$$

I $R_1 = 10$ ohms, $R_2 = 5$ ohms, and $R_3 = 15$ ohms.

What amount of voltage must flow to force 15 amperes through the circuit?

$$R_T = 5 + 10 + 15 = 30 \text{ ohms}$$

Hence, $E_T = 0.5 \times 30 = 15$ volts

What is the voltage drop across each resistance?

$E_1 = 0.5 \times 10 = 5.0$ volts
$E_2 = 0.5 \times 5 = 2.5$ volts
$E_3 = 0.5 \times 15 = 7.5$ volts

$E_T = 15$ volts

When the resistance and current are known, what is the formula to determine the voltage?

The current is 2 amperes, the resistance is 15 ohms, 10 ohms, and 30 ohms.

$E = IR \quad E = 2(15 + 10 + 30)$
$E = 2 \times 55 \quad E = 110$ volts

In the industry today most circuits are connected in parallel. In a parallel circuit the voltage is the same across each resistance (load).

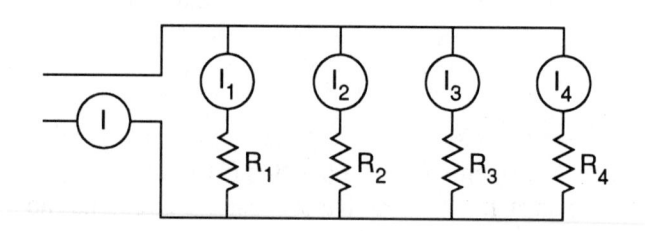

The total current is equal to the sum of all the currents.

$E = I_1R_1 = I_2R_2 = I_3R_3 = I_4R_4$ and

$I = I_1 + I_2 + I_3$

Individual Resistances =

$$I_1 = \frac{E}{R_1} \quad I_2 = \frac{E}{R_2} \quad I_3 = \frac{E}{R_3} \quad I_4 = \frac{E}{R_4}$$

Hence: $I = \frac{E}{R_1} + \frac{E}{R_2} + \frac{E}{R_3} + \frac{E}{R_4}$

or $I = E\left(\frac{1}{R_1} + \frac{1}{R_2} + \frac{1}{R_3} + \frac{1}{R_4}\right)$

and several resistances in parallel is

$$\frac{1}{R} = \frac{1}{R_1} + \frac{1}{R_2} + \frac{1}{R_3} + \frac{1}{R_4}$$

The sum of the resistances in parallel always will be smaller than the smallest resistance in the circuit where only two resistances are in parallel.

$$R = \frac{R_1 \times R_2}{R_1 + R_2} \text{ ohms}$$

Power Factor

Power factor is a phase displacement of current and voltage in an AC circuit. The cosine of the phase angle of displacement is the power factor. The cosine is multiplied by 100 and is expressed as a percentage. A cosine of 90° is 0; therefore, the power factor is 0%. If the angle of displacement were 60°, the cosine of which is .5, the power factor would be 50%. This is true whether the current leads or lags the voltage. Power is expressed in DC circuits and AC circuits that are purely resistive in nature. Where these circuits contain only resistance, P (watts) = E × I. In AC circuits that contain inductive or capacity reactances, VA (volt-ampere) = E × I. In a 60 cycle AC circuit if the voltage is 120 volts, the current is 12 amperes, and the current lags the voltage by 60°, find (a) the power factor, (b) the power and voltage-amperes (VA), and (c) the power in watts. The cosine of 60 is .5; therefore, the power factor is 50%. 120 × 12 = 1,440 volt-amperes, which is called the apparent power. 120 × 12 × .5 = 720 watts, and is called the true power. Power factor is important. There is an apparent power of 1,440 VA and a true power of 720 watts. There also are 12 amperes of line current and 6 amperes of in-phase or effective current. This means that all equipment from the source of supply to the power consumption device must be capable of handling a current of 12 amperes, which actually the device is only using the current of 6 amperes. A 50% power factor was used intentionally to make the results more pronounced. The I^2R loss is based on the 12 ampere current, while only 6 amperes are effective. Power factor can be measured by a combined use of a volt-meter, ampmeter, and watt-meter, or by the use of a power factor meter. When using the three meters, volt-ampere meter and watt-meter, all connected properly in the circuit, the readings of the three meters are taken simultaneously under the same load conditions and calculated as follows: power factor = true power (watts)/apparent power (which is volt-amperes). Power factor = W/EI.

CHAPTER 3 QUESTION REVIEW

Directions: Reduce the following fractions to their lowest terms.

1. $3/6 =$ _____
2. $4/16 =$ _____
3. $4/40 =$ _____
4. $40/100 =$ _____
5. $6/8 =$ _____
6. $15/20 =$ _____
7. $18/36 =$ _____
8. $9/15 =$ _____
9. $30/50 =$ _____
10. $4/20 =$ _____
11. $5/20 =$ _____
12. $6/24 =$ _____

Directions: Change the following mixed numbers to improper fractions.

1. $1\frac{1}{2} =$ _____
2. $3\frac{1}{4} =$ _____
3. $5\frac{3}{8} =$ _____
4. $4\frac{1}{4} =$ _____
5. $7\frac{1}{8} =$ _____
6. $2\frac{1}{6} =$ _____
7. $5\frac{3}{4} =$ _____
8. $8\frac{1}{3} =$ _____
9. $6\frac{4}{7} =$ _____
10. $3\frac{5}{8} =$ _____

36 Master Electrician's Review

Directions: Change the following improper fractions to mixed numbers.

1. $9/2 =$ _____
2. $12/5 =$ _____
3. $64/5 =$ _____
4. $26/8 =$ _____
5. $29/3 =$ _____
6. $31/2 =$ _____
7. $5/3 =$ _____
8. $13/3 =$ _____

Directions: Multiply the following whole numbers and fractions. Give the answer in proper reduced form.

1. $8 \times 1/2 =$ _____
2. $1/2 \times 3/4 =$ _____
3. $1/5 \times 1/8 =$ _____
4. $9 \times 1/2 =$ _____
5. $3/4 \times 7 =$ _____
6. $1/2 \times 5 =$ _____
7. $3/4 \times 12 =$ _____
8. $1/7 \times 15 =$ _____

Directions: Convert the following fractions to decimal equivalents.

1. $1/4 =$ _____
2. $5/8 =$ _____
3. $3/4 =$ _____
4. $13/15 =$ _____
5. $3/8 =$ _____

Directions: Resolve to whole numbers.

1. $2^4 =$ _____

2. $9^2 =$ _____

3. $50^2 =$ _____

4. $1^5 =$ _____

5. $10^3 =$ _____

Directions: Find the square root of the following numbers.

1. Square root of 25 = _____

2. Square root of 81 = _____

3. Square root of 49 = _____

4. Square root of 3136 = _____

5. Square root of 10201 = _____

6. Square root of 841 = _____

7. Square root of 6064 = _____

Directions: Solve the following problems.

1. A 20-amp load is fed with two conductors that have a *combined* resistance of 0.3 ohms. If the source voltage is 120-volts DC, the calculated voltage drop on this circuit is _____ %.

    Answer: _____

    _____

    _____

    Reference: _____

38 Master Electrician's Review

[Figure: Three-phase circuit diagram showing Phase A, Phase B, Phase C at 120V each, with three 1200 Watt Loads connected to a Neutral.]

2. Refer to the above drawing.
   I(LINE A) = I(LINE B) = I(LINE C) = 10 amps.
   The power factor in the circuit in the figure above is _____ .

   Answer: _____

   _____

   _____

   _____

   Reference: _____

3. A 120-volt branch circuit has only six 100 watt, 120-volt incandescent lighting fixtures connected to it. The current in the home run of this circuit is _____ amps.

   Answer: _____

   _____

   _____

   _____

   Reference: _____

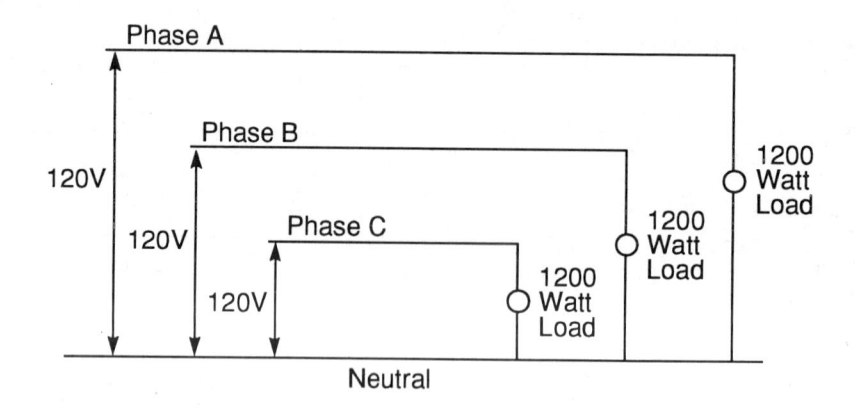

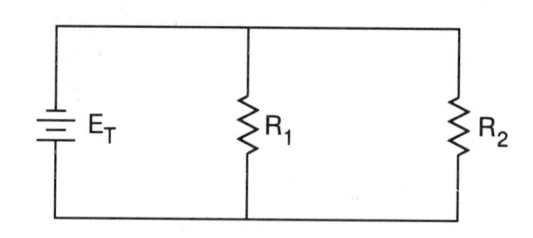

[Figure: Parallel circuit with source E_T and two resistors R_1 and R_2 in parallel.]

4. Refer to the above drawing.
   With a resistance at R_1 of 4 ohms and a resistance at R_2 of 2 ohms, the total resistance for the above circuit is _____ ohms.

   Answer: _____

   _____

   _____

   _____

   Reference: _____

5. A 1000-watt, 120-volt lamp uses electrical energy at the same rate as a/an ____ ohm resistor.

   Answer: _____

   _____

   _____

   _____

   Reference: _____

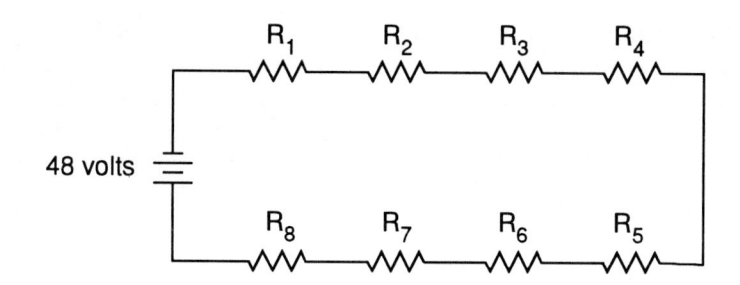

6. Eight equal resistors are connected in series to a 48-volt source. The voltage drop across each resistor is ____ volts.

   Answer: _____

   _____

   _____

   _____

   Reference: _____

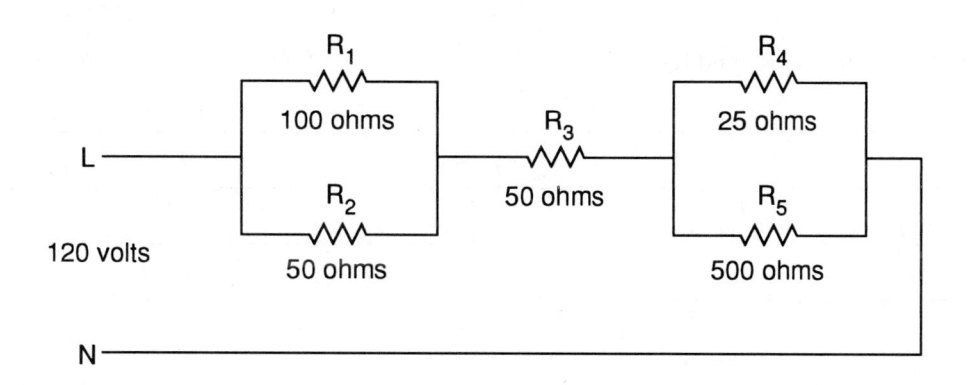

7. What is the resistance of R_1 and R_2?

   Answer: _____

   _____

   _____

   _____

   Reference: _____

8. What is the resistance of $R_3 + R_1$ and R_2?

   Answer: _____

   _____

   _____

   _____

   Reference: _____

9. What is the resistance of R_4 and R_5?

   Answer: _____

   _____

   _____

   _____

   Reference: _____

10. What is the total resistance in the circuit?

    Answer: _____

    _____

    _____

    _____

    Reference: _____

11. What is the voltage drop at R_1 and R_2?

    Answer: _____

    _____

    _____

    _____

    Reference: _____

12. What is the voltage drop at R_3?

    Answer: _____

    _____

    _____

    _____

    Reference: _____

13. What is the voltage drop at R_4 and R_5?

    Answer: _____

    _____

    _____

    _____

    Reference: _____

14. What is the total current in this circuit?

    Answer: _____

    _____

    _____

    _____

    Reference: _____

Chapter Four

INTRODUCTION TO THE *NATIONAL ELECTRICAL CODE®* (*NEC®*)

The *National Electrical Code®* is a reasonable document, developed and written in a reasonable fashion. It is a design manual, but it is not a design specification. It is not the minimum, but there are many conveniences that we take for granted that are not required by this document. There also are many safety considerations not required by this text, such as smoke detectors and other safety and detection devices that are being developed every day. The first page in *Article 90 section 90-1(c) says it is not an instruction manual for untrained persons*. You are going to find that statement to explain much of the confusion about this document. Most engineering schools, vocational schools, and apprentice programs do not have training programs on the proper use of this document as an installation and design standard. Open your 1993 *NEC®* and look up that *NEC®* Section [90-1(c)]. You will need the 1993 *National Electrical Code®* (*NEC®*) throughout your study of this manual. There are several *NEC®* handbooks available. The NFPA *National Electrical Code®* Handbook contains the exact *NEC®* text in blue print and commentary in black print. If you are using that text, remember the black commentary is only the author's opinion and may not be correct. Therefore, when studying in a classroom or for an examination, the blue text or the *NEC®* text will be that from which questions are developed.

KNOW YOUR CODE BOOK

To use the *NEC®*, we must first have a thorough understanding of Table of Contents, Index, Article 90 "The Introduction," Article 100 "Definitions," Article 110 "Requirements for Electrical Installations," and Article 300 "Wiring Methods." The rest of the code book we will refer to on an as-needed basis, but these are general Articles and we use them continually as applicable. **You must study these Articles thoroughly several times until you have a complete understanding of the content in these Articles if you expect to successfully pass your examination.**

Article 90 Introduction

Turn to Article 90. We often pass over the "Introduction" of many books that we read or study. The introduction is an important part of any book. It is the application guidelines. Read Article 90 several times. It is only about four pages but it is an important part of the document, especially for someone about to embark on the study of the *National Electrical Code®*.

Chapter 1 "General" Requirements

You must continually refer to the definitions, *NEC®* Article 100, as you apply the requirements of the *NEC®*. These terms are unique and essential to this document, which is the reason they have been included in this document.

Article 110 covers the general requirements. Section 110-3 tells us that we must select equipment and material suitable for the installation, environment, and/or application, and that it must be installed in accordance with any instructions included with its listing or labeling. Sections 110-9 and 10 provide some strong mandatory requirements on interrupting rating and circuit impedance that cannot be ignored. If the equipment is intended to break current, it must be rated to interrupt the available current at fault conditions, and the system voltage and the circuit components including the impedance shall be coordinated such that the circuit protective device will clear a fault without excessive damage. Sections 110-8 and 12 state that only recognized methods are permitted and that we are required to make our installation in a neat and professional manner. Section 110-14 gives mandatory requirements for making splices and terminations. Section 110-14(c) states that the terminations must be considered when determining the ampacity of a circuit. Read this new *NEC®* Section carefully, similar requirements have appeared in the UL Green and White books.

Section 110-16 states that we are required to carefully select the proper location for our 600 volts or less equipment (for equipment rated at over 600 volts, your code references are *NEC®* Article 110 Part B Sections 110-30 through 34) so that proper work space can be maintained. This often requires us to check the building and mechanical plans so encroachment can be avoided. To discover this encroachment in the last stages of the construction can and is often costly to one or more of the craft contractors. Other sections in Chapter 1 are equally important and must be continually referenced every time the *NEC®* is applied.

Chapter 2 Wiring and Protection

Now that we have thoroughly studied the general requirements, the next step is to study the wiring and protection requirements. Turn to Chapter 2 of the *National Electrical Code®*, Article 210, "Branch Circuits." **NOTE:** Many requirements in this Article apply only to dwellings; read the pertinent sections carefully before applying the requirements of this Article. You also will find specific branch circuit requirements in other Articles of the *NEC®*, such as Article 430 for Motors, Article 600 for Signs, and Article 517 for Health Care Facilities. Section 210-52 gives guidelines necessary to avoid the use of extension cords in dwellings. Section 210-52(a) requires outlets to be located so that no point along the floor line is more than 6 feet from an outlet, and wall spaces 2 feet wide have an outlet in them. This requirement ensures that it is not necessary to lay cords across door openings to reach an outlet. This section also points out that outlets that are part of light fixtures and appliances, located within cabinets, or above $5\frac{1}{2}$ feet cannot be counted as one of those required. Section 210-52(a) through (h) gives requirements for outlets, including the appliance outlets and the ground fault circuit protection where required. (Several of the subsections in 210-52 reference Section 210-8 that gives the requirements for GFCIs in dwellings and other locations.) Now look at Section 210-70(a) for the required lighting outlets and wall switches. Extra lighting can be added for the customer, but the required lighting will provide a safe illumination. Section 210-70(a) generally requires a switch controlled light in each room. Stairways are required to be lighted with a switch at the top and bottom of the stairs. Section 210-70(a) now requires a switch controlled lighting outlet in attics and under floor spaces where used for storage or equipment that requires servicing. This section also requires a switch controlled light at each outside exit or entrance. After we have established the branch circuit requirements, we can do the general lighting load calculations in accordance with Table 220-3(b) and footnotes. **NOTE:** Although generally extra circuits are not necessary, careful arrangement of the required branch circuits is necessary for compliance with Section 220-4(a–d). The total calculated load cannot be ascertained until all of the equipment loads are known and added to the general purpose branch circuits. Chapter 2 Articles 215 and 225 also will guide us as we select and size inside and outside feeders. **NOTE:** Several sections have been removed from Article 230 for the circuit supplying a second building on the same premises. These sections now appear in Article 225, Section 225-8. This change was made to clarify that this circuit is a feeder and not a service.

Article 230 Services is the logical next step and also the next Article. We are beginning to see some rationale for the Chapter and Article arrangement and layout. The diagram 230-1 is an excellent aid for using this Article as the sections apply to the installation. Although materials are not specified, the methods and requirements are specific. **NOTE:** When designing the service or solving questions related to the service, it is important that you properly define the components of the installation (see Article 100 Definitions). Section 230-42(a) requires the service entrance conductors to be of sufficient size to carry the loads computed in Article 220.

Article 250 Grounding. The final Article in Chapter 2 is Article 250. (Article 280, Surge arresters has not been forgotten. If you have a question related don't forget it's there.) Grounding is one important part of any project. Article 250 is specific; as you design the grounding system for this installation, you will find it to be concise and complete for most installations. However, specific grounding requirements are referenced in several other Articles of the *NEC®*. If you are researching a question related to grounding, you can find the answer in the Article specifically addressed in the question. Now that we have our branch circuitry, feeders, service, and grounding designed into the installation, it is now time to select the wiring methods and materials, which are found in Chapter 3 of the *NEC®*.

Chapter 3 Wiring Methods. As stated earlier, this Article covers general requirements that must be applied to all wiring methods. The wiring method selected must be appropriate for the environment, meet the physical protection requirements, temperature, and all other pertinent conditions that can affect the material selected. The condition or question can address the wiring method directly or can require you to select an approved method based on these factors. Be sure to study all of the "uses permitted" and the "uses not permitted" in each specific Article carefully.

For example, when selecting rigid metal conduit, you need to consider only the corrosion elements. However, when selecting nonmetallic sheath cable (Romex), many factors must be considered. Once this has been done, a wiring method can be selected from Chapter 3 based on these factors.

Temporary Wiring. Article 305 permits the use of a lower class wiring for any temporary wiring needed during the construction phase.

The conductor type is selected and sized from Article 310. The wiring method we select may need considering at the same time as the conductor type. Most installations require several wiring methods to meet all conditions, such as a method that is acceptable for dry concealed locations. For this location most methods in Chapter 3 are acceptable. We also may need to select a flexible wiring method for equipment connection and a material that can be used in wet locations and underground. There are methods acceptable in all locations; all are found in Chapter 3 including the junction, outlet, cabinets, and cutout boxes from Articles 370 and 373 *(to identify these, check Article 100 Definitions)*. Again, you are required to comply with all applicable provisions in Article 300 and in that Article covering the specific wiring method.

The requirements for switches (Article 380), switchboards and panelboards (Article 384) are also covered in Chapter 3 of the *NEC®*. Read the scope of these articles. You see that Article 380 applies to all switches, switching devices, and circuit breakers where used as switches. Article 384 the scope is not as clear, and this is an example in which the *NEC®* can be difficult to interpret. Motor control centers are not mentioned in Article 384 yet, an exception to *NEC®* Section 430-1 states: "Installation requirements for motor control centers are covered in Section 384-4." You must study the code carefully and thoroughly to successfully pass the examination the first time. Answer the following Question Review carefully. You will find it necessary to refer to each article in the *National Electrical Code®*, the Table of Contents, and the Index to solve these questions. You will find that answering these questions will prepare you for latter chapters of this book. The next eight chapters of the manual will help you gain needed understanding of the *NEC®* and give you various sample questions on each chapter of the *NEC®*.

CHAPTER 4 QUESTION REVIEW

1. Would a floating restaurant located in a river or along a dock in a harbor be covered by the *National Electrical Code®*?

   Answer: _____

   _____

   _____

   _____

   Reference: _____

2. Where in the *NEC®* would you find the requirements for grounding a community antenna television and radio distribution system?

   Answer: _____

   _____

   _____

   _____

   Reference: _____

3. Where in the *NEC®* would you find the requirements for receptacles, cord connectors, and attachment plugs?

   Answer: _____

   _____

   _____

   _____

   Reference: _____

4. All requirements for the installation and calculation of branch circuits are found in Chapter 2 of the *NEC®*? (True or False)

   Answer: _____

   _____

   _____

   _____

   Reference: _____

5. In what Article would the requirements for installing the light fixtures in a paint spray be found?

   Answer: _____

   _____

   _____

   _____

   Reference: _____

6. In what Article would you find the requirements for the installation of central heating equipment, such as a natural gas furnace, be found?

   Answer: _____

   _____

   _____

   _____

   Reference: _____

7. In what section of the *National Electrical Code®* would you find the allowable number of conductors in a conduit?

   Answer: _____

   _____

   _____

   _____

   Reference: _____

46   Master Electrician's Review

8. Which Article of the *NEC®* covers the power and lighting in mine shafts?

   Answer: _____

   _____

   _____

   Reference: _____

9. Where would you find the cover requirements for underground conductors running from building A to building B? (The conductors are Type UF direct burial conductors, the circuit is a three-phase, 2300-volt, four-wire system.)

   Answer: _____

   _____

   _____

   Reference: _____

10. The conductors in Question 9 above must terminate in a disconnecting means. What Article and Section of the *NEC®* cover the requirements for the disconnecting equipment at building B?

    Answer: _____

    _____

    _____

    Reference: _____

11. A 4-inch rigid nonmetallic conduit is being installed on the outside of a building. The length of the conduit from the junction box to the point that it enters the building is 300 feet. Where in the *National Electrical Code®* is the amount of the expansion of the rigid nonmetallic conduit found?

    Answer: _____

    _____

    _____

    Reference: _____

Chapter Four 47

12. You are going to wire a nightclub that will seat 300 people. Which article in the *NEC®* covers the wiring requirements for this nightclub?

    Answer: _____
    _____
    _____
    _____

    Reference: _____

13. Where in the *National Electrical Code®* would the requirements for grounding a portable generator be found?

    Answer: _____
    _____
    _____
    _____

    Reference: _____

14. Where in the *NEC®* would the requirements for bonding the forming shell of an inground swimming pool be found?

    Answer: _____
    _____
    _____
    _____

    Reference: _____

15. Where in the *NEC®* is the meaning of the small superscript x explained?

    Answer: _____
    _____
    _____
    _____

    Reference: _____

16. Where in the *NEC®* are the requirements for the location and installation of smoke detectors found?

    Answer: _____
    _____
    _____
    _____

    Reference: _____

48   Master Electrician's Review

17. Under what article of the *National Electrical Code®* would the wiring requirements for fire alarms be found?

    Answer: _____

    _____

    _____

    _____

    Reference: _____

18. An existing building is being remodelled. Permanent receptacles in the building are being used to supply temporary power for the construction. Must these permanent receptacles be GFCI protected?

    Answer: _____

    _____

    _____

    _____

    Reference: _____

19. You have been given the responsibility for wiring a phosphoric acid fertilizer facility. What Section of the *National Electrical Code®* covers this type of hazardous installation: (a) What class wiring is required? (b) If the hazardous dusts are not controlled and will be found on the day-to-day work schedule, what division must this area be wired in?

    Answers: _____

    _____

    _____

    _____

    Reference: _____

20. You have been asked to wire a neon sign in the bedroom of a new home you are wiring. Which section of the *National Electrical Code®* would you wire this neon sign for the bedroom?

    Answer: _____

    _____

    _____

    _____

    Reference: _____

21. You are asked to make an installation for a lawnmower repair shop where gasoline will be regularly used. The customer also will be repairing chainsaws and boat motors. Which article of the *National Electrical Code®* would you wire this facility on when gasoline is present in and around the equipment used for cleaning and testing these lawnmowers?

    Answer: _____
    _____
    _____
    _____

    Reference: _____

22. You have been asked to submit a bid for a new restaurant. After calculating the load requirements for the building, you determine that the requirements are too high. Is there an optional calculation you can use for restaurants?

    Answer: _____
    _____
    _____
    _____

    Reference: _____

23. You have been employed to wire a commercial building in electrical nonmetallic tubing. Where in the *NEC®* are the requirements for running parallel to framing members with this raceway system?

    Answer: _____
    _____
    _____
    _____

    Reference: _____

24. You have been employed to install an optional standby generator system. Which article of the *National Electrical Code®* would you make this installation in accordance with?

    Answer: _____
    _____
    _____
    _____

    Reference: _____

25. A company that you have been working with has decided to burn their garbage, using the heat derived as an energy source to drive a generator and generate power to supplement the power being bought from the utility. Which article of the *National Electrical Code*® would you make this installation under?

Answer: _____

_____

_____

_____

Reference: _____

Chapter Five

GENERAL WIRING REQUIREMENTS

The general requirements for the *National Electrical Code®* are found in Chapter 1. They include Article 100, which are the definitions, Part A—General Definitions, and Part B—the definitions for over 600 volts nominal. Article 110 gives the requirements for electrical installations, Part A—General Requirements, and Part B—the requirements for over 600 volts nominal. Article 100 contains the definitions essential to the proper application of the *National Electrical Code®*. They do not include commonly defined general terms or commonly defined technical terms from other related codes and standards. In general, only the terms used in two or more articles are defined in Article 100. Other definitions are included in the articles for which they are used but may be referenced in Article 100. Part A of Article 100 contains definitions intended to apply wherever the terms are used throughout the *NEC®*. Part B contains only the definitions applicable to installations and equipment operating at over 600 volts nominal. Electrical terms not specific to the *NEC®* may be defined in the IEEE Dictionary. Other words are defined in Webster's Dictionary. An example of a definition with a specific meaning in the *NEC®* found in Article 100 would be "accessible." Accessible is defined both for wiring methods and equipment. The definition found in Article 100 applies specifically to its use throughout the *NEC®*. Other terms or words found in Article 100 meanings generally differ from Webster's Dictionary.

Article 110 contains the general requirements for all electrical installations. It contains the approval required for installations and equipment and instructions on the examination of the equipment to judge it suitable. It includes information such as the voltages considered, conductor information, and many terms to describe the uses throughout the Code. Section 110-8 reminds us that only wiring methods recognized as suitable are included in the Code, the recognized methods of wiring permitted to be installed in any type of building or occupancy except as otherwise provided in the *NEC®*. This is important because often materials are misused and wiring methods are devised in the field that are not in

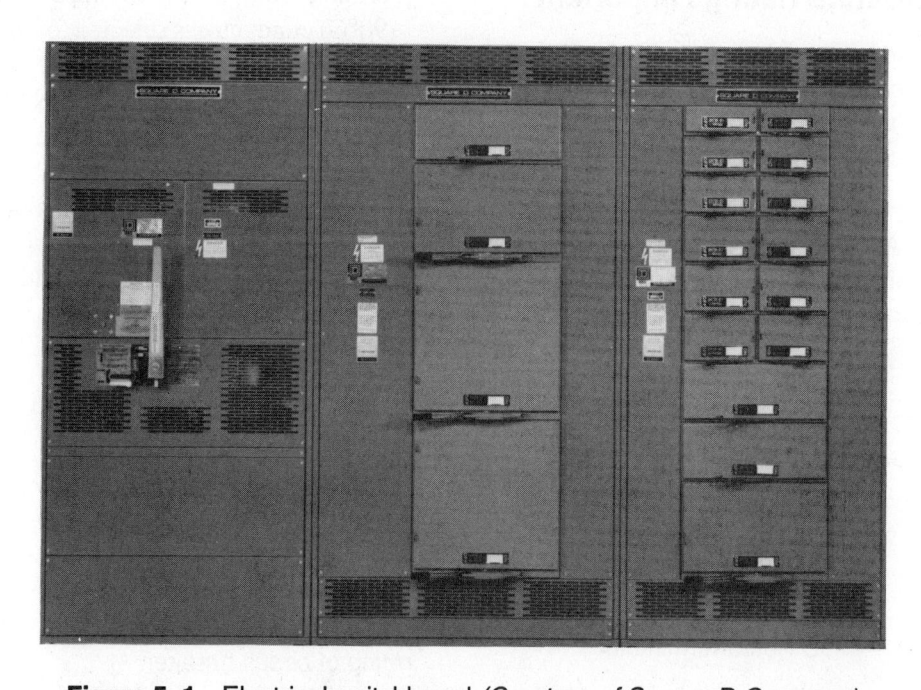

Figure 5–1 Electrical switchboard *(Courtesy of Square D Company)*

accordance with the manufacturer's instructions or design recommendations of the equipment being used. General requirements remind us that all circuitry and equipment intended to break current must be properly sized and installed so proper operation occurs. Article 110 reminds us that the proper equipment must be used where chemicals, gases, vapors, fumes, liquids, or other agents having a deteriorating effect on equipment must be considered. It reminds us that all work must be installed in a neat and professional manner and that unused openings must be closed.

Wire Temperature Termination Requirements

Wire temperature ratings and temperature termination requirements for equipment result in rejected installations. Information about this topic can be found in testing agency directories, product testing standards, and manufacturers' literature, but most do not consult these sources until it is too late.

A new section in the 1993 *National Electrical Code®*, Section 110-14(c), now provides the information contained in the standards to the inspectors, installers, and engineers. However, the appearance of the new section is likely to generate additional comments and questions, although this new section does not create any new requirements.

Why Are Temperature Ratings Important?

Conductors carry a specific temperature rating based on the type of insulation employed on the conductor. Common insulation types can be found in Table 310-13 of the *NEC®*, and corresponding ampacities can be found in Table 310-16.

Example: A 1/0 copper conductor ampacity based on different conductor insulation types:

Insulation Type	Temperature Rating	Ampacity
TW	60°C	125 amperes
THW	75°C	150 amperes
THHN	90°C	170 amperes

Although the wire size has not changed (1/0 Cu) the ampacity *has* changed due to the temperature rating of the insulation on the conductor. Higher rated insulation allows a smaller conductor to be used at the same ampacity as a larger conductor with lower rated insulation, and, as a result, the amount of copper and even the number of conduit runs needed for the job may be reduced.

One common misapplication of conductor temperature ratings occurs when the rating of the equipment termination is ignored. Conductors must be sized by considering where they will terminate and how that termination is rated. If a termination is rated for 75°C, this means the temperature at that termination may rise to 75°C when the equipment is loaded to its ampacity. If 60°C insulated conductors were employed in this example, the additional heat at the connection above the 60°C conductor insulation rating could result in failure of the conductor insulation.

When a conductor is selected to carry a specific load, the user/installer or designer must know termination ratings for the equipment involved in the circuit.

Example: Using a circuit breaker with 75°C terminations and a 150-ampere load, if a THHN (90°C) conductor is selected for the job, from Table 310-16 select a conductor that will carry the 150 amperes. Although Type THHN has a 90°C ampacity rating, the ampacity from the 75°C column must be selected because the circuit breaker termination is rated at 75°C. Looking at the table, a 1/0 copper conductor is acceptable. The installation would be as shown in Figure 5–2, with proper heat dis-

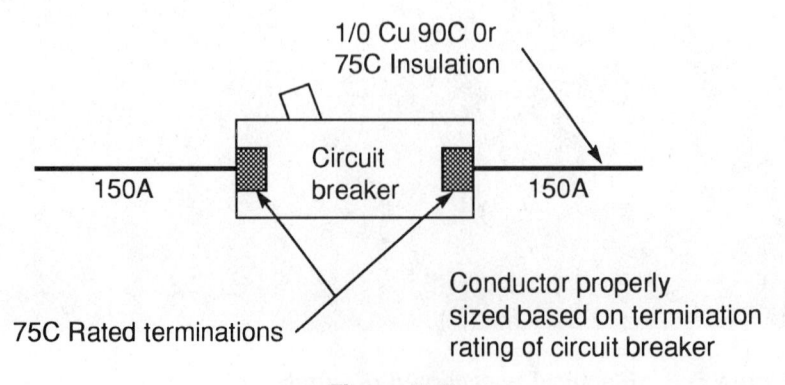

Figure 5–2

No. 1 AWG Cu
90C insulation

150A — Circuit breaker — 150A

75C Rated terminations

Conductor not properly sized based on termination rating of circuit breaker

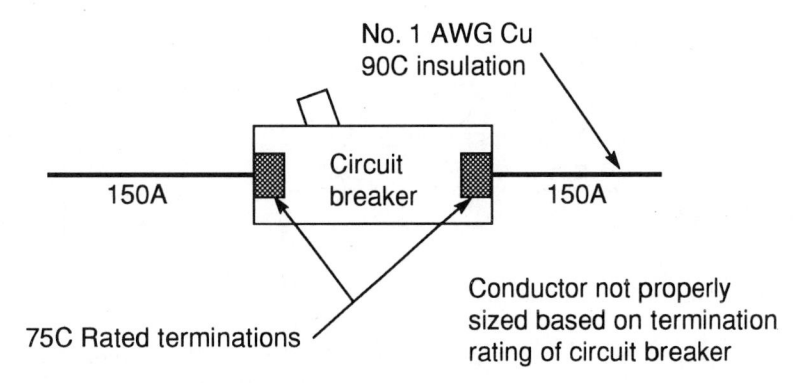

Figure 5–3

sipation at the termination and along the conductor length. If the temperature rating of the termination had not been considered, a No. 1 AWG conductor based on the 90°C ampacity may have been selected, which may have led to overheating at the termination or premature opening of the overcurrent device because of the smaller conductor size. (See Figure 5–3.)

In the example above, a conductor with a 75°C insulation type (THW, RHW, USE, and so forth) also would be acceptable because the termination is rated at 75°C. A 60°C insulation type (TW or perhaps UF) is not acceptable because the temperature at the termination could rise to a value greater than the insulation rating.

General Rules for Application

When applying equipment with conductor terminations, the following two basic rules apply:

- Rule 1 [*NEC*® Section 110-14(c)(1)]—Termination provisions for equipment rated 100A or less or equipment that is marked for No. 14 to No. 1 AWG conductors are rated for use with conductors rated 60°C. (See Figure 5–4.)

- Rule 2 [*NEC*® Section 110-14(c)(2)]—Termination provisions for equipment rated greater than 100A or equipment terminations marked for conductors larger than No. 1 AWG are for use with conductors rated 75°C. (See Figure 5–5.)

There are exceptions to the rules. Two important exceptions are:

- Exception 1—Conductors with higher temperature insulation can be terminated on lower temperature rated terminations provided the ampacity of the conductor is based on the lower rating. This is illustrated in the example where

Circuit breaker with termination rated 60C

20A receptacle with 60C terminations

20 ampere conductor based on 60C ampacity

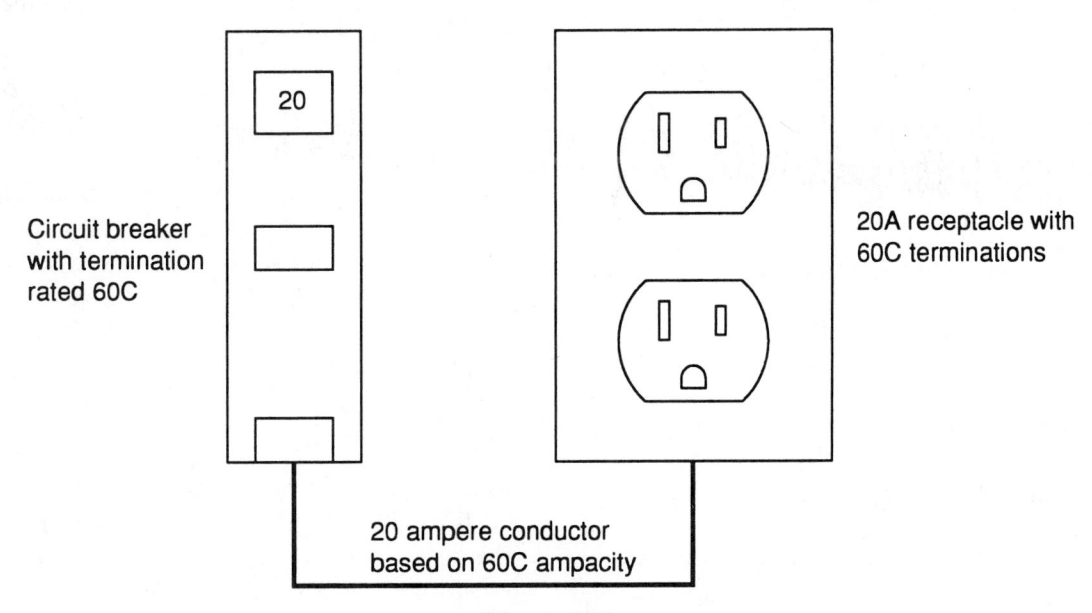

Figure 5–4

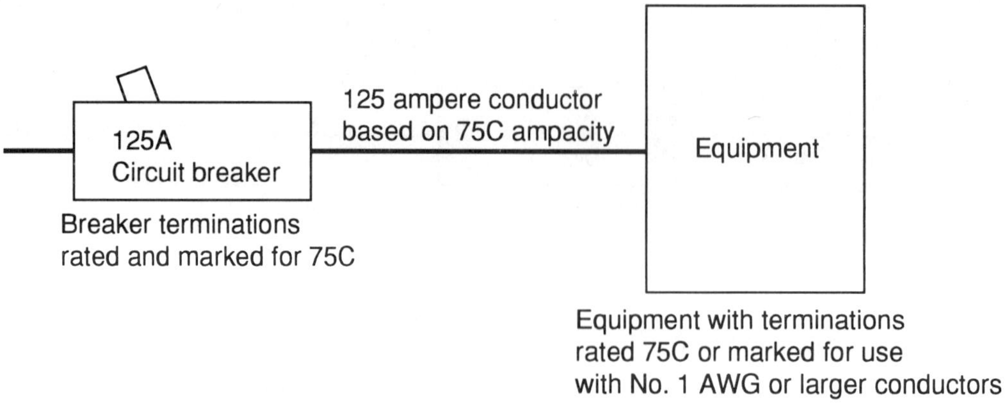

```
                        125 ampere conductor
      ┌─────────────┐   based on 75C ampacity   ┌──────────┐
──────┤    125A     ├──────────────────────────┤          │
      │Circuit breaker│                         │ Equipment│
      └─────────────┘                           │          │
      Breaker terminations                      └──────────┘
      rated and marked for 75C
                                              Equipment with terminations
                                              rated 75C or marked for use
                                              with No. 1 AWG or larger conductors
```

Figure 5–5

the THHN (90°C) conductor ampacity is based on the 75°C rating to terminate in a 75°C termination. The following table provides a quick reference of how this exception would apply to common terminations.

Conductor Insulation Versus Equipment Termination Ratings

Termination Rating	60°C	Conductor Insulation Rating 75°C	90°C
60°C	OK	OK (at 60°C ampacity)	OK (at 60°C ampacity)
75°C	No	OK	OK (at 75°C ampacity)
60/75°C	OK	OK (at 60°C or 75°C ampacity)	OK (at 60°C or 75°C ampacity)
90°C	No	No	OK*

* The equipment must have a 90°C rating to terminate 90°C wire at its 90°C ampacity.

- Exception 2—The termination can be rated for a value higher than the value permitted in the general rules if the equipment is listed and marked for the higher temperature rating.

Example: A 30-ampere safety switch could have 75°C rated terminations if the equipment was listed and identified for use at this rating.

A Word of Caution

When terminations are inside equipment, such as panelboards, motor control centers, switchboards, enclosed circuit breakers, and safety switches, it is important to note that the temperature rating identified on the equipment labeling should be followed—not the rating of the lug. It is common to use 90°C rated lugs (i.e., marked AL9CU), but the equipment rating may be only 60°C or 75°C. The use of the 90°C rated lugs in this type of equipment does not give permission for the installer to use 90°C wire at the 90°C ampacity.

The labeling of all devices and equipment should be reviewed for installation guidelines and possible restrictions.

Equipment Terminations Available Today

Remember, a conductor has two ends, and the termination on each end must be considered when applying the sizing rules.

Example: Consider a conductor that will terminate in a 75°C rated termination on a circuit breaker at one end and a 60°C rated termination on a receptacle at the other end. This circuit must be wired with a conductor that has an insulation rating of at least 75°C (due to the circuit breaker) and sized based on the ampacity of 60°C (due to the receptacle).

In electrical equipment, terminations are typically rated at 60°C, 75°C, or 60/75°C. There is no listed distribution or utilization equipment that is listed and identified for the use of 90°C wire at its 90°C ampacity. This includes distribution equipment, wiring devices, transformers motor control devices, and utilization equipment such as HVAC, motors, and light fixtures.

Installers and designers who do not know this fact have been faced with jobs that do not comply with the *National Electrical Code®*, and jobs have been turned down by the electrical inspectors.

Example: A 90°C wire can be used at its 90°C ampacity (see Figure 5–6). Note that the No. 2 90°C rated conductor does not terminate directly in the distribution equipment but in a terminal or tap box with 90°C rated terminations.

Frequently, manufacturers are asked when distribution equipment will be available with terminations that will permit 90°C conductors at the 90°C ampacity. The answer is complex and requires not only significant equipment redesign (to handle the additional heat) but also coordination of the downstream equipment where the other end of the conductor will terminate. Significant changes in the product testing/listing standards also would have to occur.

A final note about equipment is that, generally, equipment requiring the conductors to be terminated in the equiment has an insulation rating of 90°C but has an ampacity based on 75°C or 60°C. This type of equipment might include 100% rated circuit breakers, fluorescent lighting fixtures, and so on and will include a marking to indicate such a requirement. Check with the equipment manufacturer to see if any special considerations need to be considered.

Higher Rated Conductors and Derating Factors

One advantage to conductors with higher insulation ratings is noted when derating factors are applied. Derating factors can be required because of the number of conductors in a conduit, higher ambient temperatures, or possibly internal design requirements for a facility. By beginning the derating process at the conductor ampacity based on the higher insulation value, upsizing the conductors to compensate for the derating may not be required.

For the following example of this derating process, the two following points must be considered:

1. The ampacity value determined after applying the derating factors must be equal to or less than the ampacity of the conductor, based on the temperature limitations at its terminations.
2. The derated ampacity becomes the allowable ampacity of the conductor, and the conductor must be protected against overcurrent in accordance with this allowable ampacity.

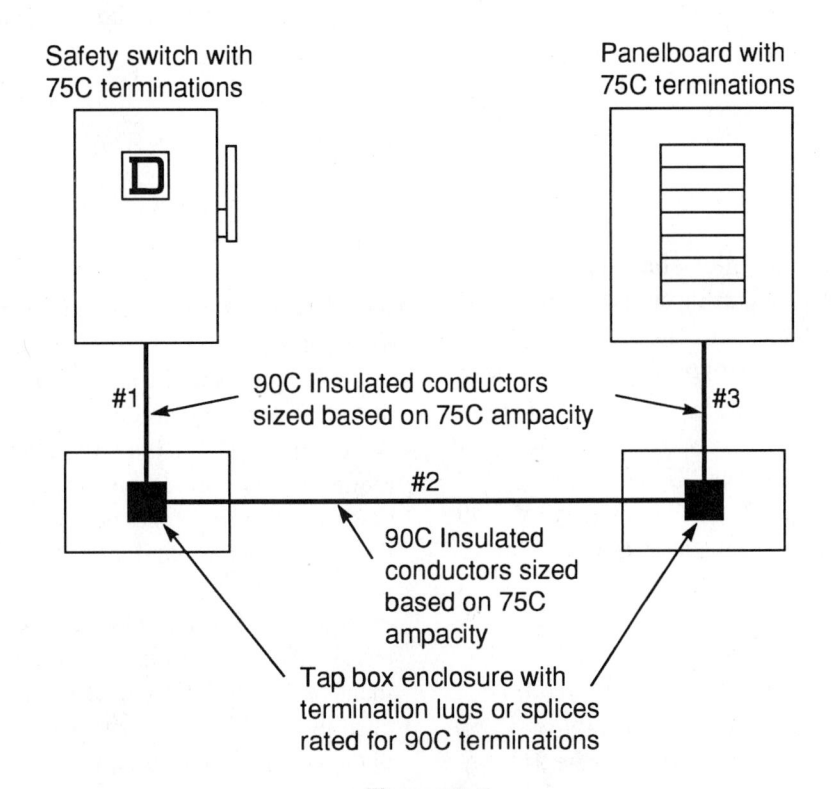

Figure 5–6

Example for the derating process: Assume a 480Y/277 VAC, 3Ø4W feeder circuit to a panelboard supplying 200 amperes of fluorescent lighting load, and assume that the conductors will be in a 40°C ambient temperature. Also, assume the conductors originate and terminate in equipment with 75°C terminations.

1. Because the phase and neutral conductors will be in the same conduit, the issue of conduit fill must be considered. Note 10(c) in the *Notes To The Ampacity Tables of 0–2000 Volts* states that the neutral must be considered to be a current-carrying conductor because it is supplying electric discharge lighting.
2. With four current-carrying conductors in the raceway, apply Note 8(a) in the *Notes To The Ampacity Tables of 0–2000 Volts*. This note requires an 80% reduction in the conductor ampacity based on four to six current-carrying conductors in the raceway.
3. The correction factors at the bottom of Table 310-16 also must be applied. An adjustment of .88 for 75°C and .91 for 90°C is required where applicable.

Now the calculations:

Using a 75°C conductor such as THWN:

300 kcmil copper has a 75°C ampacity of 285 amperes. Using the factors from above, the calculations are:

$$285 \times .80 \times .88 = 201 \text{ amperes}$$

Thus, 201 amperes is now the allowable ampacity of the 300 kcmil copper conductor for this circuit. If the derating factors for conduit fill and ambient temperatures had not been required, a 3/0 copper conductor would have met these requirements.

Using a 90°C conductor such as THHN:

250 kcmil copper has a 90°C ampacity of 290 amperes. Using the factors from above, the calculations are:

$$290 \times .8 \times .91 = 211 \text{ amperes}$$

211 amperes is less than the 75°C ampacity of a 250 kcmil copper conductor (255 amperes), so the 211 amperes would now be the allowable ampacity of the 250 kcmil conductor. If the calculation resulted in a number larger than the 75°C ampacity, the actual 75°C ampacity would have been required to be used as the allowable ampacity of the conductor. This is critical because the terminations are rated at 75°C.

Note: The primary advantage to using 90°C conductors is exemplified by this example. The conductor is permitted to be reduced by one size (300 kcmil to 250 kcmil) and still accommodate all required derating factors for the circuit.

In summary, when using 90°C wire for derating purposes, begin by derating at the 90°C ampacity. Compare the result of the calculation with the ampacity of the conductor based on the termination rating (60°C or 75°C). The smaller of the two numbers then becomes the allowable ampacity of the conductor. Note that if the load dictates the size of the required overcurrent device (i.e., continuous load × 125%), then the conductor allowable ampacity must be protected by the required overcurrent device. (See Section 240-3 that permits the conductor to be protected by the next higher standard size overcurrent device as per Section 240-6.) This may require moving to a larger conductor size and beginning the derating. In the example, if the 200 amperes of load were continuous, a 250 ampere overcurrent device would be required. The 201 or 211 amperes from the calculations would not be protected by the 250 ampere overcurrent device and, as such, we would be required to move to a larger conductor meeting these requirements to begin the derating process.

Summary

Several factors affect how the allowable ampacity of a conductor is determined. The key is not to treat the wire as a system but as a component of the total electrical system. The terminations, equipment ratings, and environment affect the ampacity assigned to the conductor. If the designer and the installer remember each rule, the installation will go smoother.

Section 110-14 is especially important in that the requirements for all electrical connections must be made in the proper manner. The 1993 Code has added text clarifications and has identified the termination requirements to meet the listing laboratories recommendations. Section 110-16 reminds us that we must have sufficient working space around electrical equipment of 600 volts nominal or less; that we must have clearances in front and above (again see chapter opener photo of switchboard, Figure 5–1); we must have headroom to work on the equipment; proper illumination;

and that all live parts must be guarded against accidental contact. Article 110 reminds us that warning signs must be placed and that electrical equipment must be protected from physical damage. Section 110-22 states that each disconnecting means for motors and appliances, and service, feeders, or branch circuits at the point where it originates must be legibly identified to indicate its purpose and that the identifying markings must be sufficiently durable to withstand the environment involved. Where circuits and fuses are installed with a series combination rating "Caution—Series Rated System," the equipment enclosure shall be legibly marked in the field to indicate the equipment has been applied with a series combination rating. Over 600 volts requirements are found in Part B of Article 110. These requirements are specific about entrance and access to working space, guarding barriers, enclosures for electrical installations (see Figure 5–1), and separation from other circuitry. Over 600 volts installations must be located in locked rooms or enclosures, except where under the observation of qualified persons at all times. Table 110-34(e) covers the elevation of unguarded live parts above working space for nominal voltages 1001 through 35kV.

NEC® 110–14(c) Temperature Limitations. The temperature rating associated with the ampacity of a conductor shall be so selected and coordinated as to not exceed the lowest temperature rating of any connected termination, conductor, or device.

(1) Termination provisions of equipment for circuits rated 100 amperes or less, or marked for Nos. 14 through 1 conductors, shall be used only for conductors rated 60°C (140°F).

Exception No. 1: Conductors with higher temperature ratings shall be permitted, provided the ampacity of such conductors is determined based on the 60°C (140°F) ampacity of the conductor size used.

Exception No. 2: Equipment termination provisions shall be permitted to be used with higher-rated conductors at the ampacity of the higher-rated conductors, provided the equipment is listed and identified for use with the higher-rated conductors.

(2) Termination provisions of equipment for circuits rated over 100 amperes, or marked for conductors larger than No. 1, shall be used only with conductors rated 75°C (167°F).

Exception No. 1: Conductors with higher temperature ratings shall be permitted, provided the ampacity of such conductors is determined based on the 75°C (167°F) ampacity of the conductor size used.

Exception No. 2: Equipment termination provisions shall be permitted to be used with the higher-rated conductors at the ampacity of the higher-rated conductors, provided the equipment is listed and identified for use with the higher-rated conductors.

(3) Separately installed pressure connectors shall be used with conductors at the ampacities not exceeding the ampacity at the listed and identified temperature rating of the conductor.

(FPN): With respect to *Sections 110–14(c)(1), (2), and (3)*, equipment markings or listing information may additionally restrict the sizing and temperature ratings of connected conductors.

(Reprinted with permission of NFPA 70-1993, the *National Electrical Code®*, Copyright © 1992, National Fire Protection Association, Quincy, MA 02269. This reprinted material is not the complete and official position of the National Fire Protection Association on the referenced subject, which is represented only by the standard in its entirety.)

CHAPTER 5 QUESTION REVIEW

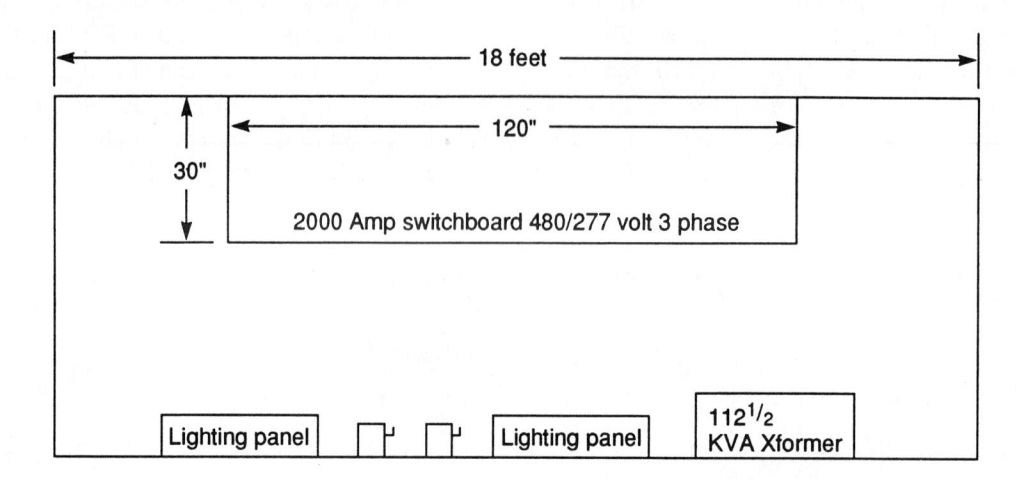

```
|←─────────────── 18 feet ───────────────→|
┌─────────────────────────────────────────┐
│     ↑  |←──────── 120" ────────→|       │
│   30"  ┌─────────────────────────┐      │
│     ↓  │ 2000 Amp switchboard    │      │
│        │ 480/277 volt 3 phase    │      │
│        └─────────────────────────┘      │
│                                         │
│  ┌──────────┐        ┌──────────┐ ┌────┐│
│  │Lighting  │ ┌┐  ┌┐ │Lighting  │ │112½││
│  │panel     │ └┘  └┘ │panel     │ │KVA ││
│  │          │        │          │ │Xfmr││
│  └──────────┘        └──────────┘ └────┘│
└─────────────────────────────────────────┘
```

This proposed electrical equipment room is to be constructed of concrete floor and concrete. Block walls will be painted.

1. In the example above, what is the space required between the switchboard and the 112½ kVA transformer?

   Answer: _____

   _____

   _____

   _____

   Reference: _____

2. In the example above, are one or two doors required for entering and leaving this room?

   Answer: _____

   _____

   _____

   _____

   Reference: _____

3. Is the electrical equipment room in the example above required to be illuminated?

   Answer: _____

   _____

   _____

   _____

   Reference: _____

4. What is the required headroom in front of the electrical equipment needed for those performing tests or maintenance on this electrical equipment?

   Answer: _____

   _____

   _____

   _____

   Reference: _____

5. In the example above, if the proposed electrical room size was increased from 10 × 18 to 18 × 18, how many doors would be required, provided that the room had no additional electrical equipment added?

   Answer: _____

   _____

   _____

   _____

   Reference: _____

6. Does the installation as shown above comply with the *NEC®*? It is a single-family dwelling. The supplementary overcurrent device and required disconnecting means are located on the dwelling wall directly behind the heat pump for convenience.

   Answer: _____

   _____

   _____

   _____

   Reference: _____

7. The working space in front of enclosed electrical equipment operating at 480 volts must have a width when facing the equipment of at least ____ inches.

   Answer: _____

   _____

   _____

   _____

   Reference: _____

8. In an electrical system, an overcurrent device shall be placed in series with each ____ conductor.

   Answer: _____

   _____

   _____

   _____

   Reference: _____

9. A conduit nipple of 18 inches in length is installed between two boxes. This nipple can be filled to a maximum of ____ % without derating the conductors.

   Answer: _____

   _____

   _____

   _____

   Reference: _____

10. A distribution panelboard must **not** be installed in a _____.

    Answer: _____

    _____

    _____

    _____

    Reference: _____

11. An unprotected cable is installed through bored holes in wood studs. The holes must be bored so the edge of the hole is at least ____ inch(es) from the nearest edge of the stud.

    Answer: _____

    _____

    _____

    _____

    Reference: _____

12. Circuit conductors run in electrical nonmetallic tubing (ENT) cannot exceed _____ volts.

    Answer: _____

    _____

    _____

    _____

    Reference: _____

13. Can liquidtight flexible nonmetallic conduit be used as a service raceway?

    Answer: _____

    _____

    _____

    _____

    Reference: _____

14. When a metal raceway is used as physical protection for a grounding electrode conductor, must this metallic raceway be bonded even though the grounding electrode conductor is bare copper? If the answer is yes, how must this metal raceway be bonded?

    Answer: _____

    _____

    _____

    _____

    Reference: _____

15. The size of the equipment grounding conductor routed with the feeder or branch circuit conductors is determined by what?

    Answer: _____

    _____

    _____

    _____

    Reference: _____

16. How many outdoor receptacle outlets are required for a one-family dwelling?

    Answer: _____

    _____

    _____

    _____

    Reference: _____

62 Master Electrician's Review

17. Are the metal parts of electrical equipment associated with a hydromassage tub required to be bonded?

    Answer: _____

    _____

    _____

    _____

    Reference: _____

18. Where rooms within dwellings are separated by railings, planters, and so forth, are receptacles required to be installed in or along these items?

    Answer: _____

    _____

    _____

    _____

    Reference: _____

19. Are the receptacle outlet spacings for room dividers, bar type counters, and fixed room dividers the same for site-built dwellings as they are for mobile homes?

    Answer: _____

    _____

    _____

    _____

    Reference: _____

20. You have been asked to install a receptacle on a rooftop for the air conditioning repair people to plug their instruments and portable tools into. In which section of the *National Electrical Code*® would this be found?

    Answer: _____

    _____

    _____

    _____

    Reference: _____

21. You have encountered a problem in making a new service installation and need to know the requirements of the meter enclosure on the outside of the building. Which article of the *National Electrical Code*® covers meter enclosures?

    Answer: _____

    _____

    _____

    _____

    Reference: _____

22. In making an installation in a home, the homeowners ask you to install a light fixture over their hydromassage bathtub. Which section of the *National Electrical Code®* governs light fixtures over hydromassage tubs?

    Answer: _____

    _____

    _____

    _____

    Reference: _____

23. As you are getting ready to install the conductors for the service drop on a large residence, you discover that all you have on the truck is a roll of No. 1 THW copper. The service is 200 amps. Realizing that No. 1 is not large enough, you parallel two No. 1s for each phase and two No. 1s for the neutral. Is this method acceptable for making this installation in accordance with the *National Electrical Code®*?

    Answer: _____

    _____

    _____

    _____

    Reference: _____

24. In Question 23, what size single conductors would be required for a 200 amp service?

    Answer: _____

    _____

    _____

    _____

    Reference: _____

25. In making an installation of over 200 feet, a concern is voltage drop. Where does one obtain the resistance for conductors to be installed in a raceway or in direct burial for the purposes of calculating voltage drop?

    Answer: _____

    _____

    _____

    _____

    Reference: _____

Chapter Six

BRANCH CIRCUITS AND FEEDERS

Branch circuits are covered in *NEC®* Article 210. Part A covers the general provisions for branch circuits, Part B covers the branch circuit ratings, and Part C gives the required outlets. Outdoor branch circuits and feeders are found in *NEC®* Article 225. *NEC®* Article 215 covers other feeders, and *NEC®* Article 220 covers the calculations for branch circuits and feeders and the optional calculations for computing feeder and service loads. However, branch circuits are found throughout the Code. Air conditioning branch circuits are found in *NEC®* Article 440, appliance branch circuits in *NEC®* Article 422, heating branch circuits in *NEC®* Article 424, and motor branch circuits in *NEC®* Article 430, and so forth. Chapters 5, 6, and 7 supplement the first four chapters of the Code. Therefore, for branch circuit studies it is necessary that you be familiar with the entire *NEC®* and use the index to quickly find the specific type branch circuits with which you are dealing. Branch circuits are defined in *NEC®* Article 100 and are defined as a branch circuit, an appliance branch circuit, a general purpose branch circuit, an individual branch circuit, a multi-wire branch circuit, and the branch circuit selection current. It is important that you read these definitions before applying the requirements in the *NEC®*, so you are properly applying the requirements. A feeder is defined in Article 100 as the circuit conductor between the service equipment or the source of a separately derived system and the final branch circuit overcurrent device. **Caution:** A feeder generally terminates in more than one overcurrent device. For instance, many lighting fixtures and other equipment contain supplementary overcurrent devices at the fixture or equipment. The conductors feeding that equipment should still be considered as the branch circuit conductors and not feeder conductors. Feeders are generally found in *NEC®* Articles 215 and 225. Calculations for sizing feeders are found in *NEC®* Article 220. Other references to feeders can be found in *NEC®* Article 364 for busways, *NEC®* Article 430 for motors, *NEC®* Article 550 for mobile homes, and *NEC®* 530 for motion picture studios. See the Index for a complete listing of where feeders can be found in the *NEC®*. See Figure 6–1.

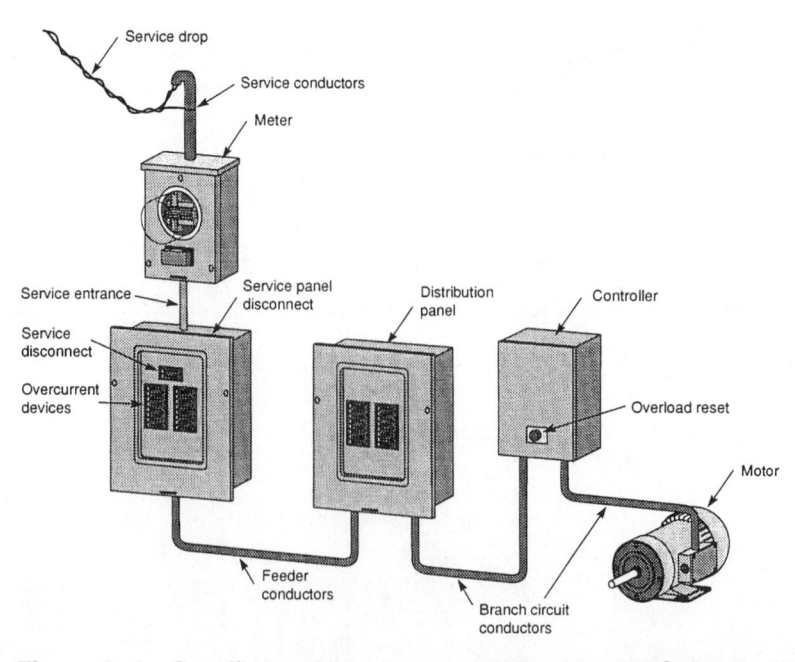

Figure 6–1 Specific conductor types as defined in *NEC®* Article 100

General Purpose Branch Circuits

The simplest form of branch circuits is general purpose branch circuits, which supply a number of outlets for lighting and appliances, and these are in 15, 20, 30, 40, and 50 ampere sizes. (See Figure 6–2.) Where conductors of higher ampacity are used for any reason, the ampere rating or the setting of the specified overcurrent device determines the circuit classification. Multi-wire branch circuits greater than 50 amps can be permitted for nonlighting outlet loads on industrial premises where maintenance and supervision indicate that a qualified person will service the equipment. A common multi-outlet branch circuit greater than 50 amperes is often found in industrial buildings using welding receptacles so the maintenance personnel can move welders throughout the plant on an as-needed basis. In these instances, you can find many receptacles on one 60 ampere or above rated circuit to supply these receptacles. Generally, no danger of overloading this circuit exists because the plant has a limited number of welding machines and personnel capable of using those machines. There are no voltage limitations on branch circuits because branch circuits can supply the equipment and motors of many varying different voltages. Section 210-6 lists the limitations for branch circuit voltages. In occupancies such as dwelling units and guest rooms of hotels, motels, and similar occupancies, the voltage is not permitted to exceed 120 volts between conductors that supply terminals of lighting fixtures and cord- and plug-connected loads 1440 VA nominal or less, or less than $\frac{1}{4}$ horsepower, and permitted to supply terminals of medium-base, screw-shell lampholders, or lampholders of other types applied within their voltage ratings, auxiliary equipment of electric discharge lamps, cord- and plug-connected, or permanently connected utilization equipment. Circuits exceeding 120 volts nominal and not exceeding 277 volts, nominal, to ground are permitted to supply

---

Example No. 6. Maximum Demand for Range Loads

Table 220-19, Column A applies to ranges not over 12 kW. The application of Note 1 to ranges over 12 kW (and not over 27 kW) and Note 2 to ranges over $8\frac{3}{4}$ kW (and not over 27 kW) is illustrated in the following examples:

A. Ranges all the same rating (Table 220-19, Note 1).
  Assume 24 ranges, each rated 16 kW.
  From Column A, the maximum demand for 24 ranges of 12 kW rating is 39 kW.
  16 kW exceeds 12 kW by 4.
  5% × 4 = 20% (5% increase for each kW in excess of 12).
  39 kW × 20% = 7.8 kW increase.
  39 + 7.8 = 46.8 kW: value to be used in selection of feeders.

B. Ranges of unequal rating (Table 220-19, Note 2).
  Assume 5 ranges, each rated 11 kW.
  2 ranges, each rated 12 kW.
  20 ranges, each rated 13.5 kW.
  3 ranges, each rated 18 kW.

  5 × 12 = 60 use 12 kW for range rated less than 12.
  2 × 12 = 24
  20 × 13.5 = 270
  3 × 18 = 54
  30   408 kW
  408 ÷ 30 = 13.6 kW (average to be used for computation).

  From Column A, the demand for 30 ranges of 12 kW rating is 15 + 30 = 45 kW.
  13.6 exceeds 12 by 1.6 (use 2).
  5% × 2 = 10% (5% increase for each kW in excess of 12).
  45 kW × 10% = 4.5 kW increase.
  45 + 4.5 = 49.5 kW: value to be used in selection of feeders.

Figure 6–2 (Reprinted with permission of NFPA 70-1993, the *National Electrical Code®*, Copyright © 1992, National Fire Protection Association, Quincy, MA 02269. This reprinted material is not the complete and official position of the National Fire Protection Association on the referenced subject, which is represented only by the standard in its entirety.)

66   Master Electrician's Review

listed electric-discharge lighting fixtures equipped with medium-base screw-shell lampholder, lighting fixtures with mogul-base screw-shell lampholders, and lampholders other than the screw-shell type applied within their voltage ratings, auxiliary equipment of electric discharge lamps, and cord- and plug-connected or permanently connected utilization equipment. Circuits exceeding 277 volts, nominal, to ground and not exceeding 600 volts, nominal, between conductors are permitted to supply auxiliary equipment of electric discharge lamps mounted in permanently installed fixtures where the fixtures are mounted in accordance with *NEC®* 210-6(d) and exceptions (see Figure 6–3). Branch circuits supplying equipment over 600 volts are required to comply with the appropriate Code sections and articles related to that equipment.

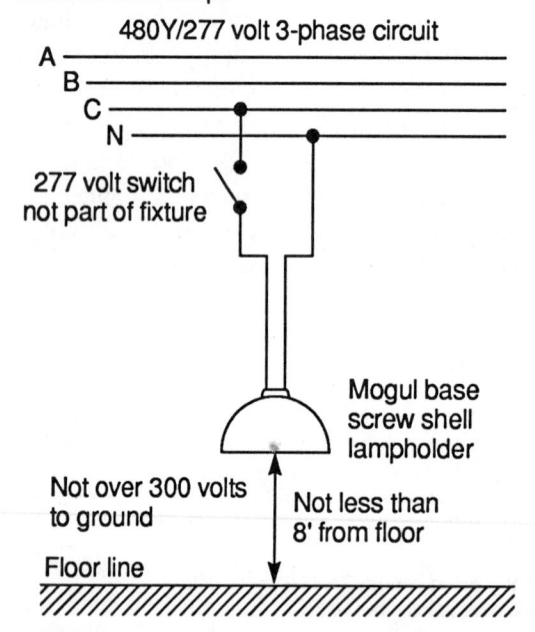

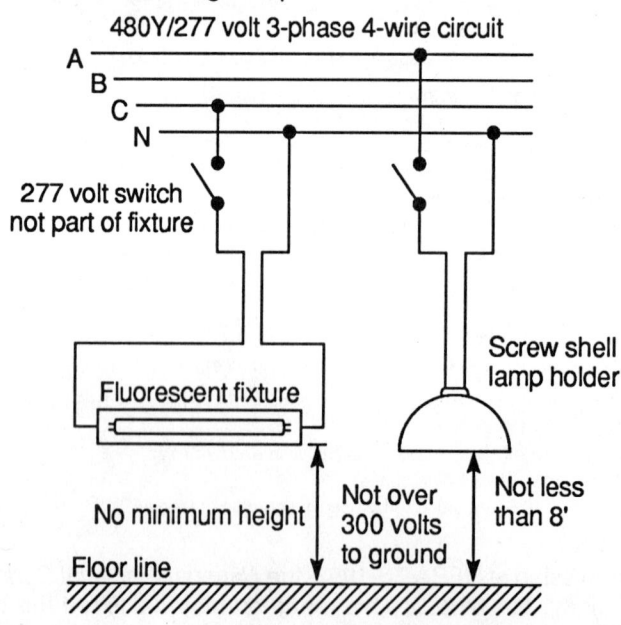

Figure 6–3 *(Courtesy of American Iron & Steel Institute)*

Ground-Fault Protection

Ground-fault circuit interrupter protection for personnel or dwelling units and other location requirements can be found in Section 210-8. However, other requirements for ground-fault circuit interrupters can be found in specific articles, such as Article 511 for commercial garages, Article 517 for health care facilities, and Article 680 for swimming pools. Section 210-52 covers the requirements and general provisions for installing receptacle outlets in dwellings and does not cover installations other than dwelling installations generally; there are some exceptions. Provisions are in the Code, in Article 220, that do cover these receptacle outlets other than dwelling installations when installed. Lighting outlets for dwellings are covered in Section 210-70. Just as for receptacle outlets, the lighting requirements are generally not specified in the *NEC®* other than for dwellings. However, the installation procedures are specified when they are installed in these installations. When calculating the loads for branch circuit conductors, one must verify whether these loads are rated as continuous duty or as noncontinuous duty. Continuous duty is defined in Article 100. Most loads in dwellings are considered to be noncontinuous loads because they would be on for less than 3 hours at any given time. An example of a continuous load would be the lighting in a commercial or industrial establishment, which would generally be energized for more than 3 hours at a time. Most branch circuit calculations are made in Article 220 (see Figure 6–4). However, specific branch circuit requirements for branch circuits supplying specific utilization equipment are also found in the appropriate article for that equipment, such as Article 440 for air conditioning equipment, and Article 430 for motors (see Figure 6–5). There also are many other specific requirements throughout the latter chapters in the *NEC®*. Several examples for calculating branch circuits are found in the latter sections of this chapter. Article 215 covers the installation requirements and minimum size and ampacity for conductors for feeders supplying branch circuit loads, as are computed in Article 220.

Example No. 1(a). One-Family Dwelling

The dwelling has a floor area of 1500 sq. ft., exclusive of unoccupied cellar, unfinished attic, and open porches. Appliances are a 12-kW range and a 5.5 kW, 240-volt dryer. Assume range and dryer kW ratings equivalent to kVA ratings in accordance with Sections 220-18 and 220-19.

Computed Load [see Section 220-10(a)]:
　General Lighting Load:
　　1500 sq. ft. at 3 volt-amperes per sq. ft = 4500 volt-amperes.

Minimum Number of Branch Circuits Required [see Section 220-4(b)]:
　General Lighting Load:
　　4500 volt-amperes ÷ 120 volts = 37.5 A:　This requires three 15 A 2-wire or two 20 A 2-wire circuits
　Small Appliance Load:　Two 2-wire 20 A circuits [see Section 220-3(b)]
　Laundry Load:　One 2-wire 20 A circuit [see Section 220-4(c)]

Minimum Size Feeder Required [see Section 220-10(a)]:

General Lighting	4500 volt-amperes
Small Appliance Load	3000 volt-amperes
Laundry	1500 volt-amperes
Total General Light and Small Appliance	9000 volt-amperes
3000 volt-amperes at 100%	3000 volt-amperes
9000 − 3000 = 6000 volt-amperes at 35%	2100 volt-amperes
Net General Lighting and Small Appliance Load	5100 volt-amperes
Range Load (see Table 220-19)	8000 volt-amperes
Dryer Load (see Table 220-18)	5500 volt-amperes
Total Load	18,600 volt-amperes

For 120/240-volt 3-wire single-phase service or feeder,
18,600 volt-amperes ÷ 240 volts = 77.5 A.
　Net computed load exceeds 10kVA. Service conductors shall be 100 amperes [see Section 230-42(b)(2)].

Figure 6–4　(Reprinted with permission of NFPA 70-1993, the *National Electrical Code®*, Copyright © 1992, National Fire Protection Association, Quincy, MA 02269. This reprinted material is not the complete and official position of the National Fire Protection Association on the referenced subject, which is represented only by the standard in its entirety.)

Neutral for Feeder and Service

Lighting and Small Appliance Load	5100 volt-amperes
Range Load 8000 volt-amperes at 70%	5600 volt-amperes
Dryer Load 5500 volt-amperes at 70%	3850 volt-amperes
Total	14,550 volt-amperes

14,550 volt-amperes ÷ 240 Volts = 60.6 amperes

Example No. 1(b). One-Family Dwelling

Same conditions as Example No. 1(a), plus addition of one 6-ampere 230-volt room air-conditioning unit and one 12-ampere 115-volt room air-conditioning unit,* one 8-ampere 115-volt rated disposal and one 10-ampere 120-volt rated dishwasher.* See Article 430 for general motors and Article 440, Part G, for air-conditioning equipment. Motors have nameplate ratings of 115 V and 230 V for use on 120 V and 240 V nominal voltage systems.

From previous Example No. 1(a), feeder current is 78 amperes (3-wire, 240 volts).

	Line A	Neutral	Line B
Amperes from Example No. 1(a)	78	61	78
One 230-V air conditioner	6	—	6
One 115-V air conditioner and 120-V dishwasher	12	12	10
One 115-V disposal	—	8	8
25% of largest motor (Section 430-24)	3	3	2
Amperes per line	99	84	104

* For feeder neutral, use largest of the two appliances for unbalance.

Example No. 2(a). Optional Calculation for One-Family Dwelling, Heating Larger than Air Conditioning

[See Section 220-30(a) and Table 220-30.]

Dwelling has a floor area of 1500 sq. ft., exclusive of unoccupied cellar, unfinished attic, and open porches. It has a 12-kW range, a 2.5-kW water heater, a 1.2-kW dishwasher, 9 kW of electric space heating installed in five rooms, a 5-kW clothes dryer, and a 6-ampere 230-volt room air-conditioning unit. Assume range, water heater, dishwasher, space heating, and clothes dryer kW ratings equivalent to kVA.

Air conditioner kVA is 6 × 230 ÷ 1000 = 1.38 kVA

1.38 kVA is less than the connected load of 9 kVA of space heating; therefore, the air conditioner load need not be included in the service calculation (see Section 220-21).

1500 sq. ft. at 3 volt-amperes	4500 volt-amperes
Two 20-ampere appliance outlet circuits at 1500 volt-amperes each	3000 volt-amperes
Laundry circuit	1500 volt-amperes
Range (at nameplate rating)	12,000 volt-amperes
Water heater	2500 volt-amperes
Dishwasher	1200 volt-amperes
Clothes dryer	5000 volt-amperes
	29,700 volt-amperes

First 10 kVA of other load at 100% = 10,000 volt-amperes
Remainder of other load at
    40% (19.7 kVA × .4) = 7900 volt-amperes
    Total of other load = 17,900 volt-amperes
9 kVA of heat at 40% (9 × .4) = 3600 volt-amperes
    Total load = 21,500 volt-amperes

Calculated load for service size
21.5 kVA = 21,500 volt-amperes
21,500 VA ÷ 240 volts = 90 amperes

Therefore, this dwelling shall be permitted to be served by a 100-ampere service.

Figure 6–4 *(continued)*

Feeder Neutral Load, per Section 220-22:

1500 sq. ft. at 3 volt-amperes	4500 volt-amperes
Three 20-amp. circuits at 1500 volt-amperes	4500 volt-amperes
Total	9000 volt-amperes
3000 volt-amperes at 100%	3000 volt-amperes
9000 VA − 3000 VA = 6000 volt-amperes at 35%	2100 volt-amperes
	5100 volt-amperes
Range — 8 kVA at 70%	5600 volt-amperes
Clothes dryer — 5 kVA at 70%	3500 volt-amperes
Dishwasher	1200 volt-amperes
Total	15,400 volt-amperes

15,400 volt-amperes ÷ 240 volts = 64.2 amperes

Example No. 2(b). Optional Calculation for One-Family Dwelling, Air Conditioning Larger than Heating

[See Section 220-30(a) and Table 220-30.]

Dwelling has a floor area of 1500 sq. ft., exclusive of unoccupied cellar, unfinished attic, and open porches. It has two 20-ampere small appliance circuits, one 20-ampere laundry circuit, two 4-kW wall-mounted ovens, one 5.1-kW counter-mounted cooking unit, a 4.5-kW water heater, a 1.2-kW dishwasher, a 5-kW combination clothes washer and dryer, six 7-ampere 230-volt room air-conditioning units, and a 1.5-kW permanently installed bathroom space heater. Assume wall-mounted ovens, counter-mounted cooking unit, water heater, dishwasher, and combination clothes washer and dryer kW ratings equivalent to kVA.

Air Conditioning kVA Calculation:
 Total amperes 6 × 7 = 42.00 amperes
 42 × 240 ÷ 1000 = 10.08 kVA of air-conditioned load; assume P.F. = 1.0

Load included at 100%:
 Air Conditioning (see below)
 Space heater (omit, see Section 220-21)

Other Load:

1500 sq. ft. at 3 volt-amperes	4500 volt-amperes
Two 20-amp. small appliance circuits at 1500 volt-amperes	3000 volt-amperes
Laundry circuit	1500 volt-amperes
Two ovens	8000 volt-amperes
One cooking unit	5100 volt-amperes
Water heater	4500 volt-amperes
Dishwasher	1200 volt-amperes
Washer/dryer	5000 volt-amperes
Total other load	32,800 volt-amperes
1st 10 kVA at 100%	10,000 volt-amperes
Remainder at 40% (22.8 kVA × .4)	9120 volt-amperes
Total other load	19,120 volt-amperes
Other load	19,120 volt-amperes
Air conditioning	10,080 volt-amperes
	29,200 volt-amperes

29,200 volt-amperes ÷ 240 volts = 122 amperes (service rating)

Feeder Neutral Load, per Section 220-22:

(It is assumed that the two 4-kVA wall-mounted ovens are supplied by one branch circuit, the 5.1-kVA counter-mounted cooking unit by a separate circuit.)

1500 sq. ft. at 3 volt-amperes	4500 volt-amperes
Three 20-amp. circuits at 1500 volt-amperes	4500 volt-amperes
Total	9000 volt-amperes
3000 volt-amperes at 100%	3000 volt-amperes
9000 VA − 3000 VA = 6000 volt-amperes at 35%	2100 volt-amperes
	5100 volt-amperes

Two 4-kVA ovens plus one 5.1-kVA cooking unit totals
 13.1 kVA
 Table 220-19 permits 55% demand factor
 13.1 kVA × .55 = 7.2 kVA feeder capacity

7200 volt-amperes × 70% for neutral load	5040 volt-amperes
Clothes washer/dryer — 5 kVA × 70% for neutral load	3500 volt-amperes
Dishwasher	1200 volt-amperes
Total	14,840 volt-amperes

14,840 volt-amperes ÷ 240 volts = 61.83, use 62 amperes

Figure 6–4 *(continued)*

Example No. 2(c). Optional Calculation for One-Family Dwelling with Heat Pump Single-Phase, 240/120-Volt Service

(See Section 220-30.)

Dwelling has a floor area of 2000 sq. ft., exclusive of unoccupied cellar, unfinished attic, and open porches. It has a 12-kW range, a 4.5-kW water heater, a 1.2-kW dishwasher, a 5-kW clothes dryer, and a 2½-ton (24-ampere) heat pump with 15 kW of back-up heat.

Heat pump kVA is 24 × 240 ÷ 1000 = 5.76 kVA. 5.76 kVA is less than 15 kVA of the back-up heat; therefore, the heat pump load need not be included in the service calculation. (See Table 220-30.)

2000 sq. ft. at 3 volt-amperes	6000 volt-amperes
Two 20-ampere appliance outlet circuits at 1500 volt-amperes each	3000 volt-amperes
Laundry circuit	1500 volt-amperes
Range (at nameplate rating)	12,000 volt-amperes
Water heater	4500 volt-amperes
Dishwasher	1200 volt-amperes
Clothes dryer	5000 volt-amperes
	33,200 volt-amperes

First 10 kVA of other load at 100% = 10,000 volt-amperes
Remainder of other load at
  40% (23,200 volt-amperes × 0.4) = 9,280 volt-amperes
    Total of other load = 19,280 volt-amperes

Heat pump and supplementary heat*

240 × 24 = 5760 volt-amperes

15-kW electric heat: 5760 volt-amperes + 15,000 volt-amperes = 20,760 volt-amperes or = 20.76 kVA

20.76 kVA at 65% = 13.49 kVA

*If supplementary heat is not on at same time as heat pump, heat pump kVA need not be added to total.

Totals:

Other load	19,280 volt-amperes
Heat pump and supplementary heat	13,490 volt-amperes
Total	32,770 volt-amperes

32.77 kVA ÷ 240 volts = 136.5 amperes

Therefore, this dwelling unit shall be permitted to be served by a 150-ampere service.

Example No. 3. Store Building

A store 50 ft. by 60 ft., or 3000 sq. ft., has 30 ft. of show window. There are a total of 80 duplex receptacles. The service is 120/240-volt, single-phase (3-wire service). Actual connected lighting load: 8500 volt-amperes.

Computed Load (Section 220-10):
Noncontinuous Loads:

Receptacle Load (Section 220-13)

80 receptacles at 180 volt-amperes = 14,400 volt-amperes

10,000 VA at 100%	10,000 volt-amperes
(14,400 − 10,000) VA at 50%	2200 volt-amperes
Total	12,200 volt-amperes

Continuous Loads:

General Lighting Load:*

3000 sq. ft. at 3 volt-amperes per sq. ft	9000 volt-amperes
Show Window Lighting Load: 30 ft. at 200 volt-amperes per ft.	6000 volt-amperes
Outside sign circuit 1200 volt-amperes [Section 600-6(c)]	1200 volt-amperes
	16,200 volt-amperes
Total noncontinuous loads plus continuous loads	28,400 volt-amperes

Minimum Number of Branch Circuits Required:

General Lighting Load: Branch circuits need only be installed to supply the actual connected load [Section 220-4(d)].

8500 volt-amperes × 1.25 = 10,625 volt-amperes

10,625 volt-amperes ÷ 240 volts = 44 amperes for 3-wire, 120/240.

The lighting load shall be permitted to be served by 2-wire or 3-wire 15- or 20-ampere circuits with combined capacity equal to 44 amperes or greater for 3-wire circuits or 88 amperes or greater for 2-wire circuits. The feeder capacity as well as the number of branch-circuit positions available for lighting circuits in the panelboard must reflect the full calculated load of 9000 volt-amperes × 1.25 = 11,250 volt-amperes.

Figure 6–4 *(concluded)*

Example No. 8. Motors, Conductors, Overload, and Short-Circuit and Ground-Fault Protection

(See Sections 240-6, 430-6, 430-7, 430-22, 430-23, 430-24, 430-32, 430-34, 430-52, and 430-62 and Tables 430-150 and 430-152.)

Determine the conductor size, the motor overload protection, the branch-circuit short-circuit and ground-fault protection and the feeder protection, for one 25-horsepower squirrel-cage induction motor (full voltage starting nameplate current 31.6 amperes, service factor 1.15, Code letter F), and two 30-horsepower wound-rotor induction motors (nameplate primary current 36.4 amperes, nameplate secondary current 65 amperes, 40°C rise), on a 460-volt, 3-phase, 60-Hertz supply.

Conductor Loads

The full-load current value used to determine the ampacity of conductors for the 25-horsepower motor is 34 amperes [Section 430-6(a) and Table 430-150]. A full-load current of 34 amperes × 1.25 = 42.5 amperes (Section 430-22). The full-load current value used to determine the ampacity of primary conductors for each 30-horsepower motor is 40 amperes [Section 430-6(a) and Table 430-150]. A full-load primary current of 40 amperes × 1.25 = 50 amperes (Section 430-22). A full-load secondary current of 65 amperes × 1.25 = 81.25 amperes [Section 430-23(a)].

The feeder ampacity will be 125 percent of 40 plus 40 plus 34, or 124 amperes (Section 430-24).

Overload and Short-Circuit and Ground-Fault Protection

Overload. Where protected by a separate overload device, the 25-horsepower motor, with nameplate current of 31.6 amperes, shall have overload protection of not over 39.5 amperes [Sections 430-6(a) and 430-32(a)(1)]. Where protected by a separate overload device, the 30-horsepower motor, with nameplate current of 36.4 amperes, shall have overload protection of not over 45.5 amperes [Sections 430-6(a) and 430-32(a)(1)]. If the overload protection is not sufficient to start the motor or to carry the load, it shall be permitted to be increased according to Section 430-34. For a motor marked "thermally protected," overload protection is provided by the thermal protector [see Sections 430-7(a)(12) and 430-32(a)(2)].

Branch-Circuit Short-Circuit and Ground-Fault. The branch circuit of the 25-horsepower motor shall have branch-circuit short-circuit and ground-fault protection of not over 300 percent of a nontime-delay fuse (Table 430-152) or 3.00 × 34 = 102 amperes. The next smaller standard size fuse is 100 amperes (Section 240-6). Since a 100-ampere fuse is adequate to carry the load, Section 430-52(a), Exception No. 1 does not apply. If a time-delay fuse is to be used, see Section 430-52(a), Exception No. 2b. Where the 100-ampere fuse will not allow the motor to run, the value for a nontime-delay fuse shall be permitted to be increased to the next larger standard size, or 110 amperes. If these fuses are not sufficient to start the motor, the value for a nontime-delay fuse shall be permitted to be increased to 400 percent [See Section 430-52(a), Exception No. 2a.]

Feeder Circuit. The maximum rating of the feeder short-circuit and ground-fault protection device is based on the sum of the largest branch-circuit protective device (100-ampere fuse) plus the sum of the full-load currents of the other motors, or 100 + 40 + 40 = 180 amperes. The nearest standard fuse that does not exceed this value is 175 amperes [Section 430-62(a)].

Figure 6–5 (Reprinted with permission of NFPA 70-1993, the *National Electrical Code®*, Copyright © 1992, National Fire Protection Association, Quincy, MA 02269. This reprinted material is not the complete and official position of the National Fire Protection Association on the referenced subject, which is represented only by the standard in its entirety.)

Outside Feeders and Branch Circuits

Article 225 covers outside branch circuits and feeders. Article 225 covers the electrical equipment and wire for the supply of utilization equipment located on or attached to the outside of public and private buildings, or run between buildings, other structures or poles on premises served. For clearances, see Figure 6–6. An example of a feeder would be those conductors being fed from an overcurrent device in the serv-

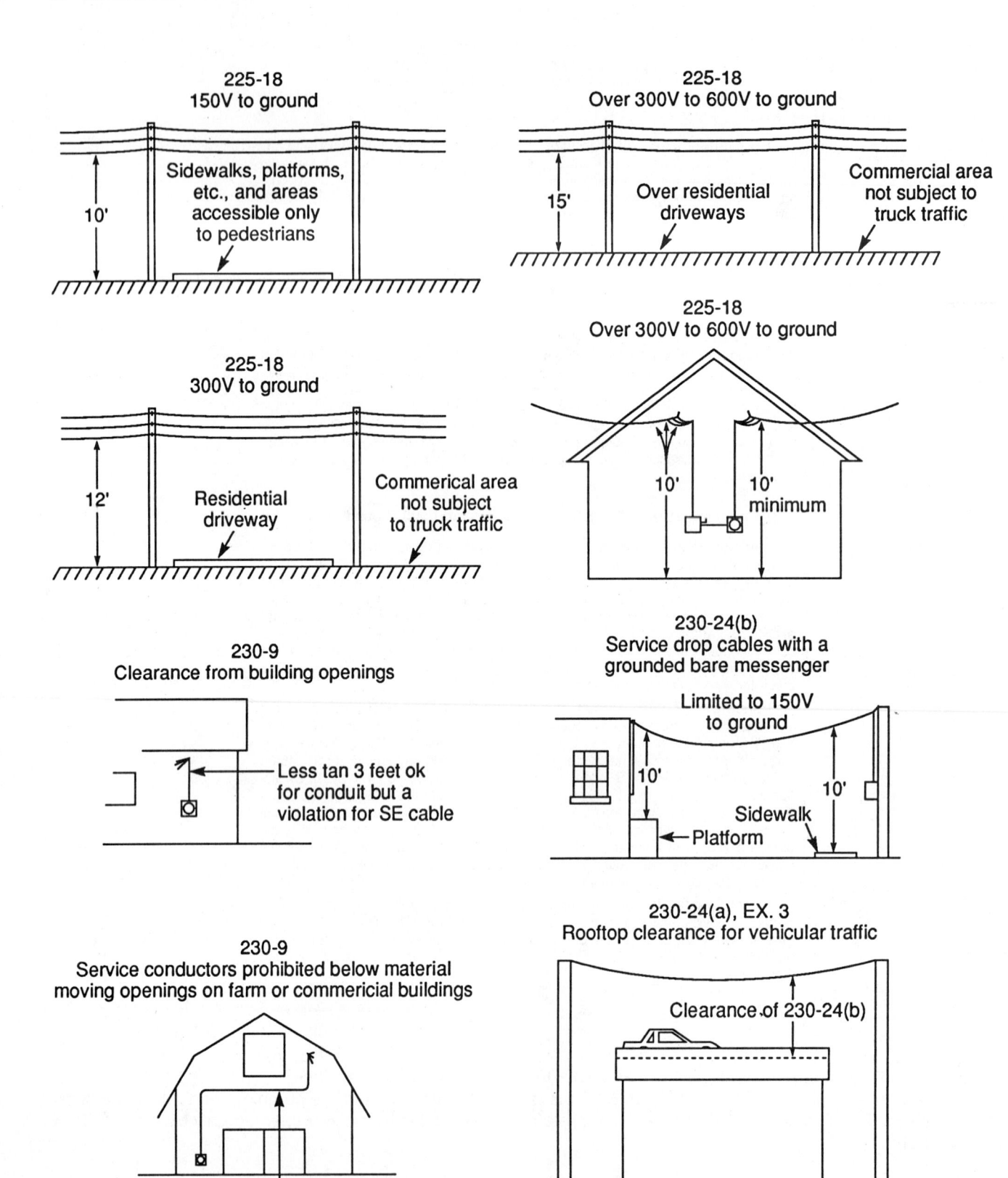

Figure 6–6 Required clearances for service and feeder conductors

ice equipment to a subpanel in another part of the building or feeding a separate building on the premises, terminating in a panel or group of overcurrent devices within the second building or in another location. A feeder also serves separately derived systems in many cases, such as a transformer where the voltage is reduced to serve lighting and receptacle branch circuit loads. The 1993 *NEC®* clarifies that the conductors feeding the second building were, in fact, feeder conductors by relocating Section 230-84 to a new section, 225-8. This new section provides that where more than one building or structure is on the same property and under single management, each building or other structure served shall be provided with a means of disconnecting all ungrounded conductors, which must be installed either inside or outside the building or structure in a readily accessible location nearest the point of entrance of the supply feeder conductors. The disconnect disconnecting these feeder conductors must be installed in accordance with Section 230-70 and 230-72 and must be suitable as service equipment. However, they are required to have overcurrent protection in accordance with Article 220 for branch circuits and Article 240 for feeders. The reference to Article 220 is necessary because residential buildings and structures such as garages and other outbuildings are permitted to be supplied with branch circuits where the loads permit. Examples for calculating feeders and outside branch circuits are shown in this chapter.

CHAPTER 6 QUESTION REVIEW

1. A 240/480 three-phase power panelboard supplies only one, 15,000-VA, 480-volt three-phase balanced resistive load. Each ungrounded conductor in the subfeeder to this power panel has a total net computed load of _____ amps.

    Answer: _____

    _____

    _____

    _____

    Reference: _____

2. Fluorescent lighting fixtures, each containing two ballasts rated 0.8 amperes each at 120 volts, are to be installed for general lighting in a store. The overcurrent protection devices are **not** listed for continuous operation at 100% of its rating. What is the maximum number of these lighting fixtures that may be permanently wired to a 20-amp, 120-volt branch circuit?

    Answer: _____

    _____

    _____

    _____

    Reference: _____

74 Master Electrician's Review

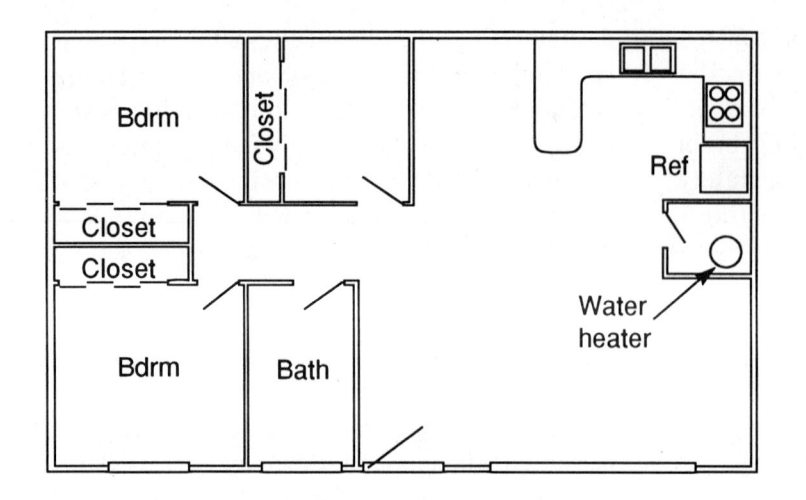

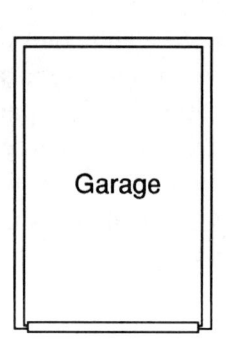

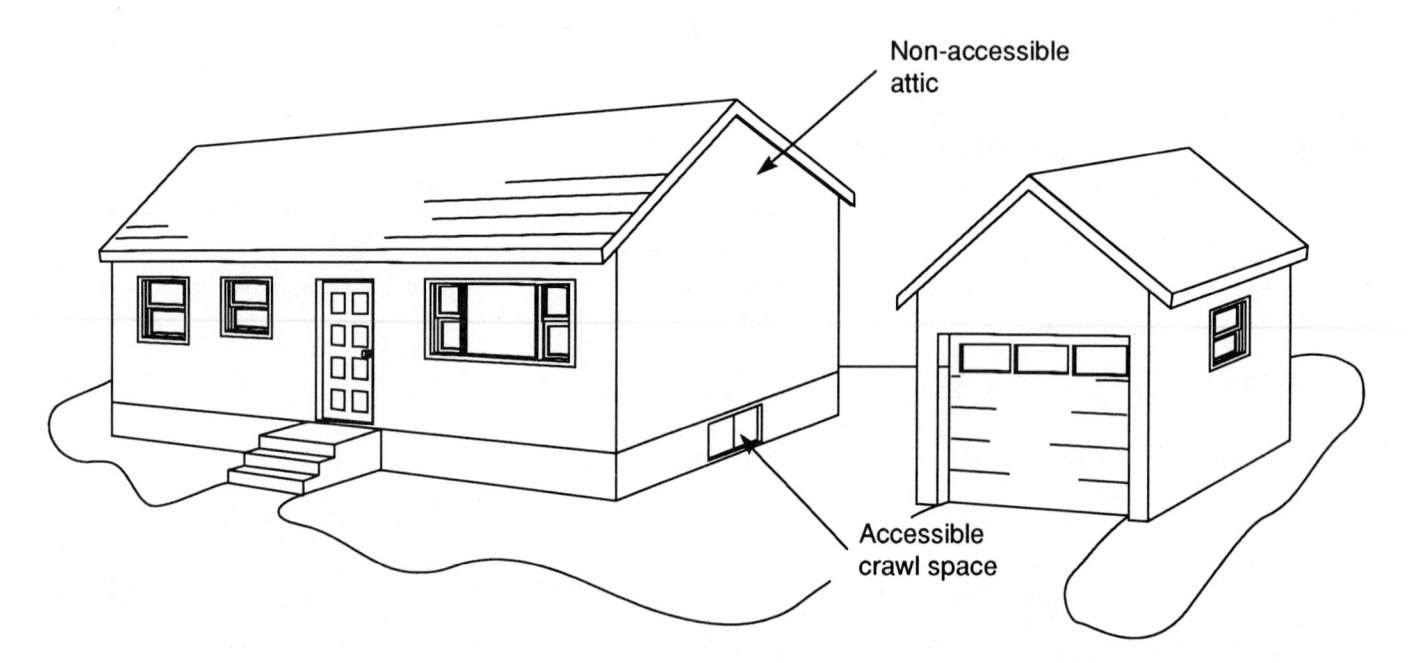

Example: You are going to build an investment single-family dwelling 28 feet by 40 feet. There will be three bedrooms and one bathroom with central natural gas heat located under the floor in the crawlspace. There will be no air conditioning. There will be an electric water heater (40 gallon, quick recovery) with two 4500 watt elements, a gas cooking range, and electric dryer. There will be a detached single-car garage. The electrical installation will be made in accordance with the 1993 *NEC*®. The attic in this dwelling is not suitable for the location of equipment or for storage.

3. How many square feet will this house contain?

   Answer: _____

   _____

   _____

   _____

   Reference: _____

4. How many 15 ampere general lighting and receptacle branch circuits are required?

   Answer: _____

   _____

   _____

   _____

   Reference: _____

5. How many 20 ampere lighting and receptacle branch circuits are required?

   Answer: _____

   _____

   _____

   _____

   Reference: _____

6. How many appliance branch circuits are required?

   Answer: _____

   _____

   _____

   _____

   Reference: _____

7. Is a special circuit required for the laundry?

   Answer: _____

   _____

   _____

   _____

   Reference: _____

8. What is the maximum standard size overcurrent device permitted to protect the electric water heater?

   Answer: _____

   _____

   _____

   _____

   Reference: _____

9. How many branch circuits are required to supply the detached garage?

   Answer: _____

   _____

   _____

   _____

   Reference: _____

10. How many outdoor receptacles are required?

    Answer: _____

    _____

    _____

    _____

    Reference: _____

11. What is the minimum number of ground-fault circuit interrupter (GFCI) devices required?

    Answer: _____

    _____

    _____

    _____

    Reference: _____

12. Are lighting outlets required at each outside door?

    Answer: _____

    _____

    _____

    _____

    Reference: _____

13. Are lighting outlets required in the attic not used for storage?

    Answer: _____

    _____

    _____

    _____

    Reference: _____

14. Is a receptacle required under the floor containing equipment requiring servicing?

    Answer: _____

    _____

    _____

    _____

    Reference: _____

15. If the answer to Question 7 above is yes, what size branch circuit would be required?

    Answer: _____

    _____

    _____

    _____

    Reference: _____

16. Would it be permissible to connect the furnace to a general lighting and receptacle branch circuit provided it does not overload the circuit?

    Answer: _____

    _____

    _____

    _____

    Reference: _____

17. Are lighting outlets required in each closet?

    Answer: _____

    _____

    _____

    _____

    Reference: _____

18. Is it permissible to connect the range to one of the required small appliance branch circuits?

    Answer: _____

    _____

    _____

    _____

    Reference: _____

19. If an outdoor receptacle is located 6 feet above the ground level, is it required to be GFCI protected?

Answer: _____

_____

_____

_____

Reference: _____

20. Could the outdoor receptacle described in Question 19 above serve as the required outdoor receptacle?

Answer: _____

_____

_____

_____

Reference: _____

21. What is the minimum size service required for this dwelling? (Calculate answers where required)
    (a) General purpose lighting and receptacle load
    (b) Appliance branch circuit load
    (c) Water heater
    (d) Laundry
    (e) Dryer
    (f) Detached garage
    (g) Furnace
    (h) Outdoor receptacle
    (i) Air conditioning
    (j) Range
    (k) Other

    (a) _____

    (b) _____

    (c) _____

    (d) _____

    (e) _____

    (f) _____

    (g) _____

    (h) _____

    (i) _____

    (j) _____

    (k) _____

Chapter Six 79

22. A branch circuit supplies a single, continuous duty, pump motor for a residential water supply. The motor circuit conductors must have an ampacity of at least ____ % of the motor full-load current rating.

   Answer: _____

   _____

   _____

   _____

   Reference: _____

23. The minimum lighting load required for the general lighting (only) of a church building having outside dimensions of 100 feet × 200 feet is ____ volt-amps.

   Answer: _____

   _____

   _____

   _____

   Reference: _____

24. For the kitchen small appliance load, a dwelling unit requires at least _____ circuit(s).

   Answer: _____

   _____

   _____

   _____

   Reference: _____

25. To lower the voltage drop in a wire, reduce its length or increase its _____

   Answer: _____

   _____

   _____

   _____

   Reference: _____

Chapter Seven

SERVICES 600 VOLTS OR LESS

NEC® Article 230 covers the requirements for services, service conductors, and equipment for the control and protection of services and their installation requirements. Parts A through G, cover services, 600 volts, nominal, or less, Part H has additional requirements for services exceeding 600 volts, nominal. It should be remembered that all of Article 230 applies to these services, and Part H is additional provisions that supplement or modify the rest of Article 230. Part H shall not apply to the equipment on the supply side of the service point. Clearances for conductors over 600 volts are found in ANSI C2, the *National Electrical Safety Code*. Section 230-2 generally limits a building or structure to one service. There are seven exceptions that permit more than one service per building or structure. It is necessary to comply with one of these exceptions any time two or more services are being installed on a building or structure. Confusing to many are the permissiveness for more than one set of service conductors permitted by Exceptions 1 and 2 of Section 230-40.

These exceptions frequently are used on strip shopping centers, condominium complexes, and apartment buildings. The rule satisfies the needs for individual supply, control, and metering of occupancies in these multi-occupancy buildings. However, it does not relieve the requirement of 230-2 for one service for each building or structure (see Figures 7–1 and 7–1A). The service conductors, as defined in Article 100, are required to go directly to the disconnecting means of that building or structure, as the disconnecting means is required to be installed at a readily accessible location either outside of the building or structure, or inside nearest the point of the entrance of the service conductors. Section 230-6 defines that conductors shall be considered outside of a building or other structure where installed under not less than two inches of concrete beneath a building or structure, where installed within a building or structure in a raceway that is encased in concrete or brick not less than two inches thick, or where installed in a transformer vault that complies with the requirements of *NEC®* Article 450, Part C. A typical service supplying a building or structure contains several carefully de-

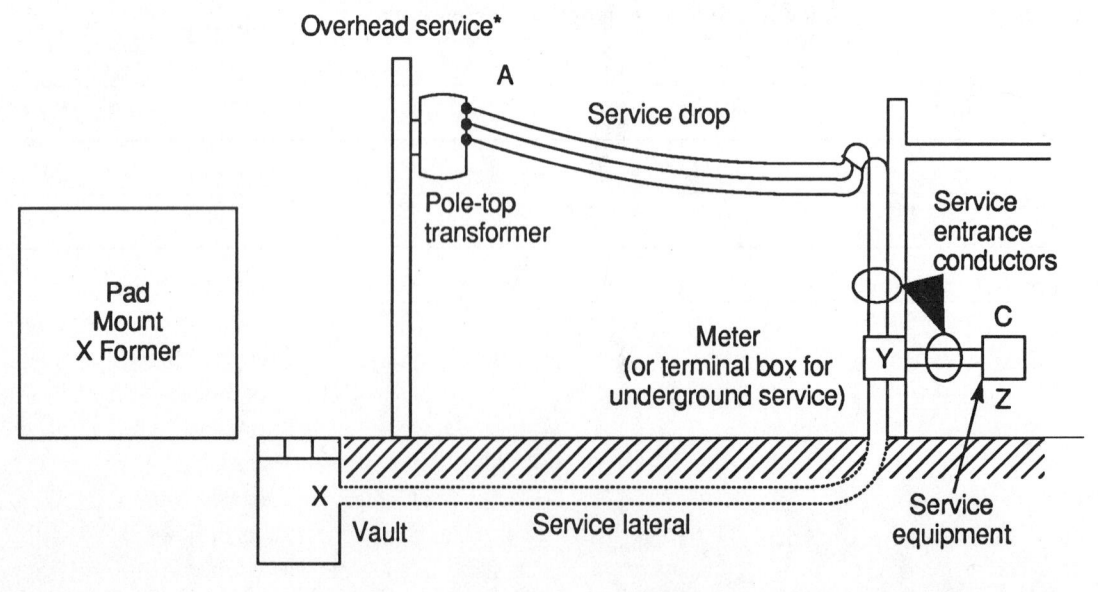

Figure 7–1 *NEC® Article 230-2* permits only one service per building. It is permissible to supply overhead or underground.

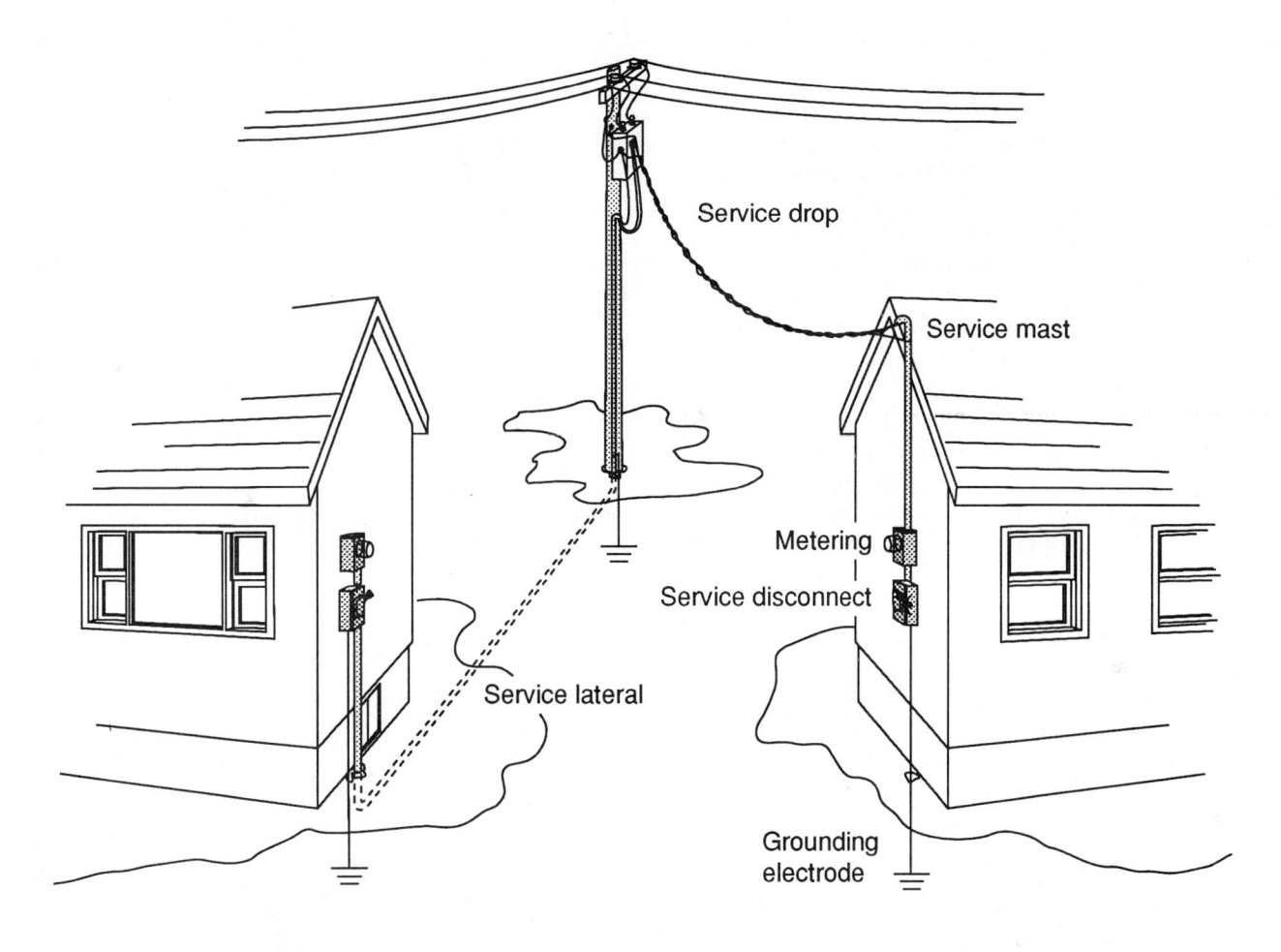

Figure 7–1A

fined parts. These parts are defined in Article 100 as service, service cable, service conductors, service drop, service entrance conductors, overhead system, service entrance conductors, underground system, service lateral, service equipment, and service point. Each service will contain several of these parts, however, not all of these parts. It is important that the components being installed in accordance with the *NEC®* be correctly defined before sizing, calculating, or installing. For instance, a typical residential service can be supplied by the serving utility either overhead or underground. When supplied overhead, it shall be supplied as a service drop. When supplied underground, it shall be supplied as a service lateral. These terms should be studied carefully in *NEC®* Article 100 so calculations and design can be made correctly. The diagram in Section 230-1 should help the reader identify and use these terms correctly. (See Figure 7–2.)

Part B of Article 230 covers the overhead service drop conductors, size and rating, and clearances required for making this installation. These conductors are generally installed by the serving utility; however, it is the responsibility of the installer to locate the service mast so these clearances will be met. Careful and close coordination between the serving utility, the customer to be served, and the installer are necessary to ensure compliance of this section of the *NEC®*. (See Figure 6–3 for clearance examples.)

Part C of Article 230 covers underground service laterals (see Figure 7–3). Care should be taken so that the service lateral is installed correctly and so that where it emerges from the ground it is amply protected by rigid or IMC steel conduit or Schedule 80 PVC to provide these conductors with physical protection against damage. Article 230, Part D covers the requirements for the service entrance conductors. These service entrance conductors, as defined in NEC Article 100, must be installed in one of the wiring methods listed in Section 230-43, and sized and rated in accordance with Section 230-42. Specific requirements for these service

Article 230 – Services

General	Part A
Overhead service drop conductors	Part B
Underground service-lateral conductors	Part C
Service-entrance conductors	Part D
Service equipment – General	Part E
Service equipment – Disconnecting means	Part F
Service equipment – Overcurrent protection	Part G
Services over 600 volts, nominal	Part H

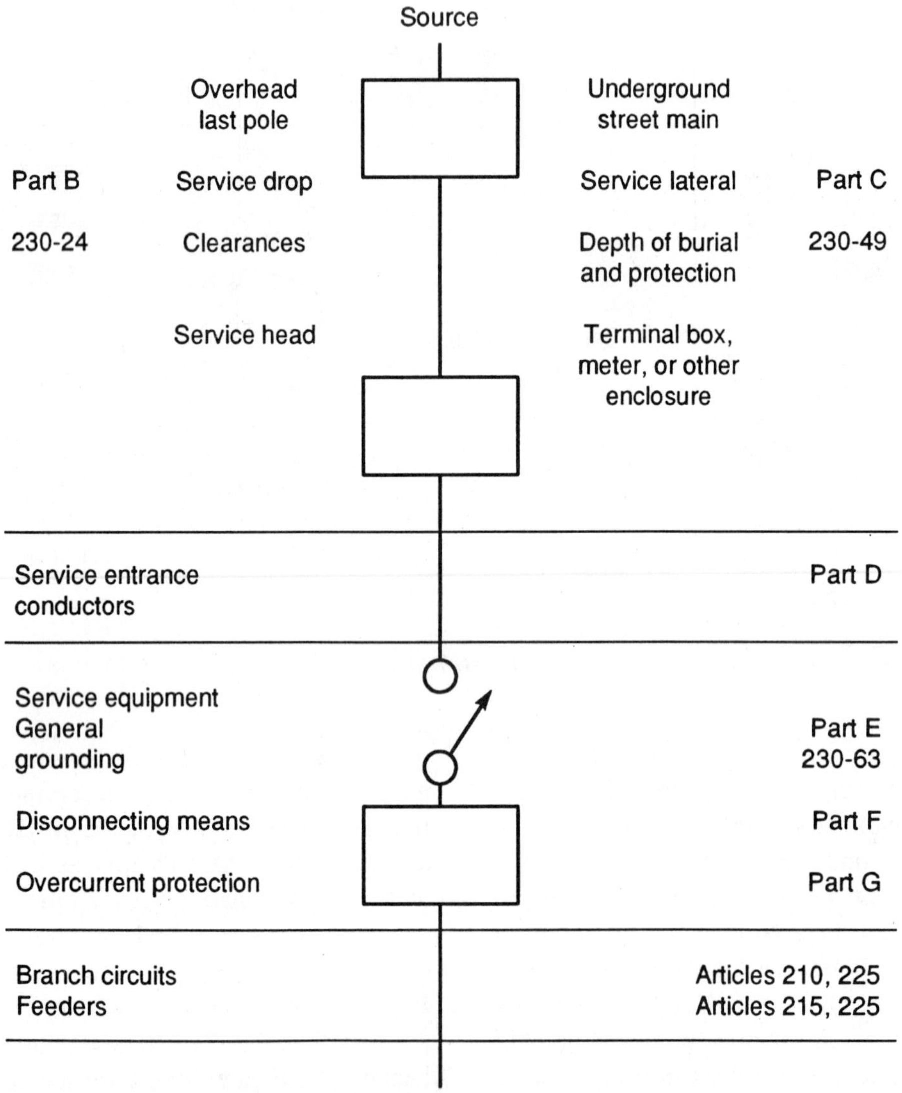

Figure 7–2 (Reprinted with permission of NFPA 70-1993, the *National Electrical Code®*, Copyright © 1992, National Fire Protection Association, Quincy, MA 02269. This reprinted material is not the complete and official position of the National Fire Protection Association on the referenced subject, which is represented only by the standard in its entirety.)

entrance conductors are covered in Article 230 Part D and must be complied with regardless of how they are routed. The general requirements for service equipment can be found in Article 230, Part E, and the disconnecting means is found in Part F.

Each service disconnecting means is required to be suitable for the prevailing equipment conditions, and when installed in hazardous locations must comply with Chapter 5 of the *National Electrical Code®*. Each service disconnecting means permitted by Section 230-

Electrical service components

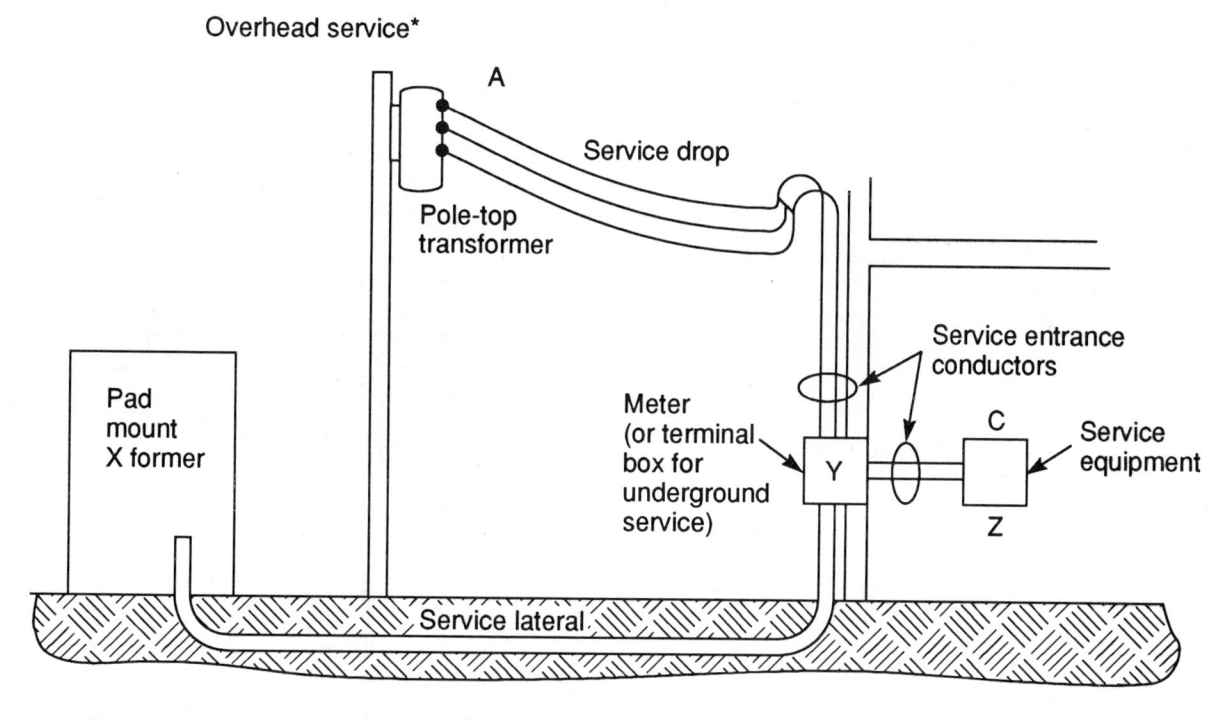

Figure 7–3 NEC® Article 230-2 permits only one service per building. It is permissible to supply overhead or underground.

2 or for each set of service entrance conductors permitted by Section 230-40, Exception 1 (see Figure 7–4) shall consist of not more than six switches or six circuit breakers mounted in a single enclosure, in a group of separate enclosures, or on a switchboard. (See Figure 7–5.) There shall be no more than six disconnects per service grouped in any one location. These requirements are critical and must be followed. The disconnecting means must be grouped; each disconnect must be marked to indicate the load served. An exception permits one of the six disconnecting means permitted where used for a water pump intended to provide fire protection to be located remotely from the other disconnecting means. In multiple-occupancy buildings, each occupant shall have access unless local building management, which supplies continuous supervision, is present. In such a case, the service disconnecting means shall be permitted to be accessible to the authorized management personnel only. Each service disconnecting means shall simultaneously disconnect the ungrounded service conductors from the premise wiring. Where the service disconnecting means does not disconnect the grounded conductor from the premise wiring, other means shall be provided for this purpose in the service equipment. A terminal or bus to which all grounded conductors can be attached by means of pressure connections shall be permitted for this purpose. Part G covers the overcurrent protection and the location of the overcurrent protection requirements for services. All solidly grounded wye services of more than 150 volts to ground but not exceeding 600 volts phase-to-phase rated 1000 amps or more must be provided with ground-fault protection of equipment. Two exceptions to this are: Exception 1 states ground-fault protection does not apply to the service disconnecting means of a continuous industrial process where a nonorderly shutdown will introduce additional or increased hazard. Exception 2 states that it does not apply to fire pumps. Specific settings and testing procedures for this ground-fault protection of equipment are covered in Section 230-95. Additional ground-fault protection of feeders where ground-fault protection has not been provided on the service is covered in both Articles 215 and 240.

Example: Where a sealed conduit contains either 3/0 AWG THWN insulated copper conductors with two of the 3/0 AWG conductors being in parallel for each phase, and two 3/0 AWG conductors being in parallel for the neutral, (a) would this installation be permitted to be terminated into a 400-ampere main service breaker where the system voltage is 120/208? (The expected loads are evenly

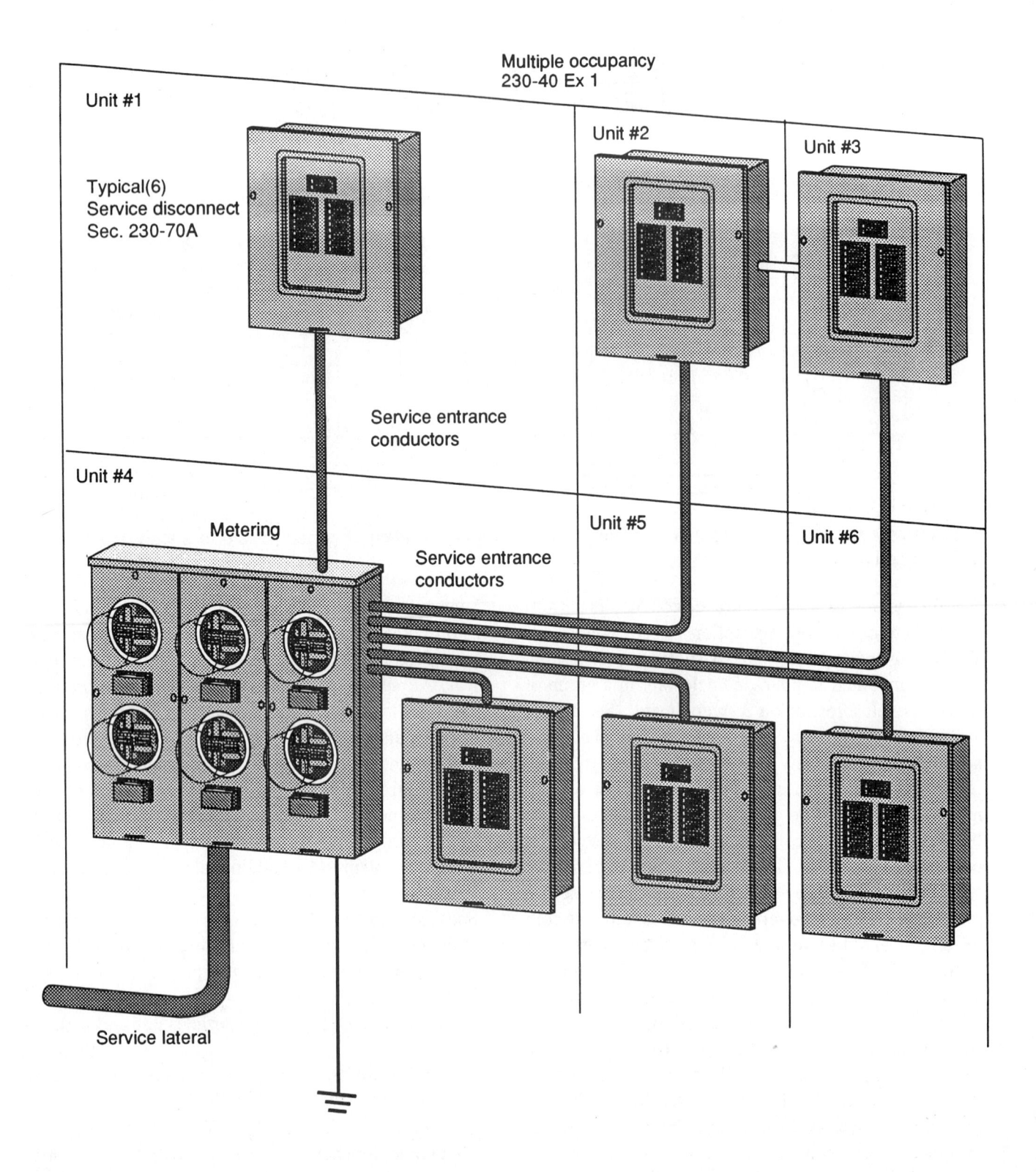

Figure 7–4

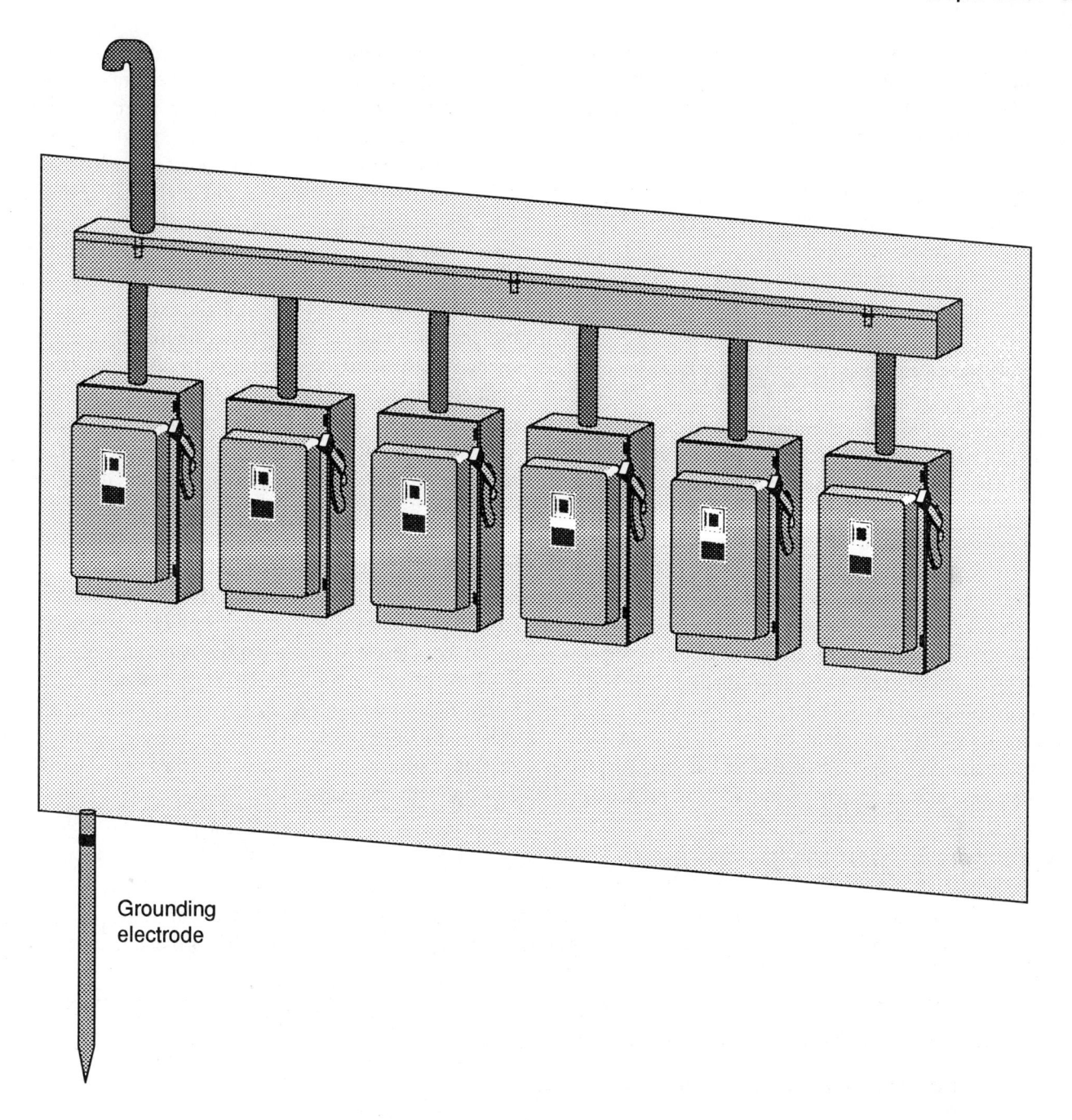

Figure 7–5

Grounding electrode

divided between fluorescent lighting, data processing equipment, and air conditioning equipment.) (b) If not, what would be the maximum ampere rating for this service?

Answer: 8 No. 3/0 THWN conductors in a single conduit, Chapter 9, Table 3B, would require a 3-inch conduit or tubing.

Table 310-16 3/0 THWN has an ampacity of 200 amperes (from 75°C column). Notes to Ampacity Tables 0–2000 Volts: Where the number of current-carrying conductors exceed three, the allowable ampacity shall be reduced as shown in the table with seven to nine conductors, a derating of 70% shall be applied. See Section 110-14(c).

Two conductors per phase:

$2 \times 200 \times 70\% = 280$ amperes.

The next higher overcurrent device per Section 240-3 would be 300 amperes. The service would be a 300-ampere service.

The 400 amperes overcurrent device would require 2-350 kcmil per phase and with eight No. 350 THWN would require the conduit or tubing to be increased to 4 inches.

For services exceeding 600 volts, nominal, Part H must be read and adhered to carefully. The definition of service-point formerly appeared in this section. The several examples below will help clarify the application of *NEC®* Article 230.

SERVICES OVER 600 VOLTS

Services over 600 volts nominal introduce an additional level of hazard. Article 230 Part H covers the requirements for services exceeding 600 volts nominal, which modify or amend the rest of Article 230. All of 230 is applicable and, in addition, Part H must be followed. Clearance requirements for over 600 volts services can be governed by ANSI C2, the *National Electrical Safety Code*, as noted in the fine print note in Section 230-200. The service entrance conductors to buildings or enclosures shall not be smaller than No. 6 unless in cable, and in cable not smaller than No. 8, and installed by the specific methods listed in Section 230-202(b). It should be noted that Article 710 and Article 110, Part B also will need to be regularly referenced in making installations over 600 volts. The requirements for support guarding and draining cables are covered in this section. Warning signs with the words "Danger. High Voltage. Keep Out." shall be posted where unauthorized people come in contact with energized parts. The disconnecting means must comply with Section 230-70 or 230-208(b) and shall simultaneously disconnect all underground conductors and shall have fault closing rating of not less than the maximum short circuit available in supply terminals, except where used switches or separate mining fuses are installed, the fuse characteristic shall be permitted to contribute to the fault closing rating of the disconnecting means. These requirements for over 600 volt services create the need for your studies to take you into new areas of the Code, such as Part B of Article 100, which covers the general requirements for over 600 volts nominal; Part B of Article 100, which gives you the unique definitions for those circuits; Part H of Article 240; Part M of Article 250 for grounding of system; and Part B of Article 300; and the requirements of 710. You must be aware that these additional requirements amend or augment the primary requirements in each article. In most instances, those requirements also apply to over 600-volt nominal services. Therefore, as you study this section, familiarize yourself with these new parts as they apply to your studies.

CHAPTER 7 QUESTION REVIEW

1. Where a circuit breaker is provided as the short-circuit protective device for service-entrance conductors exceeding 600 volts, the breaker shall have a trip setting of not more than ____ times the ampacity of the conductors.

    Answer: _____

    _____

    _____

    _____

    Reference: _____

2. An electrical service transmission line drops the voltage from 255 volts to 240 volts. The service transmission line efficiency is ____ %.

    Answer: _____

    _____

    _____

    _____

    Reference: _____

3. Where a group of buildings are served by electrical feeders from a single service drop or electrical service point in another building, is a disconnecting means required at each of the other buildings where the feeders terminate or can the feeders terminate in main lug only panels?

   Answer: _____

   _____

   _____

   _____

   Reference: _____

4. Is there a limit to the number of overcurrent devices that can be used on the secondary side of a transformer if the total of the device ratings do not exceed the allowed value for a single secondary overcurrent device?

   Answer: _____

   _____

   _____

   _____

   Reference: _____

5. Does it make a difference which opening in the service head that the service entrance conductors are brought out of? If so, why?

   Answer: _____

   _____

   _____

   _____

   Reference: _____

6. If the clearances required by Section 110-16 can be maintained in front of a panelboard or a service panel to be installed in a bathroom in a dwelling, would this installation be permitted by the *NEC®*?

   Answer: _____

   _____

   _____

   _____

   Reference: _____

7. When service entrance cable is used within a dwelling for branch circuit wiring, is the SEC type cable considered the same as Type NM nonmetallic sheath cable?

   Answer: _____

   _____

   _____

   _____

   Reference: _____

8. Can a 120/240 volt single-phase, three-wire service ground be located closer than 3 feet to a window of a dwelling?

   Answer: _____

   _____

   _____

   _____

   Reference: _____

9. The Code generally prohibits splicing a grounding electrode conductor. (a) Is this conductor permitted to be tapped? (b) If so, how are these taps sized?

   Answers: _____

   _____

   _____

   _____

   Reference: _____

10. Circuit breakers are required to open all ungrounded conductors of a circuit simultaneously. Obviously, fuses cannot be required to open all ungrounded conductors of a circuit simultaneously. Why the difference in the requirements?

    Answer: _____

    _____

    _____

    _____

    Reference: _____

11. What is a separately derived system? Where is it mentioned in the Code?

    Answer: _____
    _____
    _____
    _____

    Reference: _____

12. How many overcurrent devices are permitted to be used on the secondary of a transformer if the total of the device ratings do not exceed the allowed value for a single overcurrent device?

    Answer: _____
    _____
    _____
    _____

    Reference: _____

13. Is it necessary to identify the higher voltage-to-ground phase (hi-leg) at the disconnect of a three-phase motor where the service available is 120/240 volt, three-phase, four-wire?

    Answer: _____
    _____
    _____
    _____

    Reference: _____

14. A service panel is securely bolted to the metal frame of a building that is effectively grounded, and the panelboard's main bonding jumper is properly installed. Is this sufficient to ground the panelboard to the metal building grounding electrode as required in Section 250-81(b)?

    Answer: _____
    _____
    _____
    _____

    Reference: _____

15. What is the maximum permitted contraction or expansion for a run of Schedule 40 PVC conduit before the installation of an expansion coupling is required?

    Answer: _____
    _____
    _____
    _____

    Reference: _____

90 Master Electrician's Review

16. How can one distinguish between the mandatory rules of the *NEC®* and the explanatory rules of the *NEC®*?

    Answer: _____

    _____

    _____

    _____

    Reference: _____

17. On a recent installation, the plans called for the building steel to be used as a grounding electrode if effectively grounded. What does this term mean, and where is it used in the *NEC®*?

    Answer: _____

    _____

    _____

    _____

    Reference: _____

18. What section of the *National Electrical Code®* covers Romex?

    Answer: _____

    _____

    _____

    _____

    Reference: _____

19. What is the maximum size flexible metallic tubing available manufactured today?

    Answer: _____

    _____

    _____

    _____

    Reference: _____

20. Which article of the *National Electrical Code®* contains the requirements for fusible safety switches and other type snap switches?

    Answer: _____

    _____

    _____

    _____

    Reference: _____

21. In making an installation in a residential garage, you have determined that physical protection is needed for the nonmetallic sheath cable and that it must be installed in steel conduit. How do you size the conduit for this nonmetallic sheath cable?

    Answer: _____

    _____

    _____

    _____

    Reference: _____

22. In making an installation for a large parking garage adjacent to a hotel building, which article of the *National Electrical Code®* covers these wiring methods?

    Answer: _____

    _____

    _____

    _____

    Reference: _____

23. You have encountered a nonmetallic wireway to be installed in a wet location. Which section of the *National Electrical Code®* covers this nonmetallic wireway?

    Answer: _____

    _____

    _____

    _____

    Reference: _____

24. In attempting to calculate the load on a small office building, the receptacles are to be calculated at how many volt-amperes each?

    Answer: _____

    _____

    _____

    _____

    Reference: _____

25. In wiring a new residence, the most convenient location for the lighting and branch circuit panelboard is on a bathroom wall. Is this an acceptable wiring method? If not, what section governs this requirement?

    Answer: _____

    _____

    _____

    _____

    Reference: _____

Chapter Eight

Figure 8–1 Panelboards *(Courtesy of Square D Company)*

OVERCURRENT PROTECTION

NEC® Article 240 provides the general requirements for overcurrent protection and overcurrent protective devices not more than 600 volts, nominal. Part H of *NEC*® Article 240 covers the overcurrent protection for over 600 volts, nominal. The overcurrent protection for conductors and equipment is provided to open the circuit if the current reaches a value that will cause an excessive or dangerous temperature in conductors or conductor insulation. Sections 110-9 and 110-10 cover the requirements for the interrupting capacity and protection against fault currents. Article 240 covers the general requirements. The specific overcurrent requirements for equipment can be found in each individual article as it pertains to that specific equipment or circuitry, for example, the overcurrent protection for air conditioning and refrigeration equipment is covered in Article 440 and appliances in Article 422. A listing of the specific overcurrent requirements can be found in Section 240-2. Section 240-3 outlines the protection of conductors. This section provides the rules for the many different applications throughout the Code. Section 240-4 covers the requirements for protecting flexible cords and fixture wires, and standard overcurrent devices are listed in Section 240-6 for fuses and fixed-trip circuit breakers. See Figures 8–2 and 8–3 for examples. Adjustable trip circuit breakers are described and covered in Section 240-6(b). Part B of Article 240 specifies the location that the overcurrent device must be placed in the circuit. Part C specifies the enclosure required for enclosing the overcurrent devices. Part D covers the disconnecting requirements for disconnecting the overcurrent devices so they are accessible to the maintenance personnel performing work on those devices. Parts E and F cover fuses, fuseholders, and adapters for both plug and cartridge type fuses, and circuit breakers are covered in Part G. Article 240, Part H covers the overcurrent protection for over 600 volt, nominal, and states that feeders shall have short-circuit protection devices in each ungrounded conductor or comply with Section 230-208(d)(2) or (d)(3). The pro-

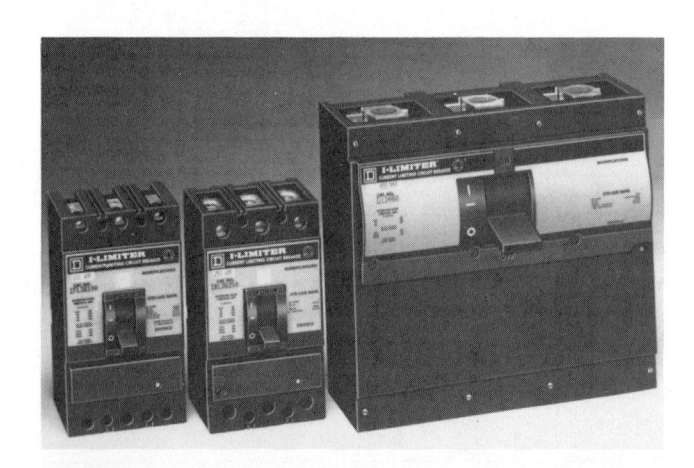

Figure 8–2 Types IFL, IKL, and ILL "I LIMITER" molded case current-limiting circuit breakers *(Courtesy of Square D Company)*

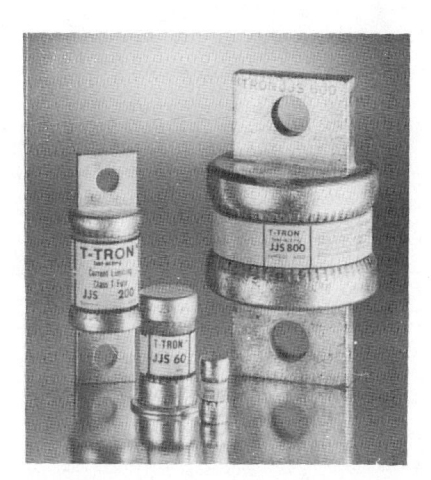

Figure 8–3 Class T current-limiting, fast-acting fuse; 200,000-ampere interrupting rating, 300 and 600 volts. Has little time delay. Generally used for protection of circuit breaker panels and for circuits that do not have high inrush loads, such as motors. When used on high-inrush loads, generally size at 300% so as to be able to override the momentary inrush current. These fuses have different dimensions compared with ordinary fuses. They will not fit in switches made for other classes of fuses. Nor will other classes of fuses fit into a Class T disconnect switch. *(Courtesy of Bussmann, Cooper Industries)*

tective device(s) shall be cpable of detecting and interrupting all values of current that can occur at their location in excess of their trip setting or melting point. In no case shall the fuse rating exceed three times the long-time trip element setting of the breaker, or six times the ampacity of the conductor. Branch circuit requirements for over 600-volt circuits are covered in Section 240-101.

Example: Provide the fuse sizes for a 10-horsepower, 230-volt, three-phase motor code lettered G, full voltage start with a service factor of 1.15.

(A) The fuse sizes using time-delay fuses for the branch circuit protection is calculated as follows: $28 \times 1.75 = 49$ amperes (Section 430-52 requires that you round down to the next lower standard size. See Section 240-6.) Therefore, install 45 ampere time-delay fuses. If the 45 ampere time-delay fuses are not sufficient to carry the load or to start the motor, a larger size time-delay fuse may be installed, but not to exceed 225%.

(B) Using time-delay fuses for motor overload protection, the size will be as follows: $28 \times 1.25 = 35$ ampere time-delay fuses.

(C) Using nontime-delay fuses, sizes for branch circuit and ground fault protection are calculated as follows: $28 \times 3 = 84$ amperes. Round down to 80 ampere nontime-delay fuses as explained in the example above. This would permit a maximum size of $28 \times 4 = 112$. Install 110 ampere nontime-delay fuses. References for making these calculations are Tables 430-150 and 152, Sections 240-6, 430-32, and 430-52.

Example: To calculate the available short-circuit current at a panelboard located 20 feet away from a transformer when the transformers are 500 kcmil copper in steel conduit. The (c) value for the conductors is 22,185. The transformer is marked 300 kVA, 208/120 volts, three-phase, four-wire. The transformer impedance is 2%. Consider the source to have an infinite amount of fault current available (often referred to as "infinite primary"). Find the transformer's full-load secondary current,

$$I_{fla} = \frac{kVA \times 1000}{E \times 1.73} = \frac{300 \times 1000}{208 \times 1.73} = 834 \text{ amperes}$$

(2) To find the transformer SCA: multiplier = 100/2 = 50.

Transformer SCA =

$I_{fla} \times \text{multiplier} = 834 \times 50 = 41{,}700$ amperes.

(3) To find the (F) factor:

$$F = \frac{1.73 \times L \times I}{C \times E_{L-L}} = \frac{1.73 \times 20 \times 41{,}700}{22{,}185 \times 208} = 0.3127$$

(4) To find the (M) multiplier:

$$M = \frac{1}{1+F} = \frac{1}{1+0.3127} = 0.76$$

(5) To find the short-circuit current of the panelboard:

$I_{sca} = \text{transformer}_{sca} \times (M) = 41{,}700 \times 0.76 = 31{,}692$ amperes.

The fault current available at the panelboard where it is located 20 feet away from the transformer is equal to 31,692 amperes.

CHAPTER 8 QUESTION REVIEW

1. Does the Code permit the use of two single-pole circuit breakers in a panelboard to serve a line-to-line connected load such as a household electrical range rated at 120/240 volts or a hot water heater rated at 240 volts?

   Answer: _____

   _____

   _____

   _____

   Reference: _____

2. Where outdoor conductors are tapped and these tapped conductors terminate into a single overcurrent device designed to limit the load so as not to exceed the ampacity of the tap conductors, how long can these tap conductors be at the maximum length?

   Answer: _____

   _____

   _____

   _____

   Reference: _____

3. Is it permissible to use 300-volt cartridge type fuses and fuse holders to protect a multi-wire, four-wire circuit such as an emergency circuit in a food store where the electrical service is 277/480 volt, three-phase, four-wire?

   Answer: _____

   _____

   _____

   _____

   Reference: _____

4. Service entrance conductors are required to be protected at their rated ampacity with exception, the exception being the next standard size overcurrent device when the service conductors are protected by a single device. When are multiple overcurrent devices supplied by a service allowed to exceed the ampacitor rating of the service entrance conductors?

   Answer: _____

   _____

   _____

   _____

   Reference: _____

5. Fixed electric space heating loads shall be computed at _____ % of the total connected load, generally.

   Answer: _____

   _____

   _____

   _____

   Reference: _____

6. A motor is to be installed in a Class I, Division 2 location. It is to be connected with 3 feet of ¾ inch liquidtight flexible metal conduit. Is it permissible to use the liquidtight flexible metal conduit as the grounding path for this installation?

   Answer: _____

   _____

   _____

   _____

   Reference: _____

7. Are feeder conductors required to be rated for 125% of the continuous load plus the noncontinuous load?

   Answer: _____

   _____

   _____

   _____

   Reference: _____

8. How many duplex receptacles can be installed on a 20-ampere branch circuit in a dwelling?

   Answer: _____

   _____

   _____

   _____

   Reference: _____

96 Master Electrician's Review

9. Table 310-16 lists the allowable ampacity of No. 12 THWN conductors as 25 amperes. Can 16 general use receptacles in an office building be connected to a 20-ampere circuit?

   Answer: _____

   _____

   _____

   _____

   Reference: _____

10. Should you use Article 310 or Appendix B to calculate underground conductor ampacities?

    Answer: _____

    _____

    _____

    _____

    Reference: _____

11. When an electric heat pump is installed with backup resistance heat in a building, (a) could one of the loads ever be considered dissimilar for the purposes of calculating the service loads? (b) Would the heat pump load have to be calculated at 125% of the service sizing?

    Answers: _____

    _____

    _____

    _____

    Reference: _____

12. Are all 20-ampere residential underground circuits required to have GFCI protection?

    Answer: _____

    _____

    _____

    _____

    Reference: _____

13. Can two circuit breakers mounted adjacent to each other in a panelboard with handle ties be used in place of a double-pole breaker to supply a 240-volt electric baseboard heater?

    Answer: _____

    _____

    _____

    _____

    Reference: _____

14. Are water heaters considered to be a continuous load so that a 125% load is required for calculating feeder and service on an installation?

    Answer: _____
    _____
    _____
    _____

    Reference: _____

15. Do the tap conductors from the secondary of a transformer always require secondary overcurrent protection? How about a 10-foot tap? A 25-foot tap?

    Answer: _____
    _____
    _____
    _____

    Reference: _____

16. Is it permitted to plug a microwave oven with a nameplate rating of 13 amperes into a 15-ampere receptacle protected on a 20-ampere circuit?

    Answer: _____
    _____
    _____
    _____

    Reference: _____

17. Can a 1400-VA load, a cord, and plug connected load, be supplied by a 240-volt circuit in a residence?

    Answer: _____
    _____
    _____
    _____

    Reference: _____

18. Three receptacles on a single yoke or strap are to be installed in a commercial or industrial facility. Are these outlets to be calculated at 180 VA or 540 VA?

    Answer: _____
    _____
    _____
    _____

    Reference: _____

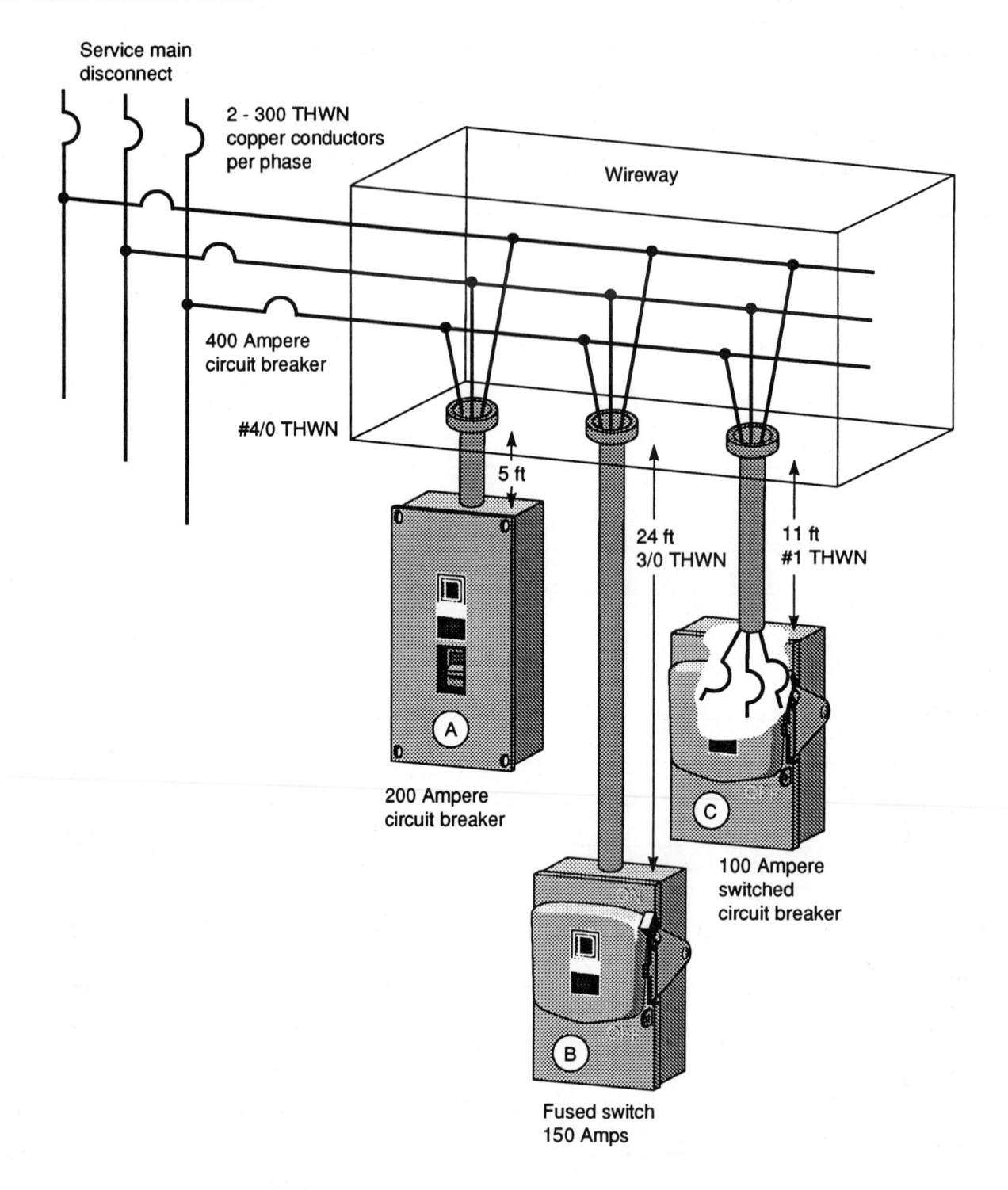

19. In the diagram above, does the circuit breaker marked A comply with the *NEC®* If so, which section?

Answer: _____

_____

_____

_____

Reference: _____

20. In the diagram above, does the switch B comply with the *NEC®*? If so, which section?

    Answer: _____

    _____

    _____

    _____

    Reference: _____

21. Does the switch circuit breaker marked C in the diagram above comply with the *NEC®*?

    Answer: _____

    _____

    _____

    _____

    Reference: _____

22. *NEC®* requires overcurrent protection where busway is reduced in size for the last 50 feet or more in other than industrial installations. (True or False)

    Answer: _____

    _____

    _____

    _____

    Reference: _____

23. A 10-foot tap is made in the field without overcurrent protection where the smaller conductor is tapped to the larger conductor. The conductors extend beyond the switchboard. The tap conductor must be:
    (A) at least $\frac{1}{10}$ the size of the conductor to which it is tapped.
    (B) at least $\frac{1}{10}$ the ampere rating of the overcurrent device that protects the conductors from which the tap is made.

    Answer: _____

    _____

    _____

    _____

    Reference: _____

24. In an extremely long feeder circuit, the circuit conductors have been increased in size to compensate for voltage drop. Is the equipment grounding conductor also required to be increased?

    Answer: _____

    _____

    _____

    _____

    Reference: _____

25. The nameplate on an air conditioning unit is marked "maximum size fuse 40 amperes." Is it permissible to install a 40-ampere circuit breaker, or must a 40-ampere fuse protection be installed?

Answer: _____

_____

_____

_____

Reference: _____

Chapter Nine

GROUNDING

NEC® Article 250 covers the general requirements and many specific requirements for grounding, such as systems, circuits, and equipment required, permitted, and/or not permitted to be grounded, circuit conductors to be grounded on grounded systems, location of the grounding connections, type and sizes of grounding and bonding conductors and electrodes, the methods of grounding and bonding, and the conditions under which guards, isolation, or insulation can substitute for grounding. Other applicable articles applying to particular cases of installations of conductors and equipment are found in *NEC®* Section 250-2. (See Figure 9–1.) The purpose of grounding can best be stated by fine print notes No. 1 and No. 2 in Section 250-1: *(FPN No. 1): Systems and circuit conductors are grounded to limit voltages due to lightning, line surges, or unintentional contact with higher voltage lines, and to stabilize the voltage to ground during normal operation. Equipment grounding conductors are bonded to the system grounded conductor to provide a low-impedance path*

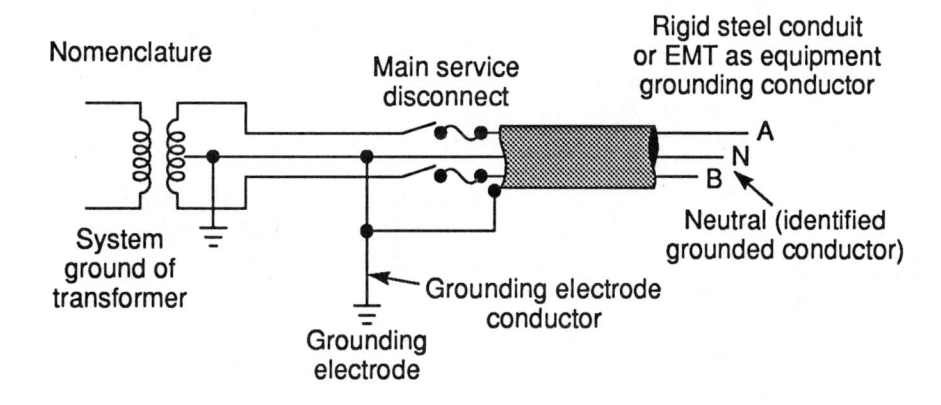

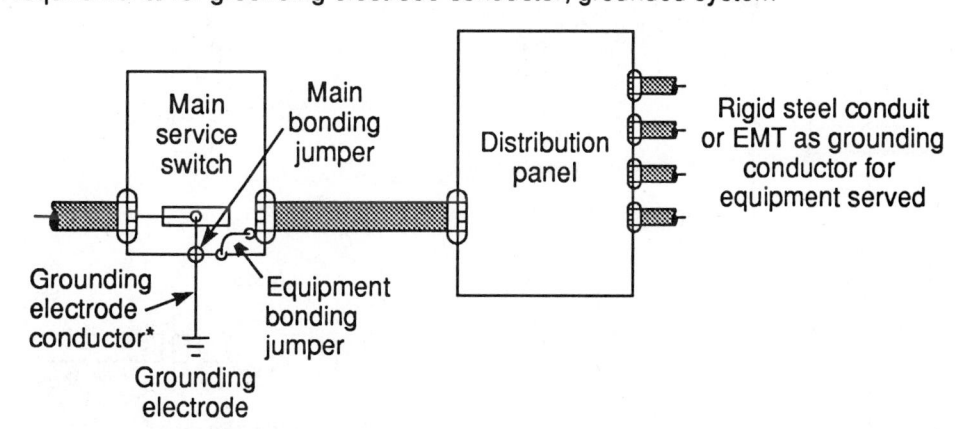

*Grounds both the identified grounded conductor and the equipment grounding conductor. It is connected to the grounded metalic cold water piping on premises, or other available grounding electrode, such as metal building frame if effectively grounded, or a made electrode.

Figure 9–1 Grounding principles *(Courtesy of American Iron & Steel Institute)*

for fault current that will facilitate the operation of overcurrent devices under ground-fault conditions. (FPN No. 2): Conductive materials enclosing electrical conductors or equipment, or forming part of such equipment, are grounded to limit the voltage to ground on these materials and bonded to facilitate the operation of overcurrent devices under ground-fault conditions. See Section 110-10. (Reprinted with permission of NFPA 70-1993, the *National Electrical Code®*, Copyright © 1992, National Fire Protection Association, Quincy, MA 02269. This reprinted material is not the complete and official position of the National Fire Protection Association on the referenced subject, which is represented only by the standard in its entirety.)

NEC® Section 250-3 for direct current systems, Section 250-5 for alternating current circuits and systems (see Figure 9–2), and Section 250-6 for portable and vehicle mounted generators outline the conditions for which these systems and circuits are to be grounded or can be grounded. Section 250-7 covers the circuits not to be grounded. Part C of Article 250 covers the locations for grounding connections. Sections 250-23 and 250-24 cover the requirements necessary for grounding buildings, single or locations with two or more buildings fed from a common service. Section 250-26 outlines the requirements for grounding a separately derived alternating-current system. Article 250, Part D covers the requirements for grounding the enclosures, Part E the equipment grounding requirements, Part F covers the methods for grounding, and Part G the bonding requirements. (See Figures 9–3 and 9–4.)

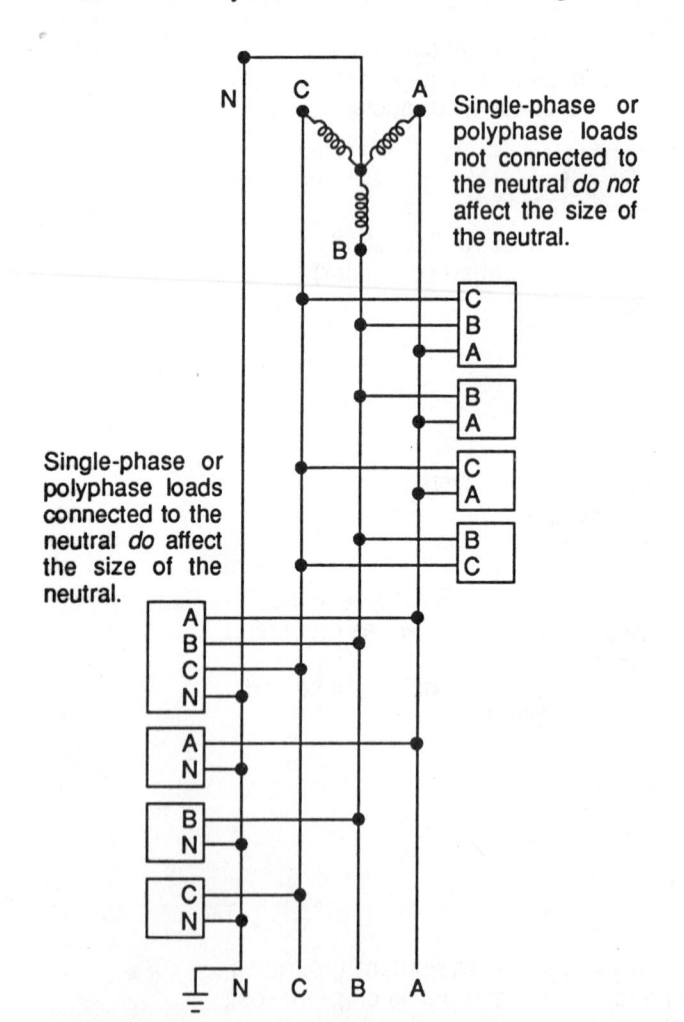

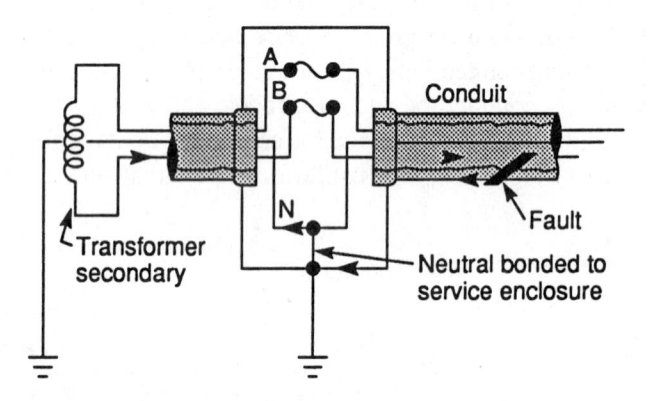

Figure 9–3 Bonding at service disconnect enclosure as required in *NEC®* Section 250-71 *(Courtesy of American Iron & Steel Institute)*

Figure 9–2 Size of neutral conductor as in *NEC®* Section 220-4(e) *(Courtesy of American Iron & Steel Institute)*

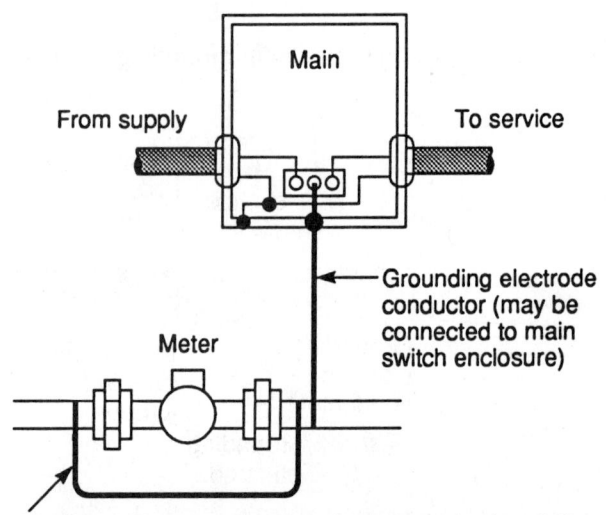

Figure 9–4 Bonding jumpers as in *NEC®* Article 250, G *(Courtesy of American Iron & Steel Institute)*

Part G clearly delineates between service equipment and other than service equipment. Part H clearly describes the grounding electrode systems (see Figure 9–4), and Part J the grounding conductors, types, sizes, and enclosures for those conductors. Most important are Table 250-94 for sizing the grounding electrode conductor for an AC system, and the Exceptions to Section 250-94 as they apply to that table. Table 250-95 lists the minimum equipment grounding conductor for grounding raceway and equipment. Section 250-95 and Exceptions provide rules for applying that table. Part K gives the conditions for the grounding conductor connections, and Part L covers instrument transformers, and so on. Part M covers systems and circuits of 1 kV (1,000 volts) or over (high voltage). Where high voltage systems are grounded, they must comply with all applicable provisions of all Article 250, Part M, which supplements or modifies Parts A through L. Sections 250-32 and 250-33 clarify a common misconception of adding an insulated equipment grounding conductor with the circuit conductors from the source, service, or subpanel to the equipment being served and terminating on the ends relieves one responsibility of installing the metallic raceways as a complete system. All metal raceway enclosures must be bonded to the equipment grounding conductor to ensure that no potential exists on that equipment, and if one of the conductors fault to the raceway would occur, that the overcurrent device will trip. The insulated conductor when installed is redundant and not required. (The redundant conductor will only carry about 3% to 5% of the current flowing, while the metal raceway will carry 95% to 97% of the current flowing.) Part E, Article 250 describes the general requirements for equipment grounding and provides a list of specific items, both the electrical and nonelectrical equipment fastened in place, and cord connected equipment in residential occupancies and other than residential occupancies. The list is specific. This part of Article 250 also covers the limited requirement for spacing lightning rods, Section 250-48 supplements the requirements in NFPA 780, which contains detailed information on grounding lightning protection systems. The types of grounding electrodes and details for the grounding electrode system are found in Sections 250-81 through 250-86. Related to those sections, you also should read Section 250-80, which is the bonding requirements for piping systems and exposed building steel, (a) covering metal water piping and (b) other metal piping, which provide additional safety to the system. As you read Section 250-81 and Exceptions, you will see that all available grounding electrodes on each premise must be used and bonded together. One electrode system must be provided where the building contains others, such as building steel, metal water pipes, and a concrete-encased electrode, all must be used and bonded together. Read these sections carefully before applying them. Section 250-54 requires common grounding electrodes and states, *"Where an ac system is connected to a grounding electrode in or at a building as specified in Sections 250-23 and 250-24, the same electrode shall be used to ground conductor enclosures and equipment in or on that building. Where separate services supply a building and are required to be connected to a grounding electrode, the same grounding electrode shall be used. Two or more grounding electrodes that are effectively bonded together shall be considered as a single grounding electrode system in this sense."* When sizing the equipment grounding conductor in accordance with Table 250-95, beware this is the minimum size conductor allowed but may not be large enough to comply with other applicable sections in Article 250, such as Section 250-51 that states, *"The path to ground from circuits, equipment, and metal enclosures for conductors shall: (1) be permanent and continuous; (2) have capacity to conduct safely any fault current likely to be imposed on it; and (3) have sufficiently low impedance to limit the voltage to ground and to facilitate the operation of the circuit protective devices in the circuit. The earth shall not be used as the sole equipment grounding conductor."* (This requirement clarifies that driving a grounding electrode at remote equipment is not an acceptable means for grounding.) An equipment grounding conductor must be run with the other conductors as stated in Section 300-3(b) from the supply to the load and sized in accordance with Table 250-95 or larger to facilitate the operation of the overcurrent device protecting that circuit. You will find after studying Article 250 that grounding is one of the most interesting sections of the *NEC®*, and is written for easy understanding. However, careful study and application are necessary to accomplish a safe design and/or installation of the electrical system.

Example: The following example illustrates how the size of the grounded (neutral) wire may be selected. A new three-phase, four-wire feeder is connected in an existing junction box to three single-phase plus grounded circuits, each consisting of two phase conductors and a grounded conductor. Two of these circuits are in one conduit, the third is in another. The single-phase circuit conductors are all

No. 2 THW copper. Determine the ampacity of each single-phase circuit and the required size of new service if electric discharge lighting ballasts are in the circuits, but the load is considered non-continuous.

Solution: Neutrals in these circuits must be counted as current-carrying conductors. (*NEC®* Note 10(b) from the 0–2000 Volt Ampacity Tables.) The ampacity of No. 2 THW conductors from Table 310-16 is 115 amperes. The circuit ampacity in the conduit with the single circuit, and no derating is required for only three conductors. Thus, the ampacity equals 115 amperes. In the conduit with two circuits, an 80% derating factor must be used, as all six wires are considered as current-carrying conductors. Each of these two circuits has a design ampacity of 80% × 115 = 92. The sum of the ampacities of three circuits is the required ampacity of the service conductors and the service neutral (Section 220-22). This value is 115 + 92 + 92 = 299. Each conductor and neutral requires not less than a 350 kcmil 75% seat conductor ampacity. Reference Article 310, Table 310-16.

CHAPTER 9 QUESTION REVIEW

1. Required grounding conductors and bonding jumpers cannot be connected solely by _____ connections.

   Answer: _____

   _____

   _____

   _____

   Reference: _____

2. The conductor permitted to bond together all isolated noncurrent carrying metal parts of an outline lighting system is at least size ____ AWG copper.

   Answer: _____

   _____

   _____

   _____

   Reference: _____

3. A connection to a concrete-encased, driven or buried grounding electrode shall be _____

   Answer: _____

   _____

   _____

   _____

   Reference: _____

4. The *grounded* conductor of an electrical branch circuit is identified by the color _____ .

   Answer: _____

   _____

   _____

   _____

   Reference: _____

   Each conduit contains
   4-500 kcmil conductors
   3-500 per phase

   Bonding jumper

   Service equipment

   Ground bus terminal

5. (Refer to the figure above.) Each of the supply side EMT conduits contain three 500 kcmil copper THW service conductors in parallel. As shown, **only one bonding jumper is used to bond all the conduits** to the grounded bus terminal.

   This bonding jumper must be at least size _____ AWG copper.

   Answer: _____

   _____

   _____

   _____

   Reference: _____

6. A rigid metal conduit contains three circuits: two 150 amp, three-phase circuits, and one 300-amp, single-phase circuit. The load side service equipment bonding jumper for this conduit must be at least size _____ AWG copper.

   Answer: _____

   _____

   _____

   _____

   Reference: _____

7. Are EMT set screw connectors approved for grounding of service raceways or are jumpers required to bond around them?

   Answer: _____

   _____

   _____

   _____

   Reference: _____

8. Where feeder conductors are paralleled and routed in separate nonmetallic raceways, is an equipment grounding conductor required in each nonmetallic raceway or only in one nonmetallic raceway?

   Answer: _____

   _____

   _____

   _____

   Reference: _____

9. Are submersible deep-well pumps required to have an equipment grounding conductor run with the other conductors and the submersible pump motor housing grounded?

   Answer: _____

   _____

   _____

   _____

   Reference: _____

10. Is it permissible to ground the secondary of a separately derived system to the grounded terminal in the service switchboard instead of the nearest effectively grounded structural steel or nearest water pipe?

    Answer: _____

    _____

    _____

    _____

    Reference: _____

11. Where installing the grounding electrode conductor in a raceway for physical protection, is the metal raceway required to be bonded to the grounding electrode conductor? If the answer is yes, is it sufficient to bond it at the service panel only or must bonding be done at all terminations?

    Answer: _____
    _____
    _____
    _____

    Reference: _____

12. Where installing the electrical hookups for RV sites in a recreational vehicle park, can a ground rod be driven at each RV site instead of pulling an equipment grounding conductor with a circuit conductor from the source?

    Answer: _____
    _____
    _____
    _____

    Reference: _____

13. You have recently installed a separately derived system, a dry type transformer in a commercial building. No effectively grounded structural steel was near the transformer location. The transformer secondary is grounded to a nearby metal water pipe. Is a supplemental grounding electrode required?

    Answer: _____
    _____
    _____
    _____

    Reference: _____

14. When bonding the noncurrent-carrying metal parts of a recreational vehicle, what is the minimum size copper or equivalent bonding conductor?

    Answer: _____
    _____
    _____
    _____

    Reference: _____

15. Does the Code require interior gas piping systems to be bonded to the grounding electrode system?

    Answer: _____
    _____
    _____
    _____

    Reference: _____

16. Two parallel runs of four 500 kcmil conductors each are run in rigid nonmetallic conduits to feed a 600-amp panel. Is an equipment grounding conductor required in each conduit? If the answer is yes, what size equipment grounding conductor is required in each conduit?

    Answer: _____

    _____

    _____

    _____

    Reference: _____

17. When terminating the equipment grounding conductor in an outlet box supplied by nonmetallic sheath cables, can the equipment grounding conductors be twisted together or must they be connected with a wire connector? (Twisting appears to make a good connection.)

    Answer: _____

    _____

    _____

    _____

    Reference: _____

18. What size aluminum equipment grounding conductor will be required to be run with the circuit conductors fed from a 60-ampere fusible switch?

    Answer: _____

    _____

    _____

    _____

    Reference: _____

19. In a run of conduit supplying a motor, it is necessary to install 4 feet of flexible metal conduit where the raceway leaves the panel to get around a column. When the conduit reaches the motor, an additional 3 feet of flexible metal conduit is employed for convenience and flexibility. Is an equipment grounding conductor required in this metal raceway?

    Answer: _____

    _____

    _____

    _____

    Reference: _____

20. When making an installation of a high-impedance grounded neutral system, how is the neutral conductor to be sized?

    Answer: _____

    _____

    _____

    _____

    Reference: _____

21. When making an installation of a hydromassage tub in a residence, how are the hydromassage bathtubs and associated electrical components to be protected?

    Answer: _____

    _____

    _____

    _____

    Reference: _____

22. Underwriters Laboratory Standard UL 67 requires that backfed plug-in type circuit breakers be secured. Does the *NEC*® require that backfed circuit breakers also be secured? If so, what section makes this requirement?

    Answer: _____

    _____

    _____

    _____

    Reference: _____

23. Where several types of grounding electrodes are available on the premises, such as an underground water line, building steel, and a concrete encased electrode, is it permissible to jump from one electrode to another with a single properly sized grounding electrode conductor?

    Answer: _____

    _____

    _____

    _____

    Reference: _____

24. Many contractors use metal 90° elbows in nonmetallic conduit underground runs for ease of pulling the conductors and to prevent burn-through. Are these 90° bends required to be bonded? If so, how can this be accomplished from a practical standpoint when they are underground?

    Answer: _____

    _____

    _____

    _____

    Reference: _____

110 Master Electrician's Review

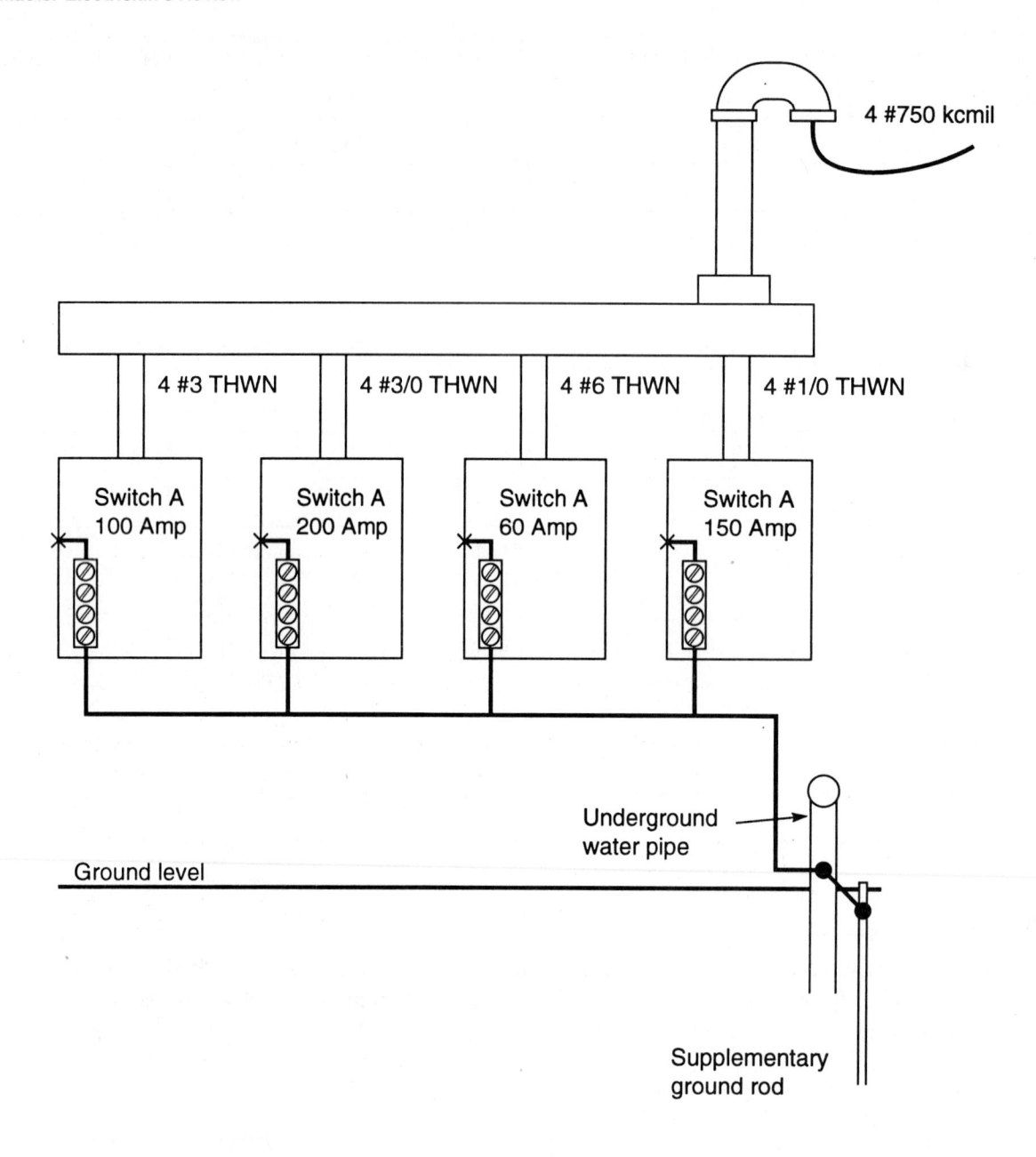

Figure: Service entrance with 4 #750 kcmil conductors feeding a gutter/trough, which distributes to four switches:
- *Switch A 100 Amp — 4 #3 THWN*
- *Switch A 200 Amp — 4 #3/0 THWN*
- *Switch A 60 Amp — 4 #6 THWN*
- *Switch A 150 Amp — 4 #1/0 THWN*

Grounding electrode conductor runs to underground water pipe with supplementary ground rod.

25. In the figure above, what is the required size of the grounding electrode conductor and the main grounding electrode conductor to Switch A, Switch B, Switch C, and Switch D?

Answer: _____

_____

_____

_____

Reference: _____

Chapter Ten

WIRING METHODS

The general requirements for wiring methods are found in Article 300. As you may recall in earlier chapters of this book, several times you were told you must know the contents of Article 300 thoroughly, as it is used in some application for every installation.

The general wiring methods in Article 300 include requirements that apply to the specific wiring methods, such as nonmetallic sheath cable that is covered in Article 336 (see Figure 10–2), as specific support requirements and installation criteria, such as found in Section 300-4(a), (b), (c), and (d) and the Exceptions. These requirements are mandatory and must be adhered to when making an installation of any wiring method and are in Chapter 3 as they apply. Chapter 3 also covers burial depths (in Table 300-5 for conductors under 600 volts) and many conditions that apply to underground installations, such as splices, taps, and the back fill material used so as not to damage the raceway or cable that you have selected as a wiring method for your installation. When a wiring method is installed in earth that is not suitable as physical protection, select fill material should be used such as sand or soil without rocks. This article also covers such things as seals and protection against corrosion, indoor wet locations, raceways exposed to different temperatures, and expansion joints. This chapter discusses much about the expansion joint requirements for nonmetallic raceways. However, Section 300-7 covers all wiring methods and where they are necessary to compensate for expansion or contraction they are required (see Figures 10–3, 10–4, and

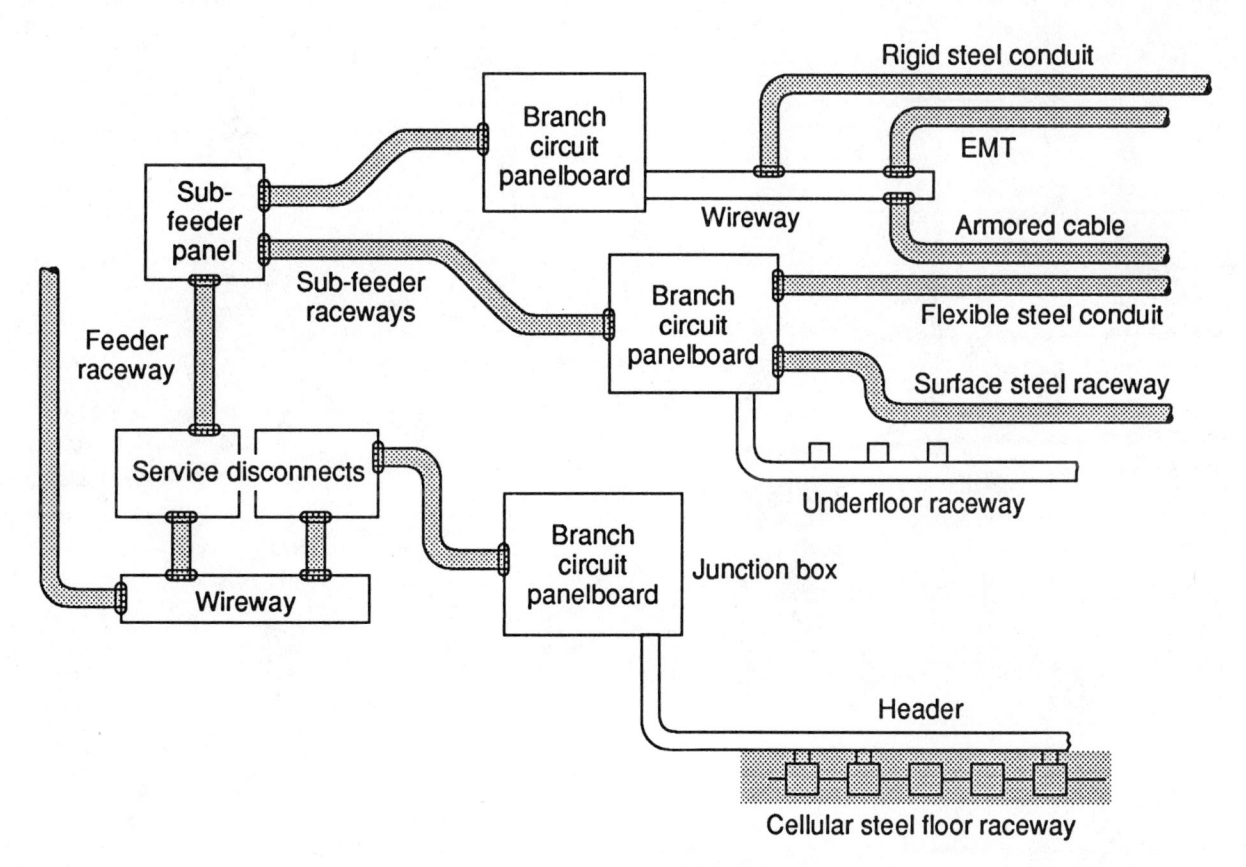

Figure 10–1A Service raceway 600 volts or less per *NEC® Article 230*, over 600 volts per *NEC® Article 710*. **NOTE:** Location of overcurrent devices not shown. *(Courtesy of American Iron & Steel Institute)*

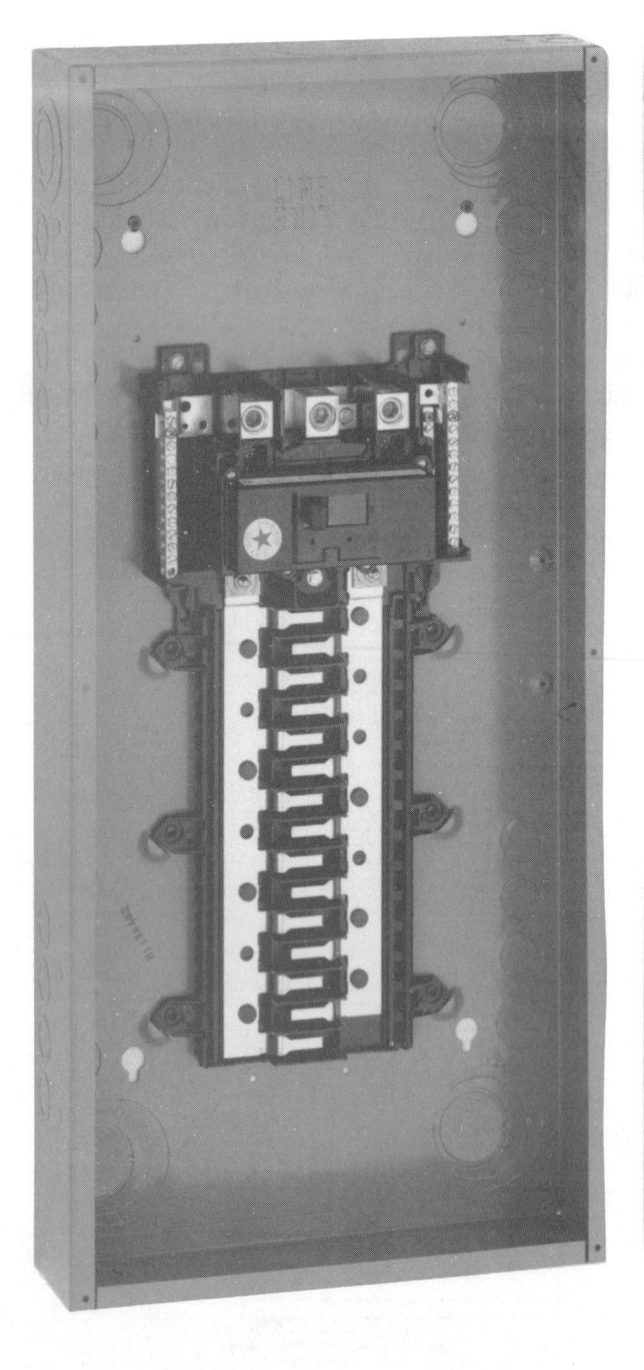

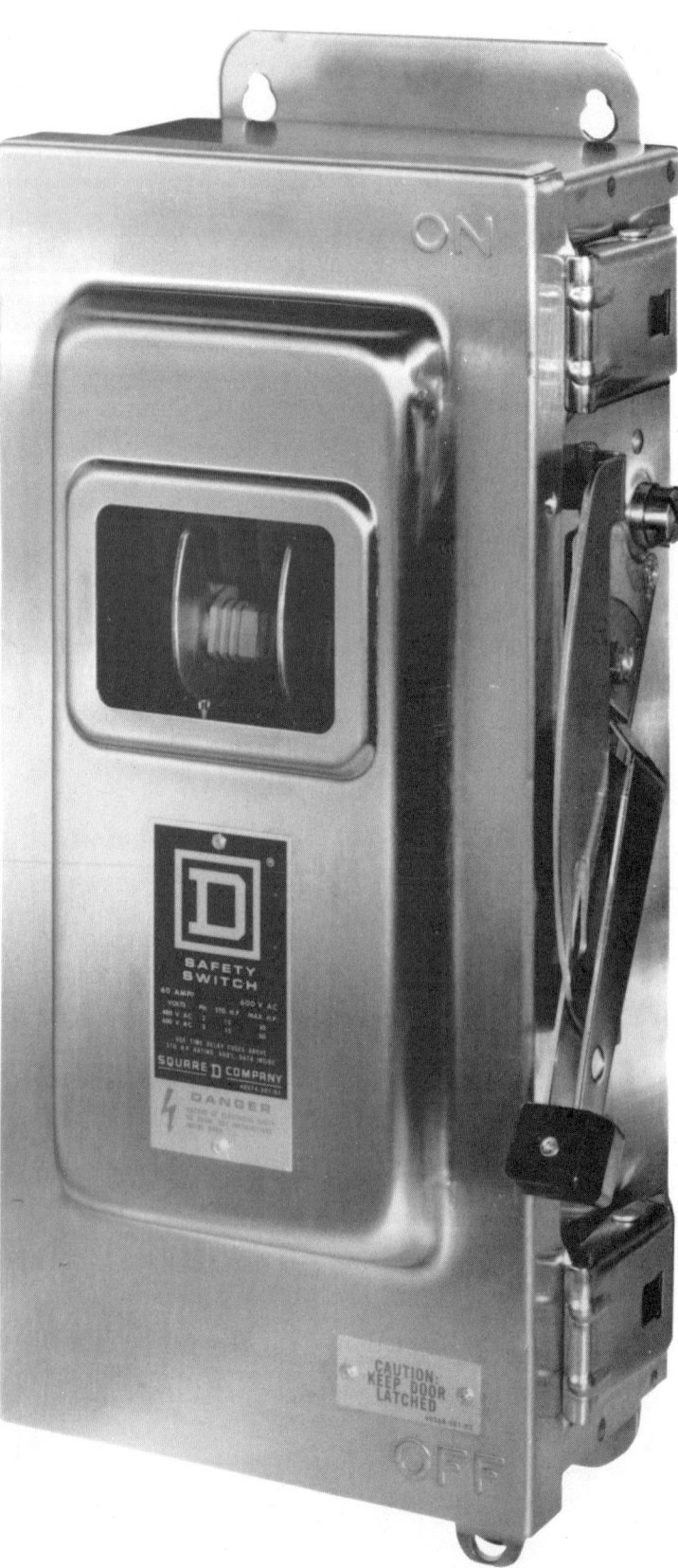

Figure 10–1B *(Courtesy of Square D Company)*

Figure 10–1C *(Courtesy of Square D Company)*

300-11(a)
Securing and Supporting

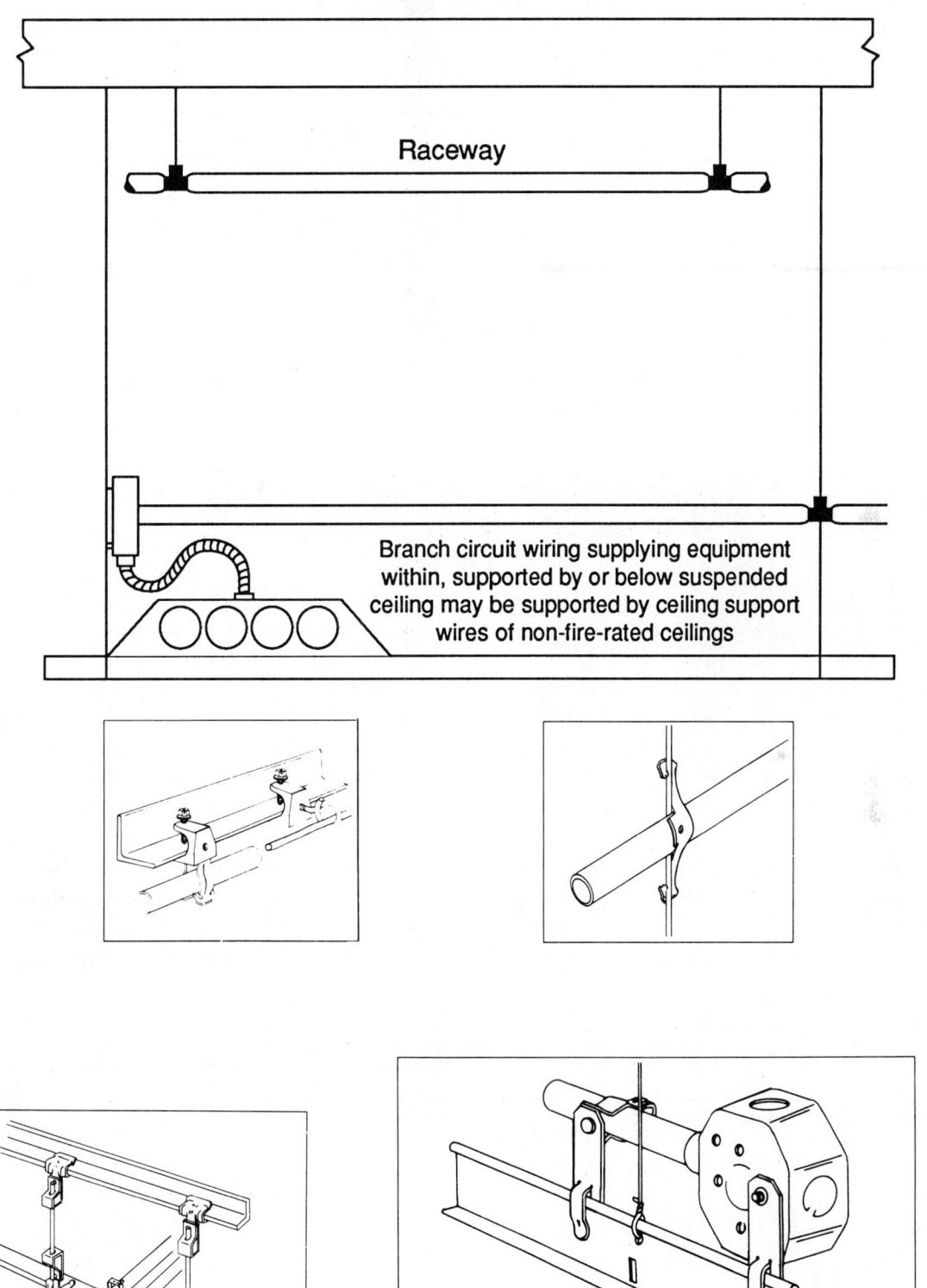

Raceway

Branch circuit wiring supplying equipment within, supported by or below suspended ceiling may be supported by ceiling support wires of non-fire-rated ceilings

Figure 10–2

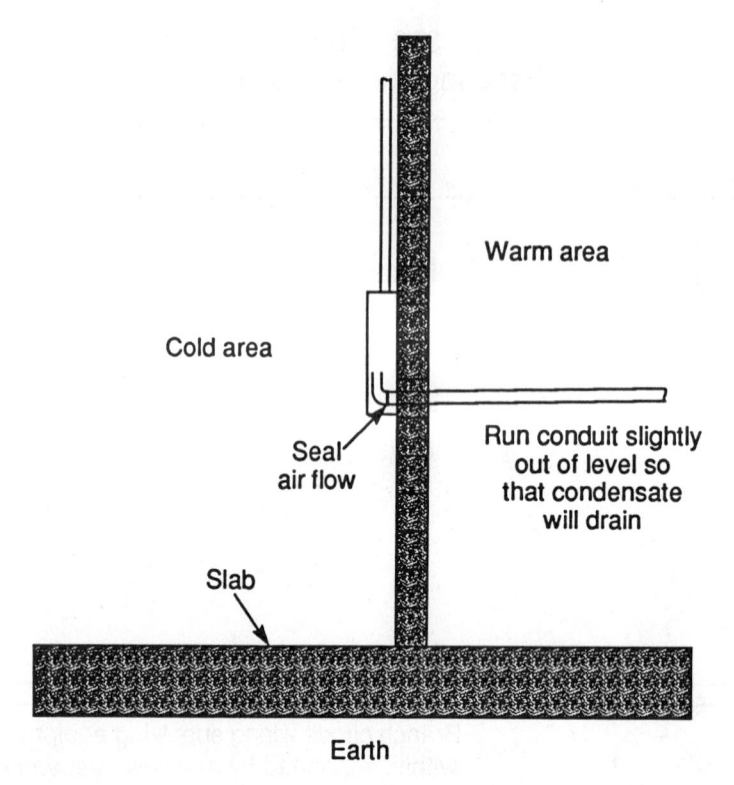

Figure 10–3 Seals are required to comply with *NEC® Section 300-7* where the raceway is subjected to more than one temperature such as shown. The system enters the building from outside.

10–5). The electrical and mechanical continuity for metallic systems, both the raceway and the enclosures 300-11 require that these methods, raceways, cable assemblies, boxes, cabinets, and fittings all be securely fastened where supported on support wires above suspended ceilings.

For branch circuit wiring, read this section carefully. The wiring systems for feeders or services are not permitted to be secured in this manner; only certain branch circuits are permitted. Specifically, the junction box out of the box section in Article 370 places more stringent requirements for supporting these box systems. All requirements must be followed. Article 300 also tells us where a box or fitting is required and where a box only is permitted. Article 300-15(c) states that all fittings and connectors must be designed for the specific wiring method for which they are used and listed for that purpose. Look up the definition for a branch circuit listed in Article 100 now. An important requirement for all wiring methods is found in Section 300-21.

Table 10. Expansion Characteristics of PVC Rigid Nonmetallic Conduit
Coefficient of Thermal Expansion = 3.38×10^{-5} in./in./°F

Temperature Change in Degrees F	Length Change in Inches per 100 ft. of PVC Conduit	Temperature Change in Degrees F	Length Change in Inches per 100 ft. of PVC Conduit	Temperature Change in Degrees F	Length Change in Inches per 100 ft. of PVC Conduit	Temperature Change in Degrees F	Length Change in Inches per 100 ft. of PVC Conduit
5	0.2	55	2.2	105	4.2	155	6.3
10	0.4	60	2.4	110	4.5	160	6.5
15	0.6	65	2.6	115	4.7	165	6.7
20	0.8	70	2.8	120	4.9	170	6.9
25	1.0	75	3.0	125	5.1	175	7.1
30	1.2	80	3.2	130	5.3	180	7.3
35	1.4	85	3.4	135	5.5	185	7.5
40	1.6	90	3.6	140	5.7	190	7.7
45	1.8	95	3.8	145	5.9	195	7.9
50	2.0	100	4.1	150	6.1	200	8.1

Figure 10–4 (Reprinted with permission of NFPA 70-1993, the *National Electrical Code®*, Copyright © 1992, National Fire Protection Association, Quincy, MA 02269. This reprinted material is not the complete and official position of the National Fire Protection Association on the referenced subject, which is represented only by the standard in its entirety.)

Example

380 ft. of conduit is to be installed on the outside of a building exposed to the sun in a single straight run. It is expected that the conduit will vary in temperature from 0°F in the winter to 140°F in the summer (this includes the 30°F for radiant heating from the sun). The installation is to be made at a conduit temperature of 90°F. From the table, a 140°F temperature change will cause a 5.7 in. length change in 100 ft. of conduit. The total change for this example is 5.7" × 3.8 = 21.67" which should be rounded to 22". The number of expansion couplings will be 22 ÷ coupling range (6" for E945, 2" for E955). If the E945 coupling is used, the number will be 22 ÷ 6 = 3.67 which should be rounded to 4. The coupling should be placed at 95 ft. intervals (380 ÷ 4). The proper piston setting at the time of installation is calculated as explained above.

$$0 = \left[\frac{140 - 90}{140}\right] 6.0 = 2.1 \text{ in.}$$

Insert the piston into the barrel to the maximum depth. Place a mark on the piston at the end of the barrel. To properly set the piston, pull the piston out of the barrel to correspond to the 2.1 in. calculated above.

See the drawing below.

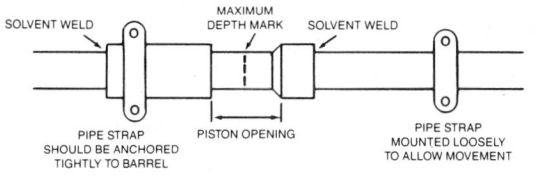

Figure 10–5 Rigid nonmetallic conduit. For steel raceway, 21.7 × .05 − 1.08 inches total expansion on expansion fitting would not generally be required. For aluminum raceway, 21.7 × 0.1 − 2.17 inches an expansion fitting may be required. *(Courtesy of Carlon, A Lamson & Sessions Company)*

The requirement limits the spread of fire or products of combustion and requires that the wiring system be installed so as to minimize the spread of fire and products of combustion; that all penetrations through floors, walls, and ceilings be sealed in an improved manner to accomplish this purpose. Section 300-22, which are the provisions for the installation and uses of electrical wiring and equipment in ducts, plenums, and other air handling spaces, states these sections are safety concerns in determining the wiring method to be used in electrical installations; these should be studied carefully before making the final selections. Part B of Article 300 is the requirements for installations over 600 volts nominal. Section 300-37 covers the underground installations and references the Table and Section in Article 710, which are the general wiring requirements for installations over 600 volts. Other general articles are found in Chapter 3 relating to wiring methods. Article 305 covers the temporary wiring requirements needed to provide power for the workers making the installation and constructing the facility during the construction phases of a job. Article 305 also covers other types of temporary wiring, such as temporary wiring concerned with Christmas tree lighting, trade fairs, emergencies and tests, experimental and development work. Be aware of Article 305. Article 310 is the conductors to be used for general wiring. In Article 310 you will find all requirements related to conductors for general wiring. They do not apply to conductors that form an integral part of equipment, such as motors, motor controllers, and similar equipment, or conductors specifically provided for elsewhere in the Code. An example would be Article 400 and Article 402 for flexible cords and fixture wires. Other sections dealing with conductors would be Chapter 7 of the *NEC®* and Chapter 8 of the *NEC®*, which deal with limited voltage wire, such as in Article 725, 760, and so on. Article 310 must be used in every installation. The requirements in 310-1 through 310-15 are general requirements for the use and installation of conductors. The tables apply as applicable. **Note:** When using the tables, read each heading carefully. Read the columns carefully. Read the ambient temperature correction factors below each column, and remember the notes (footnotes) under each table are part of those tables and are mandatory requirements. This is important when studying for a test because many test questions will require you to study these additional parts of the table to get the correct answer. For example, the obelisk note under each table applies to general wiring methods. However, notice that it says unless specifically otherwise provided for or permitted in this Code. Motors, for instance, generally are not required to comply with this obelisk note because Article 430 has specific requirements for branch circuit conductors for motors. Study these tables carefully. The next important part of Article 310 is the notes to the Ampacity Tables 0–2000 Volts. Notes 1 through 11 provide language that can be found in many test questions. They are important. Note 3 has less restrictive ampacity requirements for dwelling services and feeders. **Note:** These only apply to single-phase, 120/240 volt, three-wire systems, and do not apply to any other type of system. Also, note the neutral conductor is permitted to be a maximum of two AWG sizes smaller than ungrounded conductors, provided they meet the requirements of Section 215-2, 220-22, and 230-42. This is an important sentence in using this table. The derating adjustment factors in Note 8 also will likely be referenced in most examinations for calculating the allowable ampacity of conductors in a raceway or cable.

116 Master Electrician's Review

NEC Section 318-9(a)(1)

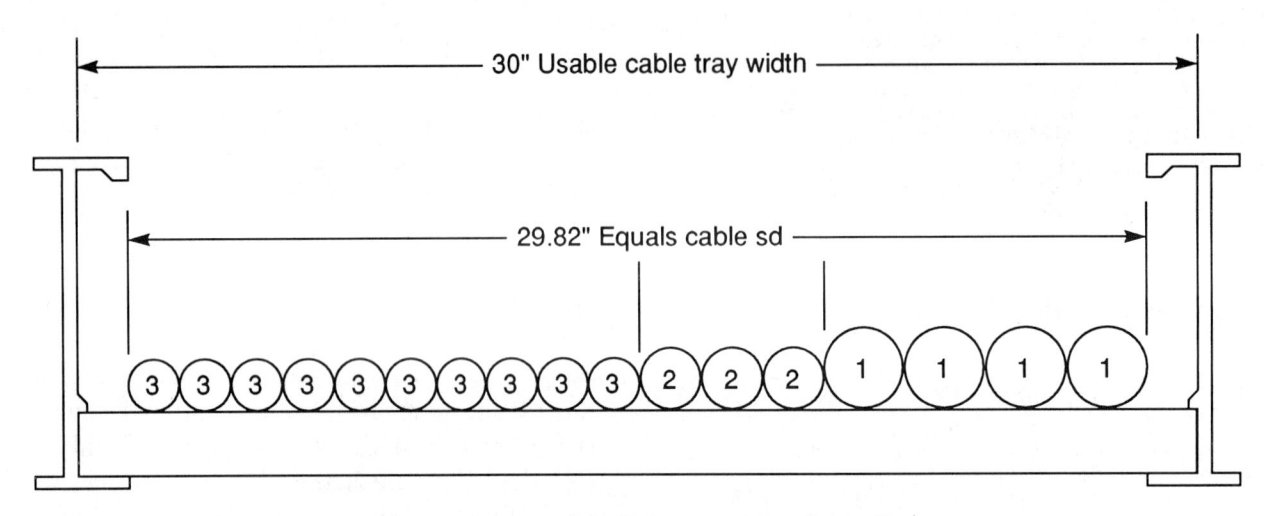

Cross Section of the Cables and the Cable Tray

(Courtesy of B-Line Systems)

Example: *NEC®* Section 318-9(a)(1)

Width selection for cable tray containing 600-volt multiconductor cables, sizes No. 4/0 AWG and larger only. Cable installation is limited to a single layer. The sum of the cable diameters (Sd) must be equal to or less than the usable cable tray width.

Cable tray width is obtained as follows:

A—Width required for No. 4/0 AWG and larger multiconductor cables:

Item Number	List Cable Sizes	(D) List Cable Outside Diameter	(N) List Number of Cables	Multiply (D) × (N) Subtotal of Sum of Cable Diameters (Sd)
1.	3/C–No. 500 kcmil	2.25 inches	4	9.04 inches
2.	3/C–No. 250 kcmil	1.76 inches	3	5.28 inches
3.	3/C–No. 4/0 AWG	1.55 inches	10	15.50 inches

The sum of the diameters (Sd) of all cables (add total Sd for items 1, 2, and 3):

9.04 inches + 5.78 inches + 15.50 inches = 29.82 inches (Sd)

A cable tray with a usable width of 30 inches is required. For approximately 15% more a 36-inch wide cable tray could be purchased that would provide for some future cable additions.

NOTES:
1. Cable sizes used in this example are a random selection.
2. Cables—copper conductor with cross linked polyethylene insulation and a PVC jacket. (These cables could be ordered with or without an equipment grounding conductor.)
3. Total cable weight per foot for this installation:
   61.4 lbs/ft (without equipment grounding conductors)
   69.9 lbs/ft (with equipment grounding conductors)

This load can be supported by a load symbol "B" cable tray–75 lbs/ft.

Example: *NEC®* Section 318-9(a)(2)

Width selection for cable tray containing 600-volt multiconductor cables, sizes No. 3/0 AWG and smaller. Cable tray allowance fill areas are listed in Column 1 of Table 318-9.

Cable tray is obtained as follows:

Item Number	List Cable Sizes	(A) List Cable Cross Sectional Areas	(N) List Number of Cables	Multiply (A) × (N) Total of Cross Sectional Area for Each Item
1.	3/C–No. 500 kcmil	2.25 inches	4	9.04 inches
2.	3/C–No. 250 kcmil	1.76 inches	3	5.28 inches
3.	3/C–No. 4/0 AWG	1.55 inches	10	15.50 inches

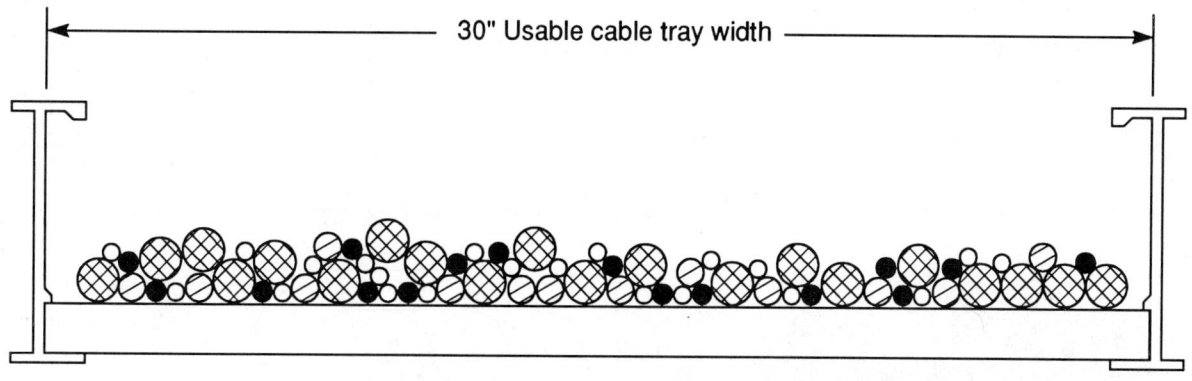

Cross Section of the Cables and the Cable Tray

(Courtesy of B-Line Systems)

METHOD 1.

The sum of the total areas for items 1, 2, 3, and 4:

3.34 sq in + 3.04 sq in + 6.02 sq in + 16.00 sq in = 28.40 sq inches

From Table 318-9, Column 1, a 30-inch wide tray with an allowable fill area of 35 sq inches must be used. The 30-inch cable tray has the capacity for additional future cables (6.60 sq inches additional allowable fill area can be used).

METHOD 2.

The sum of the total areas for items 1, 2, 3, and 4 multiplied by (6 sq in / 7 sq in) = cable tray width required:

3.34 sq in + 3.04 sq in + 6.02 sq in + 16.00 sq in = 28.40 sq inches

$$\frac{28.40 \text{ sq inches} \times 6 \text{ sq inches}}{7 \text{ sq inches}} = 24.34 \text{ inch cable tray width required}$$

Use a 30-inch cable tray.

NOTES:
1. The cable sizes used in this example are a random selection.
2. Cables—copper conductors with cross linked polyethylene insulation and a PVC jacket. These cables could be ordered with or without an equipment grounding conductor.
3. Total cable weight per foot for this installation: 31.9 lbs/ft. (Cables in this example do not contain equipment grounding conductors.) This load can be supported by a load symbol "A" cable tray–50 lbs/ft.

Example: NEC Section 318-9(a)(3)

Width selection for cable tray containing 600-volt multiconductor cables, sizes No. 4/0 AWG and larger (single layer required) and No. 3/0 AWG and smaller. These two groups of cables must have dedicated areas in the cable tray.

Cable tray width is obtained as follows:

A—Width required for No. 4/0 AWG and larger multiconductor cables:

Item Number	List Cable Sizes	(D) List Cable Outside Diameter	(N) List Number of Cables	Multiply (D) × (N) Subtotal of Sum of Cable Diameters (Sd)
1.	3/C–No. 500 kcmil	2.26 inches	3	6.78 inches
2.	3/C–No. 4/0 AWG	1.55 inches	4	6.20 inches

Total cable tray width required for items 1 and 2:

6.76 inches + 6.20 inches = 12.98 inches

B—Width required for No. 3/0 AWG and smaller multiconductor cables:

Item Number	List Cable Sizes	(A) List Cable Cross Sectional Areas	(N) List Number of Cables	Multiply (A) × (N) Total of Cross Sectional Area for Each Item

Total cable tray width required for items 3, 4, and 5:

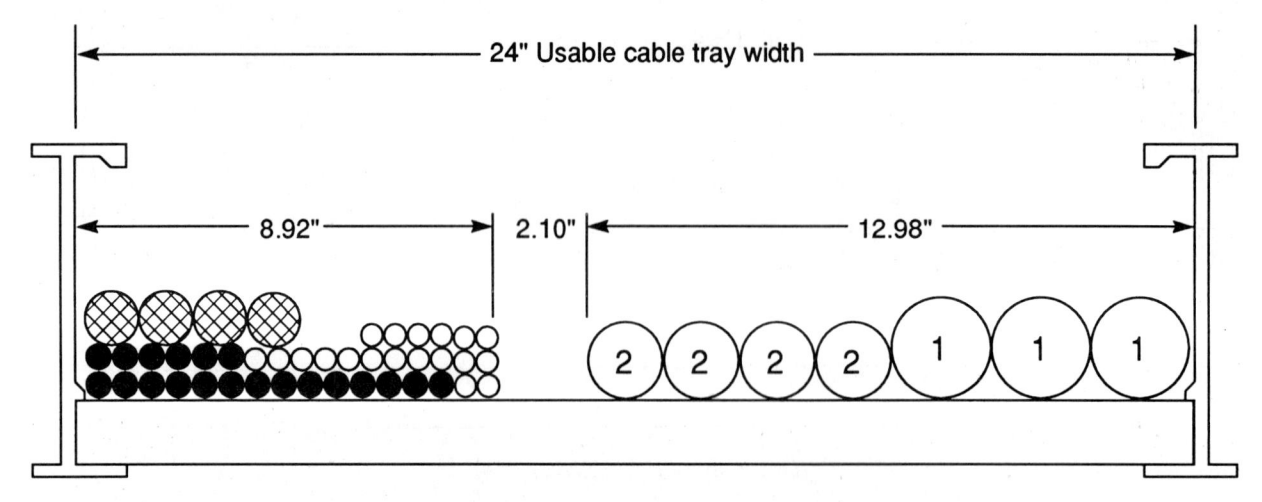

Cross Section of the Cables and the Cable Tray

(Courtesy of B-Line Systems)

(3.20 sq in + 4.00 sq in + 3.20 sq in) (6 sq in / 7 sq in)1= (10.4 sq in) (6 sq in / 7 sq in)1 = 8.92 inches

Actual cable tray width is A "Width" (12.98 inches) + B "Width" (8.92 inches) = 20.88 inches

A 24-inch wide cable tray is required. The 24-inch cable tray has the capacity for additional future cables (3.1 inches or 3.6 sq inches allowable fill can be used).

NOTES:
1. This ratio is the inside width of the cable tray in inches divided by its maximum fill area in square inches from Table 318-9, Column 1.
2. The cable sizes used in this example are a random selection.
3. Cables—copper conductor with cross linked polyethylene insulation and a PVC jacket.
4. Total cable weight per foot for this installation:
   40.2 lbs/ft. (Cables in this example do not contain equipment grounding conductors.) This load can be supported by a load symbol "A" cable try–50 lbs/ft.

Example: *NEC®* Section 318-9(b)
50% of the cable tray usable cross-sectional area can contain type PLTC cables

4 inches × 6 inches × .500 = 12 square inches allowable fill area

2/C-No. 16 AWG 300-volt shielded instrumentation cable O.D. = .224 inches.

Cross-Sectional Area = 0.04 square inches.

$$\left(\frac{12 \text{ sq in}}{0.04 \text{ sq in cable}}\right) = 300 \text{ cables can be installed in this cable tray.}$$

$$\left(\frac{300 \text{ cables}}{26 \text{ cables / row}}\right) = 11.54 \text{ rows can be installed in this cable tray.}$$

NOTES:
1. The cable sizes used in this example are a random selection.
2. Cable—copper conductors with PVC insulation, aluminum/mylar shielding, and PVC jacket.

Following these general application articles 300, 305, and 310, the remaining articles in Chapter 3 are specific articles. Article 318 covers cable trays (again see Figure 10–4). Cable trays are a support system and not a wiring method.

Note: As you begin to study this part of Chapter 3, refer to the definitions in Article 100 for raceway. A fine print note will alert you to types of raceways. There is a distinct difference between raceways and cables and other wiring methods. Study that definition or refer to it

← 6" Usable cable tray width →

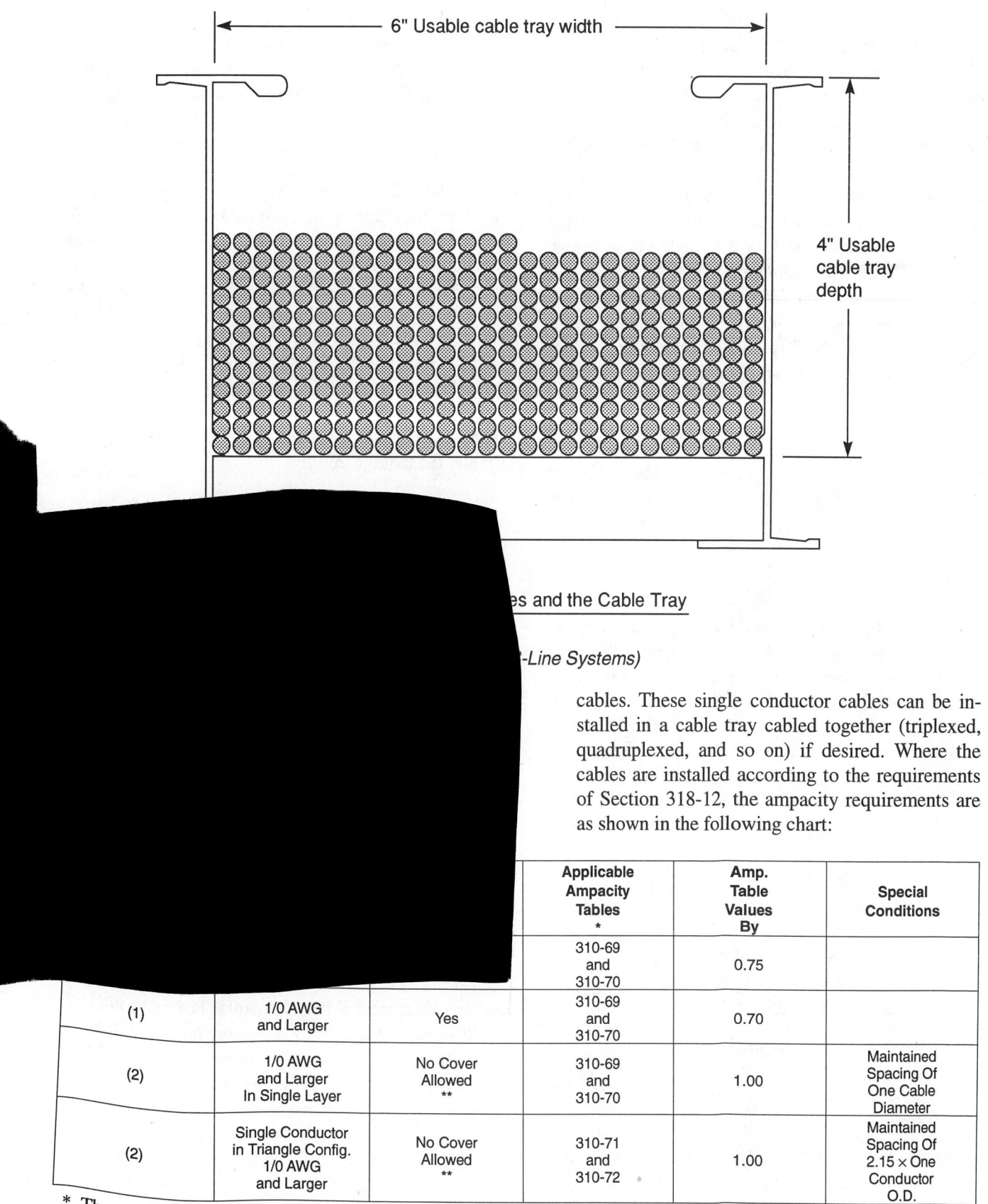

4" Usable cable tray depth

...es and the Cable Tray

(...-Line Systems)

cables. These single conductor cables can be installed in a cable tray cabled together (triplexed, quadruplexed, and so on) if desired. Where the cables are installed according to the requirements of Section 318-12, the ampacity requirements are as shown in the following chart:

			Applicable Ampacity Tables *	Amp. Table Values By	Special Conditions
			310-69 and 310-70	0.75	
(1)	1/0 AWG and Larger	Yes	310-69 and 310-70	0.70	
(2)	1/0 AWG and Larger In Single Layer	No Cover Allowed **	310-69 and 310-70	1.00	Maintained Spacing Of One Cable Diameter
(2)	Single Conductor in Triangle Config. 1/0 AWG and Larger	No Cover Allowed **	310-71 and 310-72	1.00	Maintained Spacing Of 2.15 × One Conductor O.D.

* The ambient ampacity correction factors must be used.
** At a specific position where it is determined that the tray cables require mechanical protection, a single cable tray cover of six feet or less in length can be installed.

As needed, refer to the articles from 320 through 365 for the specific types of raceways and cables. In addition to the references for raceways in Chapter 3, you will find references related to Chapter 9 tables and examples, such as the reference found in Section 345-7, the numbers of conductors in conduit. When sizing the number of conductors in a raceway, conduit, tubing, and so on, it is necessary to refer to the permitted percent fills specified in the tables in Chapter 9, Table 1 for percentage fill and Table 4 for the conduit dimensions. Similar references are found in Section 346-6 for rigid metal conduit, Section 347-11 for rigid nonmetallic conduit, Section 348-6 for electrical nonmetallic conduit, and Section 350-7 for flexible metallic conduit. Other similar references are found for other raceway types. The number of conductors allowed in a given raceway must be calculated in square inches, based on the insulation and wire size listed in Tables 5-A and 5-B of Chapter 9. For bare conductors, Chapter 8 dimensions are used. Where conductors of all the same size are installed in a single raceway, Table 3 applies generally. The following example will help you learn to apply Chapter 9.

Example: What size PVC rigid nonmetallic conduit is required for a multiple branch circuit supplying a cooler tower located adjacent to a hospital containing three No. 8 THWN stranded copper conductors, three No. 4 THWN copper conductors, four No. 1/0 THW copper conductors, and one No. 2 bare copper conductors?

Answer: _____

References:
　Rigid nonmetallic conduit—Article 347.
　Number of conductors—Section 347-11, Ref. Chapter 9
　Chapter 9, Notes 1, 2, Table 1 and Notes. Read carefully Notes 4 and 5.
　Table 1—permits a 40% fill.

Note: Table 4 dimensions and percent are of conduit and tubing is in square inches. Therefore, select square inch dimensions from Table 5 and 8 for the correct calculations.

Table 5—3 No. 8 THWN = .0373 × 3 = .1119 sq inches
　　　　　3 No. 4 THWN = .0845 × 3 = .2535 sq inches
　　　　　4 No. 1/0 THW = .2367 × 4 = .9468 sq inches
Table 8—1 No. 2 BARE = .067 × 1 = .067 sq inches
　　　　　　　　　　　　Total　1.3792 sq inches

Table 4—Not lead covered over two conductors in a raceway 40% 2 inches raceway is 1.34 square inches. 2½ raceway is 1.92. Therefore, two inches is not large enough, and it is necessary to install 2½ inch rigid nonmetallic conduit for this installation.

Article 370 covers outlet boxes, pull boxes, junction boxes, conduit bodies, and fittings. A conduit body is what is normally termed a Tee, LB, and so forth or as a condulet. Use Article 370 carefully, and as you determine the number of conductors for a given box, the sizing requirements are found in Section 370-16. **Note:** Most boxes with the integral type clamps, such as nonmetallic outlet boxes, are clamps and a deduction must be added. If uncertain, check with the manufacturer. In a test, normally the test material will tell you whether there are clamps to be calculated. Study 370-16(a) and (b) and the tables carefully.

Example: What size outlet box with integral clamps is required to accommodate a 15-ampere snap switch to control lighting and a 20-ampere duplex receptacle? The switch has a two-wire w/GRD No. 14 nonmetallic sheathed cables feeding it with another No. 14 NM two-wire w/GRD feeding the light fixture. The receptacle is fed with a two-wire w/GRD No. 12 nonmetallic sheath cable, and a two-wire w/GRD nonmetallic sheath cable goes to feed additional duplex receptacles in the same area.

Solution: Minimum size of box in cubic inches:

Based on largest conductor in the box:
  Cable clamps = No. 12, 2.25 × 1 = 2.25

Based on size conductor connected:
  Conductors to receptacle = No. 12, 2.25 × 4 = 9.0
  Conductors to switch = No. 14, 2 × 4 = 8.0
  Receptacle = No. 12, 2.25 × 2 = 4.5

Based on size conductor connected:
  Switch = No. 14, 2 × 2 = 4.0

Based on longest conductor in the box:
  Equipment grounding
    conductors = 2.25
    ─────────
    30.0 cubic inches

In addition, in most examinations will include sizing pull boxes and junction boxes. These requirements are found in 370-28.

Example: Four 3 inch rigid steel conduits or EMT, straight pull.

Three 2 inch rigid steel conduits or EMT, angle pull.

Recommended minimum spacing for 3 inch conduit, 4¾ inch; for 2 inch conduit, 3⅜ inch; and for 3 inch to 2 inch conduits, 4 inch.

For depth of box (A):
  Three 2 inch conduits, angle pull:
    A = (6 × 2 inch) + 2 inch + 2 inch = 16 inch

Spacing for two rows of conduit:
  Edge of box to center of 3 inch conduit = 2⅜ inch
  Center to center, 3 inch to 2 inch conduit = 4
  Minimum distance, D = 8½
        Total   14⅞ inch

Required depth: Use 16 inch

For length of box (B):
  Four 3 inch conduits, straight pull:
    B = 8 × 3 inch = 24 inch
  2 inch conduit, angle pull: B = A = 16 inch

Required length: Use 24 inch

For width of box (C):
  Recommended spacing for 3 inch conduits = 4¾ inch
    C = 4 × 4¾ inch = 19 inch

Required width: Use 19 inch

Angle pull distance (D):
    D = 6 × 2 inch = 12 inch

Required distance: Use 12 inch

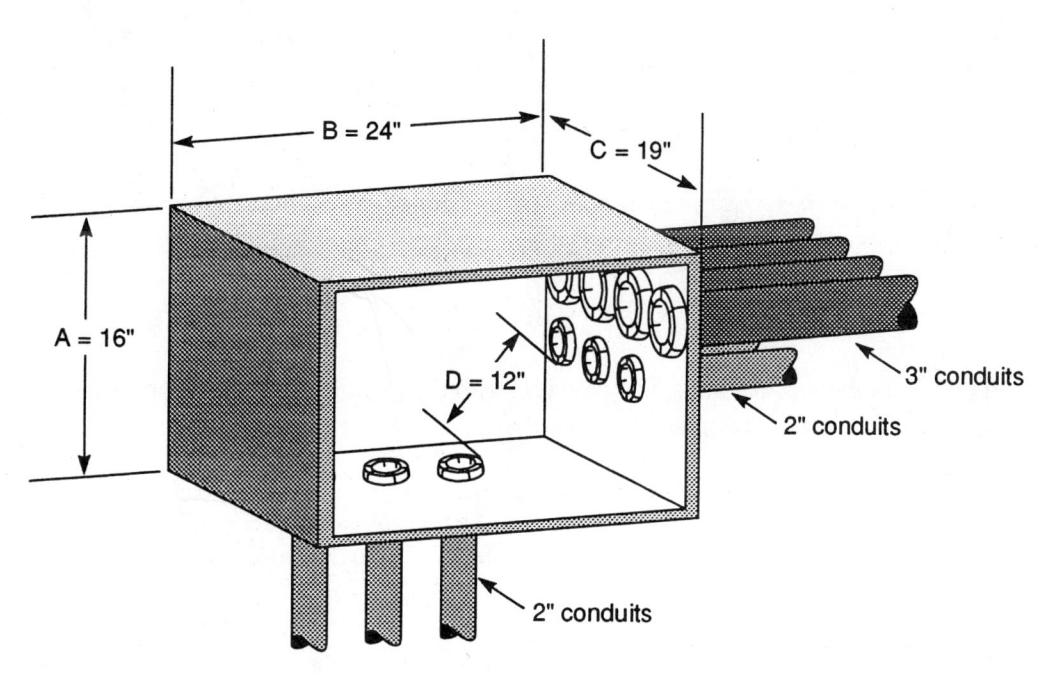

Pull and junction boxes. Minimum size as required in *NEC®* Section 370-18 *(Courtesy of American Iron & Steel Institute)*

This section also should be studied, and you must know where to find it quickly so you can determine the size, outlet, or junction box as required. Article 373 covers cabinet, cut-out boxes, and meter socket enclosures. Most panelboards, switches, disconnecting switches, conductors, and so on are enclosed in a cabinet or cut-out box. When determining the requirements for cabinet and cut-out boxes, refer to Article 373. Article 374 covers auxiliary gutters. Auxiliary gutters are not raceways or cables. An auxiliary gutter is an extension used to supplement wiring spaces in meter enclosures, motor control centers, distribution centers, switchboards, and similar wiring systems, and can enclose conductors or bus bars but is not permitted for switches, overcurrent devices, appliances, or other similar equipment. Although similar in appearance to a wireway, an auxiliary gutter has a different application and use. Be aware of Article 374 for auxiliary gutters. Article 380 covers switches. In this article you can find requirements for all types of switches and some installation requirements, such as accessibility and grouping. Article 384 covers switchboards and panelboards. This article undoubtedly will be included in most examinations and includes many requirements for panelboards and switches. Of particular importance is Section 384-4, an installation requirement. You are urged to mark in your code book in Section 110-16 to also refer to Section 384-4 for the dedicating working space above and below equipment. This section also will be in most examinations. You need not try to memorize this section as it is specific requirements for this equipment, and your proficiency in using this as a reference document will adequately guide you through these wiring method articles. Be familiar with the article names as listed in the Table of Contents and it will assist you as you use this book for reference or for a tool in the field and will increase your proficiency in preparing for an examination.

CHAPTER 10 QUESTION REVIEW

1. When installing Type SE service entrance cable in a building from the service panel to a dryer outlet, what are the support requirements for this cable?

   Answer: _____

   _____

   _____

   _____

   Reference: _____

2. When installing conduit bodies that are legibly marked with their cubic inch capacity, how can one determine the number of conductors and splices and taps permitted?

   Answer: _____

   _____

   _____

   _____

   Reference: _____

3. Is ⅜ liquidtight flexible metal conduit permitted for enclosing motor leads in accordance with Section 430-145(b)?

   Answer: _____

   _____

   _____

   _____

   Reference: _____

4. Structural bar joists are often spaced at up to 5 foot intervals. Support is difficult to achieve when installing metal raceways on these bar joists. Is it permitted to mount an outlet box on a bar joist and extend IMC (Intermediate Metal Conduit) 5 feet to the first support?

   Answer: _____

   _____

   _____

   _____

   Reference: _____

5. When installing electrical nonmetallic tubing (ENT), is it permissible to install this material through metal studs without securing the ENT every 3 feet as required by the Code?

   Answer: _____

   _____

   _____

   _____

   Reference: _____

6. What is the support requirements for Type MC cable?

   Answer: _____

   _____

   _____

   _____

   Reference: _____

7. What are the support requirements and methods of securing Type NM cable (12/2 with ground Romex)?

   Answer: _____

   _____

   _____

   _____

   Reference: _____

8. Can nonmetallic wireways be installed in hazardous Class 1 or Class 2 locations?

   Answer: _____

   _____

   _____

   _____

   Reference: _____

9. What wiring methods are permitted for installing a branch circuit routed through the bar joists of a metal building for a 4,160 volt, three-phase motor?

   Answer: _____

   _____

   _____

   _____

   Reference: _____

10. A contract to install a branch circuit under a bank president's desk to serve a calculator is awarded to a contractor. The bank has asked that there be no drilled holes in the walls. Can the conductors be laid on the floor under the carpet? If so, what type of wiring method should be used?

    Answer: _____

    _____

    _____

    _____

    Reference: _____

11. When installing wireway metallic or nonmetallic, either type, how do you derate the conductors when exceeding the number permitted?

    Answer: _____

    _____

    _____

    _____

    Reference: _____

12. When installing Type AC cable on wood studs, routing them parallel to the framing members, what distance clearance must be maintained where the Type AC cable is likely to be penetrated by screws or nails?

    Answer: _____

    _____

    _____

    _____

    Reference: _____

13. Is it necessary to support EMT (electrical metallic tubing) within 3 feet of each coupling or fitting?

    Answer: _____

    _____

    _____

    _____

    Reference: _____

14. How do the installation requirements for intermediate metal conduit (IMC) and that of rigid metal conduit (RMC) differ? What is permitted by one wiring method that is not permitted by the other?

    Answer: _____

    _____

    _____

    _____

    Reference: _____

15. What is the minimum size conductor that can be run in parallel, generally? (Do not consider exceptions.)

    Answer: _____

    _____

    _____

    _____

    Reference: _____

16. Where installing electrical nonmetallic tubing for the branch circuit conductors in a five-story building, does the material from which the walls, floors, and ceilings need to be considered?

    Answer: _____

    _____

    _____

    _____

    Reference: _____

17. Christmas tree lights have been installed on the local courthouse. The mayor suggested that we leave them up throughout the year. Is this permitted? If so, how long are they permitted to remain installed?

Answer: _____

_____

_____

_____

Reference: _____

18. When installing a run of conduit, it is necessary to install a piece of liquidtight flexible metal conduit 6 feet long through a plenum. Is this permitted by the Code? If not, how much liquidtight flexible metal conduit can you install in a plenum?

Answer: _____

_____

_____

_____

Reference: _____

19. Where making an installation of rigid nonmetallic conduit, Schedule 80, along the outside of a building to feed outdoor floodlights and studying the Code, you find that in Section 300-7(b) expansion joints must be provided where necessary. How much expansion or contraction must rigid nonmetallic conduit have before an expansion fitting is required?

Answer: _____

_____

_____

_____

Reference: _____

20. A column panel is a narrow panel that is installed inside or flush with an I-beam. It is primarily used in warehouses and areas where the panel is subjected to moving traffic and warehousing storage, such as forklifts and palletizing. When installing a column type panel, an auxiliary gutter is installed above the panel, and the neutral connections are made up at the top of the auxiliary gutter and not brought down into the panelboard. Is this an acceptable wiring method?

Answer: _____

_____

_____

_____

Reference: _____

21. When wiring a single-family dwelling to save space, a 2 inch rigid conduit is installed from the top of the panelboard into the attic space and nonmetallic sheath cables are pulled within that 2 inch conduit. Is this a permitted wiring method?

    Answer: _____

    _____

    _____

    _____

    Reference: _____

22. A small lake cabin is to be wired. This cabin will contain wood heating, no air conditioning, and only six single-pole branch circuits for lighting and receptacle outlets. Using only six switches in this panel, are you therefore exempt from using a main circuit breaker or a single disconnecting switch as required in Chapter 2 for services?

    Answer: _____

    _____

    _____

    _____

    Reference: _____

23. A location in a small commercial office building appears to be dangerous. The light switches located on the wall are installed in a standard three-gang switch box. Each light switch controls a row of fixtures, and each is fed from a different phase in the panelboard. The light switch on the right is fed from B-phase of a 480-volt, 277-volt panelboard. The center switch is fed from B-phase, and the switch on the right is fed from C-phase, making the potential between each switch at 480 volts. This appears to be dangerous, but you can find no section in the Code that prohibits it.

    Answer: _____

    _____

    _____

    _____

    Reference: _____

24. Two new circuits are to be added in an existing old house with knob-and-tube wiring. The owner has asked that this extension be made with the same wiring method that is now in the house, concealed knob-and-tube. The attic space has had 6 inches of blown insulation installed; the wires are covered by the insulation. The owner has stated that you should pull the insulation back, make the extension, and then relocate the insulation over the wiring. Is this wiring method still acceptable in the Code?

    Answer: _____

    _____

    _____

    _____

    Reference: _____

25. What is the minimum size single conductor cable permitted within cable trays to be used as ungrounded circuit conductors?

Answer: _____

_____

_____

_____

Reference: _____

Chapter Eleven

GENERAL USE UTILIZATION EQUIPMENT

The equipment covered in Chapter 4 is generally found in most facilities—residential, commercial, and industrial. Chapter 4 begins with two articles about wire. One might think that these articles are misplaced and should be found in Chapter 3. However, flexible cords and cables and fixture wires are usually associated with general purpose equipment, such as flexible cords for appliances, motors, and fixture wiring is usually associated with lighting fixtures. Chapter 4 covers all types, lighting fixtures, appliances, fixed electric space heating equipment, motors, transformers, air conditioning and refrigeration equipment, and several special use types of general equipment, such as phase converters often used in rural farm type installations, and capacitors now generally found in many types of equipment that improves the efficiency of many services as a power factor correction. Generators are often used to supplement or provide emergency power or standby power. As you study Chapter 4 of the *NEC®*, be aware that the requirements in 410 cover not only lighting fixtures but also cover receptacles, cord connectors, and attachment caps. (See Figure 11–1.) However, Article 410 does not cover signs. Article 422, which covers the appliances, generally has specific requirements for branch circuits, installation of appliances, and the control and protection of appliances. Space heating, motors, air conditioning and refrigeration equipment articles also have general provisions and installation provisions within those articles. Study them carefully for correct application. Many examination questions are derived from Chapter 4.

Figure 11–1 *(Courtesy of TayMac Corporation)*

410-57 (b) Exception
REVISION

LANGUAGE

410-57. Receptacles in Damp or Wet Locations.

(b) Wet Locations

A receptacle installed outdoors where exposed to weather or in other wet locations shall be in a weatherproof enclosure, the integrity of which is not affected when the receptacle is in use (attachment plug cap inserted).

Exception: An enclosure that is weatherproof only when a self-enclosing receptacle cover is closed shall be permitted to be used for a receptacle installed outdoors where the receptacle is provided for use with portable tools or other portable equipment normally connected to the outlet only when attended.

PURPOSE

To assure receptacles remain weatherproof when equipment is used in wet locations.

APPLICATION

The main rule would apply to equipment, such as outdoor vending machines, pool pump motors and low voltage lighting. The exception would apply to equipment, such as portable lawn tools or other portable equipment, that is used in wet locations.

WET LOCATIONS

"Std Wp cover"

"Raintite" while in use

Meets exception

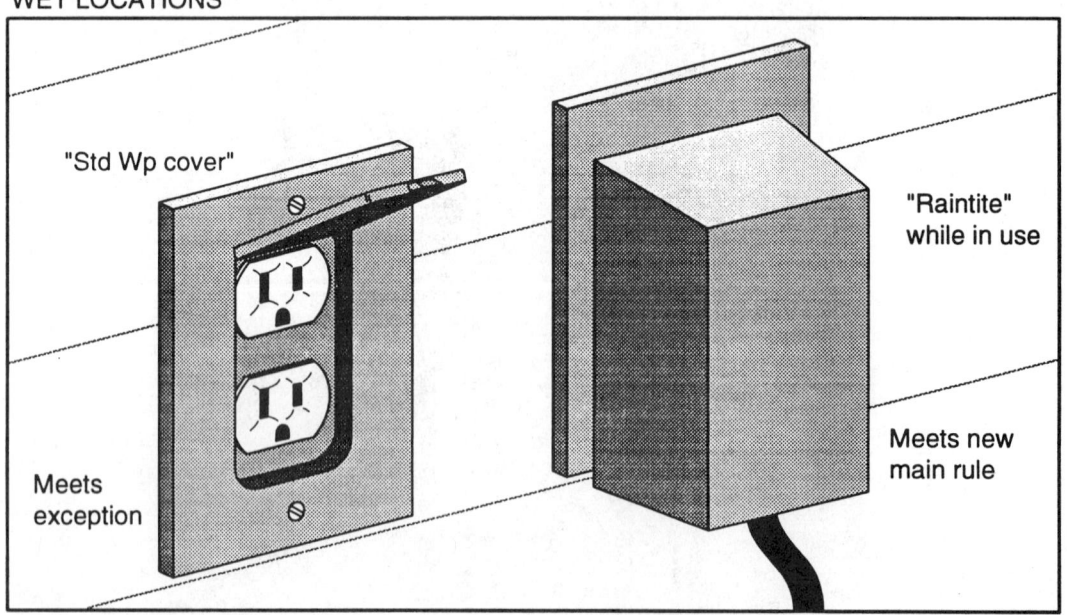

Meets new main rule

CHAPTER 11 QUESTION REVIEW

THREE-PHASE MOTOR AS FOLLOWS:
 Type: squirrel cage, high reactance
 Voltage: 460 volts, three-phase
 HP: 30
 Current: nameplate is the same as NEC® current Tables 430-147-150
 Duty: continuous
 Start: high reactance
 Code letter: none
 Service factor: none

Refer to the three-phase motor above for Questions 1–3.

1. What NEC® section requires that the Tables 430-147-150 be used instead of actual current rating marked on motor nameplate?

   Answer: _____

   _____

   _____

   _____

   Reference: _____

2. What size inverse time circuit breaker is needed to provide the maximum branch circuit, short-circuit, ground-fault protection?

   Answer: _____

   _____

   _____

   _____

   Reference: _____

3. A dual element (time delay) fuse can be sized a maximum of _____ % when used as branch circuit protection.

   Answer: _____

   _____

   _____

   _____

   Reference: _____

132 Master Electrician's Review

4. A 120-volt single-phase store sign panel supplies only the sign as follows: Hours-on 24 hours a day, three ¼ horsepower motors, 24-0.80 amp ballasts. Each ungrounded conductor in the subfeeder to the sign panel must be at least No. ____ AWG *aluminum* THW.

   Answer: _____

   _____

   _____

   _____

   Reference: _____

5. Where small conductors are tapped to larger conductors as permitted in the *NEC®*, the tap cannot be larger than the _____ size of the conductor being tapped.

   Answer: _____

   _____

   _____

   _____

   Reference: _____

In Questions 6 through 9 calculate the ampere rating of the fuse size needed for protection of the following copper conductors where no motors are connected:

6. No. 12 THHN _____ ampere rated fuse

   Answer: _____

   _____

   _____

   _____

   Reference: _____

7. No. 8 TW _____ ampere rated fuse

   Answer: _____

   _____

   _____

   _____

   Reference: _____

8. No. 3 THW _____ ampere rated fuse

   Answer: _____

   _____

   _____

   _____

   Reference: _____

9. No. 3/0 THWN _____ ampere rated fuse

   Answer: _____

   _____

   _____

   _____

   Reference: _____

10. For a transformer that has no secondary overcurrent protection, the primary overcurrent protection shall not exceed _____% of the primary current.

    Answer: _____

    _____

    _____

    _____

    Reference: _____

11. The maximum branch circuit and ground fault protection using time-delay fuses for typical motors are sized at _____% of the motor's full-load current.

    Answer: _____

    _____

    _____

    _____

    Reference: _____

12. A motor pulls 22 amperes at 230 volts, and the feeder circuit is 150 feet in length. If a No. 10 copper wire is desired, what would be the voltage drop?

    Answer: _____

    _____

    _____

    _____

    Reference: _____

134 Master Electrician's Review

13. What is the percentage of voltage drop on the circuit described in Question 12?

    Answer: _____

    Reference: _____

14. What is the correct size conductor to maintain 3% or less voltage drop in Question 12?

    Answer: _____

    Reference: _____

15. Overload protection (heaters, relays, time delay fuses, thermal overloads) for typical motors having a service factor of 1.15 is generally sized at ____ percent of the motor's full-load current.

    Answer: _____

    Reference: _____

16. A motor controller that is installed with the expectation of its being submerged in water occasionally for short periods, shall be installed in a rated enclosure type No. ____.

    Answer: _____

    Reference: _____

17. Where recessed high-intensity discharge fixtures are installed _____ and operated by remote ballasts, both the fixture and the ballast require thermal protection.

    Answer: _____

    Reference: _____

18. Any pipe or duct system foreign to the electrical installation must **not** enter a transformer vault. The _____ piping is **not** considered foreign to the vault.

    Answer: _____

    _____

    _____

    _____

    Reference: _____

19. A 10 kVA dry type transformer rated at 480 volts can be installed on a building column and shall **not** be required to be _____.

    Answer: _____

    _____

    _____

    _____

    Reference: _____

20. A capacitor is located indoors. It must be enclosed in a vault if it contains more than a minimum of _____ gallon(s) of flammable liquid.

    Answer: _____

    _____

    _____

    _____

    Reference: _____

21. A single-phase hermetic refrigerant motor-compressor has a rated load current of 24 amps and a branch circuit selection current of 30 amps. The branch circuit conductors are copper, type TW. They operate at 80° Fahrenheit, and they are the only conductors in the conduit to this compressor. The smallest possible branch circuit conductors must be least size _____ AWG.

    Answer: _____

    _____

    _____

    _____

    Reference: _____

136 Master Electrician's Review

22. A 50-horsepower, three-phase, squirrel-cage motor with full-voltage reactor starting has no code letter. The calculated size of the fuse to protect the branch circuit of the motor would be sufficient for the starting current of the motor. The nontime delay fuse to protect the circuit of this motor shall be sized at ____ % of full-load current.

Answer: _____

_____

_____

_____

Reference: _____

23. A dry type transformer is to be installed indoors. If rated more than ____ kilovolt-amps, the transformer must be installed in a fire-resistant transformer room.

Answer: _____

_____

_____

_____

Reference: _____

24. An electric resistance heater is rated for 2400 watts at 240 volts. What power is consumed when the heater is operated at 120 volts?

Answer: _____

_____

_____

_____

Reference: _____

25. A 230-volt single-phase circuit has 10 kilowatts of load and 50 amps of current. The power factor is ____ %.

Answer: _____

_____

_____

_____

Reference: _____

Chapter Twelve

SPECIAL EQUIPMENT AND OCCUPANCIES

As you begin to study Chapter 12, remember Chapter 11 covered general equipment, equipment that is found in most every common installation, equipment that is for general use, such as light fixtures, receptacles, motors, transformers, and so forth. In this chapter, we note that we are studying special occupancies, special equipment, and special conditions. These rules apply to these items and are not generally found in all buildings. They are special. Study carefully Article 90, Chapter 1, and Article 300 of the text. Article 90-3 outlines the use of the *NEC®* and states that Chapters 1 through 4 apply generally and Chapters 5, 6, and 7 amend, supplement, or modify Chapters 1 through 4.

Special occupancies include such things as hazardous areas, bulk storage plants, service stations, paint spray booths, commercial garages, and many more hazardous occupancies. They also include occupancies, such as hospitals, mobile homes, and agriculture buildings. On the other hand, Chapter 6 deals with special equipment, such as signs, not found in every building but still common equipment. Swimming pools, welding equipment, electric welders, elevators, dumbwaiters, office furnishings, and many more types are covered. Chapter 7 deals with special conditions. These are conditions that are special and not general conditions, such as emergency systems, and legally and optional standby systems, over 600 volt systems, optical fiber cables, and raceways. Finally, in Chapter 7 is Article 780, which deals with the intelligent buildings, such as Smart House, the closed loop systems that offer a totally different concept to the safe use of electricity as we have known it in the past. Study the questions carefully. Remember as you study these chapters, they modify, supplement, or amend Chapters 1 through 4 of the *NEC®*.

CHAPTER 12 QUESTION REVIEW

1. Name three types of optical fiber cable.

   Answer: _____

   _____

   _____

   _____

   Reference: _____

2. Fixed wiring over a Class I location in a commercial garage cannot be enclosed in nonmetallic sheathed cables. (True or False)

   Answer: _____

   _____

   _____

   _____

   Reference: _____

138 Master Electrician's Review

3. Are seals required for outdoor propane-dispensing unit located 50 feet from the office where the branch circuit supplying the unit originates? If so, where are they to be placed?

   Answer: _____

   _____

   _____

   _____

   Reference: _____

4. Emergency electrical systems are those systems legally required and classed as emergency by Article 700 *NEC®*? (True or False)

   Answer: _____

   _____

   _____

   _____

   Reference: _____

5. In a commercial garage, Class ____ Division ____ wiring methods must be met for an electrical outlet installed 12 inches above the floor if there is no mechanical ventilation?

   Answer: _____

   _____

   _____

   _____

   Reference: _____

6. According to the *National Electrical Code®*, transfer equipment for emergency systems shall be designed and installed to prevent _____ of normal and emergency sources of power.

   Answer: _____

   _____

   _____

   _____

   Reference: _____

7. The sealing compound in a completed conduit seal for a Class I location must be at least ____ inch thick.

   Answer: _____

   _____

   _____

   _____

   Reference: _____

8. Bare conductors are field-connected to fixed terminals of different phases on an outdoor, 13.8 kV circuit. According to the *National Electrical Code®*, the air separation between these conductors shall be a minimum of _____ inches.

   Answer: _____

   _____

   _____

   _____

   Reference: _____

9. In a hospital's general care areas, the number of receptacles required to be in a patient-bed location is a minimum of

   Answer: _____

   _____

   _____

   _____

   Reference: _____

10. A 120-volt single-phase store sign panel supplies only the sign as follows: Hours-on 24 hours a day, three ⅓ horsepower motors, 36-0.80 amp ballasts. Each ungrounded conductor in the subfeeder to the sign panel must be at least No. _____ AWG copper THWN.

    Answer: _____

    _____

    _____

    _____

    Reference: _____

11. Portable gas tube signs for interior use can be connected with a supply cord having a maximum length of _____ meters. (feet)

    Answer: _____

    _____

    _____

    _____

    Reference: _____

12. Electric discharge tubing shall be a _____ and _____ so as to **not** to cause steady overvoltage on the transformer.

    Answer: _____

    _____

    _____

    _____

    Reference: _____

Answer the following questions about Type S fuses True or False.

13. The S indicates "size rejection."

    Answer: _____

    _____

    _____

    _____

    Reference: _____

14. A 15-ampere Type S fuse will fit into a 20-ampere Type S adapter.

    Answer: _____

    _____

    _____

    _____

    Reference: _____

15. A 20-ampere Type S fuse will fit into a 20-ampere Type S adapter.

    Answer: _____

    _____

    _____

    _____

    Reference: _____

16. A 20-ampere Type S fuse will fit into a 15-ampere Type S adapter.

    Answer: _____

    _____

    _____

    _____

    Reference: _____

17. A 30-ampere Type S fuse will fit into a 20-ampere Type S adapter.

    Answer: _____

    _____

    _____

    _____

    Reference: _____

18. A 25-ampere Type S fuse will fit into a 30-ampere Type S adapter.

    Answer: _____

    _____

    _____

    _____

    Reference: _____

19. A $6\frac{1}{2}$ ampere Type S adapter will accept a $6\frac{1}{4}$ Type S fuse.

    Answer: _____

    _____

    _____

    _____

    Reference: _____

20. A $6\frac{1}{4}$ Type S adapter will accept a 3 amp, 4, $4\frac{1}{2}$, 5, $5\frac{6}{10}$, and $6\frac{1}{4}$ ampere Type S fuse.

    Answer: _____

    _____

    _____

    _____

    Reference: _____

21. The *NEC®* requires that the available fault current be marked when series rated systems are installed. (a) Who is required to install that marking? (b) What must the marking state? (c) Who is responsible for providing this information?

    Answers: _____

    _____

    _____

    _____

    Reference: _____

142 Master Electrician's Review

22. A 1½ horsepower single-phase motor which has an efficiency of 80%, operates at 230 volts, and has an input current of _____ amps. (1 hp = 746 watts.)

   Answer: _____

   _____

   _____

   _____

   Reference: _____

23. A 30-horsepower wound-rotor induction motor with no code letter is to be installed with 460-volt, three-phase, alternating current. Disregarding all the exceptions, the nontime delay fuse for short-circuit protection of the motor branch circuit must be rated at a maximum _____ amps.

   Answer: _____

   _____

   _____

   _____

   Reference: _____

24. The following 480-volt, three-phase, three-wire, intermittent use equipment is in a commercial kitchen: two 5,000 watt water heaters, four 3,000 watt fryers, and two 6,000 watt ovens. Each ungrounded conductor in the feeder circuit for this kitchen equipment must be sized to carry a minimum computed load of _____ amps.

   Answer: _____

   _____

   _____

   _____

   Reference: _____

25. A 240/480 3-phase power panelboard supplies only one, 15,000-VA, 480-volt, three-phase balanced resistive load. Each ungrounded conductor in the subfeeder to this power panel has a total net computed load of _____ amps.

   Answer: _____

   _____

   _____

   _____

   Reference: _____

Chapter Thirteen

COMMUNICATION INSTALLATIONS

This chapter deals primarily with Chapter 8 of the *NEC®*, which covers communications systems. Chapter 8 is independent of the other chapters, except where they are specifically referenced therein. Chapter 8 of the *NEC®* stands alone, and therefore, none of the rules in the previous chapters apply unless specifically referenced. For instance, an often asked question is "What is the burial depth of communication lines on residential property?" Many electricians reference *NEC®* Table 300-5, which gives the burial depth for conductors, cables, and raceways. However, Chapter 8 does not reference Table 300-5, and therefore the table cannot be applied. There are no burial depth requirements for these communication cables because the *NEC®* is primarily a safety document. Communications generally pose no safety hazards by interruption of service. Therefore, the installers often bury these cables for service and economic reasons. However, they pose no safety hazards so the *NEC®* does not set any requirements. Local requirements or amendments may apply. Chapter 8 is unique. Study it carefully. Study the scope of each article carefully to be sure that you are answering a question from the right section of this chapter.

CHAPTER 13 QUESTION REVIEW

1. Does the Code address grounding CATV service to a dwelling unit?

    Answer: _____

    _____

    _____

    _____

    Reference: _____

2. After making an installation of a satellite dish in a local residence, the inspector informs you that you must bury the cable from the satellite dish to the house in accordance with Table 300-5 of the *NEC®*. Was that inspector correct?

    Answer: _____

    _____

    _____

    _____

    Reference: _____

3. Does the Code require grounding the metal sheath of CATV cable entering a dwelling?

   Answer: _____

   _____

   _____

   _____

   Reference: _____

4. According to the *National Electrical Code*®, open conductors for communication equipment on a building shall be separated at least ____ feet from *lightning* conductors.

   Answer: _____

   _____

   _____

   _____

   Reference: _____

5. In a highly hazardous location where propane is used, the customer states that the wiring to the control circuit is of sufficient low voltage that a spark cannot be generated from this circuit. The devices and wiring have a control drawing number on them. What type wiring is most likely employed in this installation where there is not enough voltage to generate a spark sufficient to cause an explosion?

   Answer: _____

   _____

   _____

   _____

   Reference: _____

6. A motor control center is to be installed in a local factory. Does the *National Electrical Code*® cover this type of installation? If so, which article is applicable?

   Answer: _____

   _____

   _____

   _____

   Reference: _____

7. You have been asked to install a small transformer that will transform 480 volts to 240/120 volts. The inspector states this must be wired as a separately derived system. What is a separately derived system? How do you wire this transformer?

   Answer: _____

   _____

   _____

   _____

   Reference: _____

8. You recently encountered an orange receptacle. You were told this was an isolated ground receptacle. Where can this be found in the *National Electrical Code®*?

   Answer: _____

   _____

   _____

   _____

9. You recently encountered a rental storage facility that is an area in a building to store old movie films. Where would the wiring requirements for this area be found in the *NEC®*?

   Answer: _____

   _____

   _____

   _____

10. The inspector informed you that a bathroom area was defined in the *NEC®*. You looked under the definitions in Article 100 and could not find it. Where is it?

    Answer: _____

    _____

    _____

    _____

    Reference: _____

11. In wiring a large church, an installation of a pipe organ was to be required. Are pipe organs covered in the *NEC®*?

    Answer: _____

    _____

    _____

    _____

    Reference: _____

146 Master Electrician's Review

12. In making an installation, it was noted the clearance between two large switchboards was rather tight, only about 3 feet. With the doors open there was not room to pass. Are there any requirements in the *National Electrical Code®* to cover the requirements for working clearances?

    Answer: _____

    _____

    _____

    _____

    Reference: _____

13. After completing the wiring of a new home, the inspector came and told you to cover the panelboard, that if the sheetrockers got their spackling material or paint on the panel it was a code violation. Is this correct?

    Answer: _____

    _____

    _____

    _____

    Reference: _____

14. You recently mounted a new service on a new mobile home on a privately owned lot. The inspector turned it down and said this was not permitted. Which section of the *National Electrical Code®* did you violate?

    Answer: _____

    _____

    _____

    _____

    Reference: _____

15. What is an optional standby system?

    Answer: _____

    _____

    _____

    _____

    Reference: _____

16. Is color coding required in the *NEC®*? You know that the neutral conductor is required to be white and the grounding conductor is required to be green, but what about the other ungrounded conductors?

    Answer: _____

    _____

    _____

    _____

    Reference: _____

17. You recently installed No. 14 THHN conductors to supply a general purpose branch circuit supplying duplex receptacles in an office area. You connected the No. 14 Type THHN to a 20-ampere breaker. Table 310-16 allows 25 amperes on a No. 14 conductor. Was this in accordance with the *NEC®*?

    Answer: _____

    _____

    _____

    _____

    Reference: _____

18. In making an installation to an outdoor ballfield, you proposed a common neutral to be run with multiple branch circuit conductors. How many branch circuit conductors can be supplied with a common neutral?

    Answer: _____

    _____

    _____

    _____

    Reference: _____

19. You have been asked to wire a dairy barn. Are there any specific requirements in the *NEC®* for this type installation?

    Answer: _____

    _____

    _____

    _____

    Reference: _____

148 Master Electrician's Review

20. You recently observed a local electrician installing wiring under a carpet in a bank. Is this wiring method permitted? If so, which article covers wiring under carpets?

    Answer: _____

    _____

    _____

    _____

    Reference: _____

21. More buildings are increasingly requiring lightning arrestors to be installed on the service. Which section of the *National Electrical Code*® covers this installation?

    Answer: _____

    _____

    _____

    _____

    Reference: _____

22. Are temporary wiring methods covered in the *National Electrical Code*®? You regularly install a temporary meter loop to supply power to the construction crews until the permanent power system has been installed. Where is this covered in the *NEC*®?

    Answer: _____

    _____

    _____

    _____

    Reference: _____

23. A 480-volt, three-phase circuit with a rated load of 150 KW draws 210 amperes. The power factor is _____ percent.

    Answer: _____

    _____

    _____

    _____

    Reference: _____

24. A _____ horsepower, 208-volts, three-phase motor operating at 91% efficiency has an input current of 141 amperes (HP = 746 watts).

    Answer: _____

    _____

    _____

    _____

    Reference: _____

25. A feeder supplying a second building requires a maximum conductor resistance of 0.02 ohms. The minimum size conductor must have a resistance of no more than _____ ohms per 1000 feet. The total length of the circuit conductor is 190 feet.

    Answer: _____

    _____

    _____

    _____

    Reference: _____

Chapter Fourteen

ELECTRICAL AND *NEC*® QUESTION REVIEW

Now you are about to begin the final chapter of this book, Chapter 14. You should have a good understanding of the Code arrangement and its content. You should now understand that the chapters are arranged in the order in which you use them in most common installations, from Article 90, Introduction; Chapter 1, general requirements; Chapter 2, wiring and protection; Chapter 3, wiring methods and materials; Chapter 4, general use equipment; Chapters 5, 6, and 7, special occupancies, uses and equipment; and Chapter 8, communication systems. The final chapter in the *NEC*® is Chapter 9, which has the tables and examples. Furthermore, you should now understand this arrangement and know the articles contained in each chapter. For instance, you should understand that the article covering rigid metal conduit would automatically be found in Chapter 3, because rigid metal conduit is a wiring method. You should understand that the conditions for installation of a light fixture would be found in Chapter 4, because light fixtures are general use equipment. Or you should understand that hospital installation requirements would be found in Chapter 5, because a hospital is a special occupancy and those are found in Chapter 5, such as mobile homes, hazardous materials and spray booths, commercial garages, and so on. You should automatically remember that a sign would be found in Chapter 6, special equipment because although signs are common, they are not general equipment found in all types of installations. Signs are special equipment and are found in Chapter 6. These things must be understood before your taking the examination. It is necessary so that you might use the *National Electrical Code*® efficiently as a reference document. Memorization is not recommended. However, with a good, clear understanding, one can refer to the Table of Contents in the front of the book and quickly find the article number if you have an understanding of where that article should be found in the Code. You should then be able to go to the Index and find the proper reference so you might quickly look up the question and prove the answer. Once this has been done, you should have no problem passing any competency exam given throughout the country that is based on your knowledge of the *National Electrical Code*®.

With that competency achieved and your understanding of the *National Electrical Code*® arrangement ensured, the next important task you must improve is the understanding of what each question is asking. You must read the question through, then determine exactly what the question is asking. You will find key words, and with those key words you can simplify the question generally so it can be quickly researched using the Table of Contents, the Index, and your knowledge of the Code arrangement. As an example, in the following question, select the key words and simplify the question:

> "May Type NM cable be used for a 120/140-volt branch circuit as temporary wiring in a building under construction where the cable is supported on insulators at intervals of not more than 10 feet?"

Now, you may rephrase the question slightly different in your mind, or you may look for the key words. NM cable is a key word, temporary wiring is a key, and supported on insulators at intervals of not more than 10 feet is a key. So,

> "Is NM cable permitted as temporary wiring where supported on insulators."

Now, with the three key words in mind, we must remember that they are not asking where this wiring method is acceptable, they are telling you that the installation would be temporary wiring. If we go to the Table of Contents in the front of the *NEC*®, we find that temporary wiring is found in Article 305. As we turn to Article 305 and we look in the scope under 305-1, we find out that yes, lesser methods are acceptable. As we research that article quickly, we find in Section 305-4(c) that this open wiring cable is an acceptable method, provided that it is supported on insulators at intervals of not more than 10 feet, so the answer would be yes. References would be Article 305, Sections 305-1 and 305-4(c). As you see in this case when selecting the key words or phrases, using the Table of Contents was sufficient and it was not necessary to go to the Index.

Let's look at another example. The question is:

"Is liquidtight flexible nonmetallic conduit permitted to be used in circuits in excess of 600 volts?"

As we look at that question, we see that liquidtight flexible nonmetallic conduit is the key phrase. Then we look at permitted use as a key phrase and 600 volts as the final key phrase.

Again, we can go to the Table of Contents and find that liquidtight flexible nonmetallic conduit is found in Article 351, Part B. As we turn to Article 351, Part B, we see uses permitted in Section 351-23(a), uses not permitted in Section 351-23(b). Immediately as we scan down the uses permitted and the uses not permitted, we find that in Section 351-23(b)(4) that liquidtight flexible nonmetallic conduit is not permitted for circuits in excess of 600 volts. However, you will note there is an exception for electrical signs, so the answer is: "Yes, for electrical signs by exception. No, generally."

If you will take each question and break it down so you are looking at the key phrase or key word, you will find it much easier to reference the correct Code article and section, and you will not waste the valuable time needed to complete your examination. All questions will not be as easy as these two examples to identify the key phrases, but in all questions there are key phrases or key words and with the practice you will receive during the following exercises in Chapter 14, you should greatly improve your competency and speed when preparing for an examination, which will allow you ample time to get through the examination and still have some time to do additional research on the tough questions. What you will find by following these examples is that there will be fewer tough questions in the exam. Good luck!

PRACTICE EXAM 1

1. Are all 125-volt, 15- and 20-ampere receptacles in the service area of a commercial garage required to be protected by a GFCI?

   Answer: _____

   _____

   _____

   _____

   Reference: _____

2. A lighting fixture is installed over a hydromassage bathtub. Is GFCI protection required for the fixture?

   Answer: _____

   _____

   _____

   _____

   Reference: _____

3. A nongrounding type receptacle is to be replaced with a GFCI receptacle because a grounding means does not exist within the box. Can you supply downstream receptacles from the GFCI? If yes, are these downstream receptacles required to be two-wire, nongrounding type?

   Answer: _____

   _____

   _____

   _____

   Reference: _____

152 Master Electrician's Review

4. A kitchen range has a built-in 125-volt receptacle and is within 6 feet of the sink. Does the Code require such a receptacle to be protected with a GFCI?

   Answer: _____

   _____

   _____

   _____

   Reference: _____

5. Does the 1993 *NEC®* prohibit the use of nonmetallic outlet boxes with metal raceways?

   Answer: _____

   _____

   _____

   _____

   Reference: _____

6. Does the 1993 *NEC®* permit lighting fixtures to be installed on trees?

   Answer: _____

   _____

   _____

   _____

   Reference: _____

7. Electrical nonmetallic tubing (ENT) can be used concealed within walls, ceilings, and floors where the walls, ceilings, and floors provide a thermal barrier of material that has at least

   Answer: _____

   _____

   _____

   _____

   Reference: _____

8. Does the Code require any extra protection for electrical nonmetallic tubing (ENT) when it is installed through openings in metal studs?

   Answer: _____

   _____

   _____

   _____

   Reference: _____

9. What are the strapping requirements for ENT when installed in metal studs?

   Answer: _____
   _____
   _____
   _____

   Reference: _____

10. Does the Code require any extra protection for type NM cable when it is installed through openings in metal studs?

    Answer: _____
    _____
    _____
    _____

    Reference: _____

11. When UF cable is used for interior wiring, are the conductors required to be rated at 90°C?

    Answer: _____
    _____
    _____
    _____

    Reference: _____

12. Must "hospital grade" receptacles be installed throughout hospitals?

    Answer: _____
    _____
    _____
    _____

    Reference: _____

13. When railings or lattice work are used for room dividers, does the Code require receptacles to be installed as if they were solid walls?

    Answer: _____
    _____
    _____
    _____

    Reference: _____

14. In an apartment project, multi-wire branch circuits are routed through outlet boxes to supply receptacles. Can the neutral conductor "continuity be assured" by two connections to the screw terminals of the receptacle?

    Answer: _____

    _____

    _____

    _____

    Reference: _____

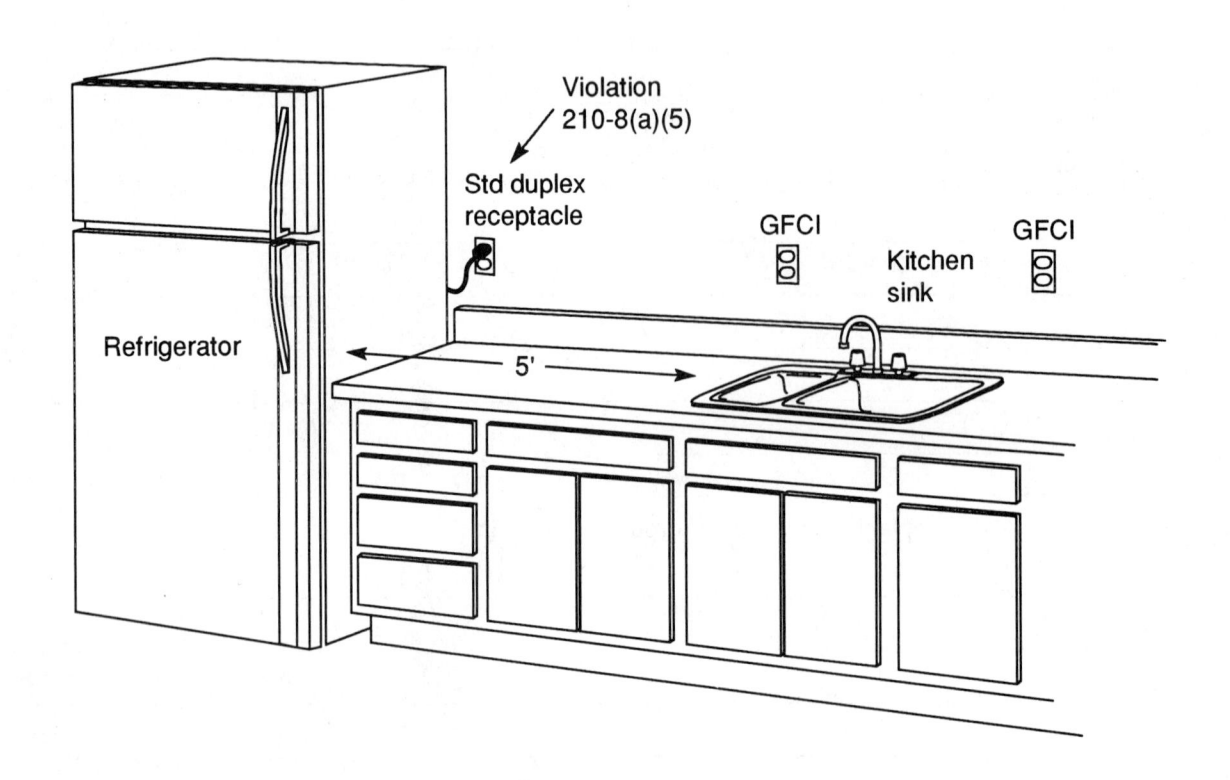

[Illustration: Kitchen with refrigerator plugged into "Std duplex receptacle" labeled "Violation 210-8(a)(5)", 5' from kitchen sink with GFCI receptacles on either side.]

15. A small appliance circuit receptacle is above the counter next to the refrigerator. The refrigerator is plugged into this receptacle that is 5 feet from the kitchen sink. Is the receptacle required to have GFCI protection?

    Answer: _____

    _____

    _____

    _____

    Reference: _____

16. Does the Code require outdoor receptacles to be installed on balconies of high-rise apartment buildings?

    Answer: _____
    _____
    _____
    _____

    Reference: _____

17. The Code did require that switches and circuit breakers used as switches be installed so the center of the grip of the operating handle when in its highest position was no more than 6 feet above the floor or working platform. Is this still the case?

    Answer: _____
    _____
    _____
    _____

    Reference: _____

18. When can lighting fixtures be used as raceways for circuit conductors?

    Answer: _____
    _____
    _____
    _____

    Reference: _____

19. In what areas of a hospital are hospital grade receptacles required?

    Answer: _____
    _____
    _____
    _____

    Reference: _____

20. Is it a requirement that receptacles over a wet bar be spaced according to cabinet or floor line requirements?

    Answer: _____
    _____
    _____
    _____

    Reference: _____

156 Master Electrician's Review

21. A weatherproof receptacle cover must be weatherproof when in use (attachment cap inserted) or shall have a _____ feature.

    Answer: _____

    _____

    _____

    _____

    Reference: _____

22. Would electrical nonmetallic tubing be an acceptable wiring method for a wet-niche lighting fixture or some portion of the circuit?

    Answer: _____

    _____

    _____

    _____

    Reference: _____

23. Can nonmetallic sheathed cable be used to supply a recessed fluorescent fixture?

    Answer: _____

    _____

    _____

    _____

    Reference: _____

24. Do conductors for festoon lighting have to be rubber covered?

    Answer: _____

    _____

    _____

    _____

    Reference: _____

25. Can an electric discharge fixture such as a fluorescent strip be used as a raceway for circuit conductors?

    Answer: _____

    _____

    _____

    _____

    Reference: _____

PRACTICE EXAM 2

1. All fixtures located in agriculture buildings are to be listed as dusttight or watertight?

   Answer: _____
   _____
   _____
   _____

   Reference: _____

2. When installing AC cable horizontally through metal studs, are insulators needed?

   Answer: _____
   _____
   _____
   _____

   Reference: _____

3. What types of cable assemblies are acceptable for use as feeders to a floating building?

   Answer: _____
   _____
   _____
   _____

   Reference: _____

4. Would nonmetallic sheathed cable be acceptable to wire a pool associated motor?

   Answer: _____
   _____
   _____
   _____

   Reference: _____

5. Would liquidtight flexible conduit be acceptable for enclosing conductors from an adjacent weatherproof box to an approved swimming pool junction box?

   Answer: _____
   _____
   _____
   _____

   Reference: _____

6. Is the cord and plug connection of an electric water heater acceptable under the *NEC*®? If so, would a standard dryer cord be acceptable (240 V with neutral)? (Assume ampacities, and so on, were acceptable under the *NEC*®.)

   Answer: _____

   _____

   _____

   _____

   Reference: _____

7. Does the Code permit motor controllers and disconnects to be used as junction boxes or wireways?

   Answer: _____

   _____

   _____

   _____

   Reference: _____

8. In determining the locked-rotor kVA per horsepower of dual voltage electric motor, which voltage must be considered?

   Answer: _____

   _____

   _____

   _____

   Reference: _____

9. Would electrical nonmetallic tubing be acceptable to enclose the insulated conductors for a panelboard (not part of service equipment) feeding pool associated equipment?

   Answer: _____

   _____

   _____

   _____

   Reference: _____

10. Does the Code permit the installation of 15-ampere rated receptacles on the 20-ampere small appliance branch circuits required in a dwelling?

    Answer: _____

    _____

    _____

    _____

    Reference: _____

11. Does the Code permit the installation of a 15-ampere rated duplex receptacle on a 20-ampere separate circuit for a microwave oven with a nameplate reading of 13 amperes?

    Answer: _____

    _____

    _____

    _____

    Reference: _____

12. What does the term "conduit in free air" mean at the top of Table 310-17?

    Answer: _____

    _____

    _____

    _____

    Reference: _____

13. Has the 1993 *National Electrical Code®* restricted nonmetallic conduit (PVC) from being installed in cold weather areas?

    Answer: _____

    _____

    _____

    _____

    Reference: _____

14. All temporary wiring 15A and 20A, 125 V receptacles not a part of the permanent building structure need to have GFCI protection. Does this mean that if a permanently installed receptacle is installed and energized that no GFCI protection is required even though this outlet is to be used by construction personnel?

    Answer: _____

    _____

    _____

    _____

    Reference: _____

15. In the 1993 Code would a GFCI protected receptacle be required on a wet bar in a dwelling?

    Answer: _____

    _____

    _____

    _____

    Reference: _____

16. Must the garbage disposal receptacle located under the sink be GFCI protected because it is within 6 feet of the sink?

    Answer: _____
    _____
    _____
    _____

    Reference: _____

17. What is the maximum number of duplex receptacles permitted on a two-wire, 15-amp circuit in a dwelling?

    Answer: _____
    _____
    _____
    _____

    Reference: _____

18. How many receptacles are required at the patient's bed location in a critical care facility?

    Answer: _____
    _____
    _____
    _____

    Reference: _____

19. A ¼ horsepower circulating pump is to be located in a closet space housing a gas fired water heater in a large single-family residence. Can the circulating pump be cord and plug connected?

    Answer: _____
    _____
    _____
    _____

    Reference: _____

20. In lieu of switching a ceiling light, the Code permits switching a receptacle. Can the switched receptacle be counted as one of the receptacles required for the 12-foot rule?

    Answer: _____
    _____
    _____
    _____

    Reference: _____

21. A laundry room in a single family residence has a receptacle outlet on one wall and a second receptacle on the opposite wall. Can these two outlets be fed by the same laundry branch outlet?

    Answer: _____

    _____

    _____

    _____

    Reference: _____

22. Antique stores and others are selling hanging fixtures without a grounding conductor. Are these fixtures in violation of the *NEC*®?

    Answer: _____

    _____

    _____

    _____

    Reference: _____

23. Can 6 feet of flexible metal conduit be installed between "rigid nonmetallic" raceways used for service entrance conductors?

    Answer: _____

    _____

    _____

    _____

    Reference: _____

24. Many kitchen and bathroom sinks, tubs, and showers are provided with short sections of metal pipe (6 inches to 2 feet) for connection to nonmetallic pipe systems. Is it required to bond the several short sections of metal pipe to the service equipment, etc.?

    Answer: _____

    _____

    _____

    _____

    Reference: _____

162   Master Electrician's Review

25. Is there a limit on the number of extension boxes that can be joined together?

    Answer: _____
    _____
    _____
    _____

    Reference: _____

PRACTICE EXAM 3

1. A switch box was installed in a floor of a dwelling. A receptacle outlet with a standard plate was installed on the box. Does this standard outlet box and receptacle installed in the floor comply with the Code?

   Answer: _____
   _____
   _____
   _____

   Reference: _____

2. A 4 inch × 2⅛ inch square NM box is marked with the number of No. 14, 12, and 10 conductors that can be installed in it. Does this mean that No. 8 conductors are not permitted in the box?

   Answer: _____
   _____
   _____
   _____

   Reference: _____

3. Are outdoor receptacle outlets required at each dwelling unit of a multi-family dwelling?

   Answer: _____
   _____
   _____
   _____

   Reference: _____

4. Do portable lamps wired with flexible cord require a polarized attachment plug?

   Answer: _____
   _____
   _____
   _____

   Reference: _____

5. When figuring receptacles for a bedroom in a residence, must you include the space behind the door?

   Answer: _____

   _____

   _____

   _____

   Reference: _____

6. Are metal boxes required for splices in temporary wiring?

   Answer: _____

   _____

   _____

   _____

   Reference: _____

7. What is the maximum number of No. 12 conductors permitted in a 4 inch × 1½ inch deep octagon box?

   Answer: _____

   _____

   _____

   _____

   Reference: _____

8. When a junction box contains a combination of conductors, such as size No. 14 and No. 12 wires, what Table is used to determine the size of the box?

   Answer: _____

   _____

   _____

   _____

   Reference: _____

9. Can you use threaded intermediate metal conduit in a Class 1, Division 1 location?

   Answer: _____

   _____

   _____

   _____

   Reference: _____

10. Does the Code consider EMT as conduit?

    Answer: _____
    _____
    _____
    _____

    Reference: _____

11. Can suspended fluorescent fixtures be connected together with unsupported EMT between them?

    Answer: _____
    _____
    _____
    _____

    Reference: _____

12. Must lay-in fluorescent lighting fixtures be fastened to the grid tee-bars of a suspended ceiling system?

    Answer: _____
    _____
    _____
    _____

    Reference: _____

13. Can fixtures marked "suitable for damp locations" be used in wet locations?

    Answer: _____
    _____
    _____
    _____

    Reference: _____

14. A UL listed fluorescent fixture is equipped with a Class "P" ballast. Does the Code permit such a fixture to be surface mounted on a combustible low density cellulose fiberboard ceiling?

    Answer: _____
    _____
    _____
    _____

    Reference: _____

15. Are all recessed incandescent fixtures required to have thermal protection?

    Answer: _____

    _____

    _____

    _____

    Reference: _____

16. Is it necessary to identify the high leg of a 240/120 volt system at the motor disconnect of a three-phase motor?

    Answer: _____

    _____

    _____

    _____

    Reference: _____

17. Is a GFCI protection required for a receptacle located on a food preparation island that is less than 6 feet from the countertop sink?

    Answer: _____

    _____

    _____

    _____

    Reference: _____

18. Can conductors that pass through a panelboard also be spliced in the panelboard?

    Answer: _____

    _____

    _____

    _____

    Reference: _____

19. Are the enclosures for mercury vapor lighting fixtures installed 8 feet above grade on poles outdoors required to be grounded?

    Answer: _____

    _____

    _____

    _____

    Reference: _____

20. Would an exit light with an emergency pack capable of supplying more than 1½ hours of power have to be connected to an emergency circuit ahead of the main disconnect?

Answer: _____

_____

_____

_____

Reference: _____

21. A grounded 20-amp receptacle is installed in a residential garage to serve an air compressor. Would this receptacle require GFCI protection?

Answer: _____

_____

_____

_____

Reference: _____

22. If additional receptacles are installed in a bathroom, not adjacent to the wash basin, are they required to be GFCI protected?

Answer: _____

_____

_____

_____

Reference: _____

23. Must a receptacle in the kitchen installed for a lighting fixture be GFCI protected?

Answer: _____

_____

_____

_____

Reference: _____

24. When installing recessed fixtures (hi-hats) in the ceiling, does Section 370-20 apply to the fixture housing?

Answer: _____

_____

_____

_____

Reference: _____

25. What are the support requirements for service entrance cables?

   Answer: _____

   _____

   _____

   _____

   Reference: _____

PRACTICE EXAM 4

1. An underground cable is routed from the service equipment in the building underground and up a pole in a commercial parking lot to supply an overhead light. What is the minimum burial depth for the cable below the base of the pole?

   Answer: _____

   _____

   _____

   _____

   Reference: _____

2. Does the Code permit 15-ampere duplex receptacles to be installed on a 20-ampere branch circuit?

   Answer: _____

   _____

   _____

   _____

   Reference: _____

3. We are designing a new hotel with 250 guest rooms. It is our intent to install permanent beds, desks, and dressers in each guest room, bolted to the wall. If receptacles are located in accordance with Section 210-52, the receptacles will be located behind the headboard of the bed and behind the dresser. Are the receptacles required behind these pieces of furniture? Would the number of required receptacles be less if the furniture was not bolted to the wall?

   Answer: _____

   _____

   _____

   _____

   Reference: _____

4. Can a paddle fan be supported only by an outlet box if it weighs less than 50 pounds?

   Answer: _____

   _____

   _____

   _____

   Reference: _____

5. When can lighting fixtures be used as raceways for circuit conductors?

   Answer: _____

   _____

   _____

   _____

   Reference: _____

6. Do recessed incandescent lighting fixtures require thermal protection?

   Answer: _____

   _____

   _____

   _____

   Reference: _____

7. Are medium base lampholders in industrial occupancies permitted on 277-volt lighting circuits?

   Answer: _____

   _____

   _____

   _____

   Reference: _____

8. Can a battery powered (unit equipment) emergency light be supplied directly from a circuit in a panelboard that supplies normal lighting in the area that will be illuminated by the battery light under emergency conditions?

   Answer: _____

   _____

   _____

   _____

   Reference: _____

9. A single transfer switch is permitted to serve one or more branches of the essential electrical system in a small hospital. What constitutes a small hospital?

   Answer: _____

   _____

   _____

   _____

   Reference: _____

10. Would a three-phase, four-wire, 1200-ampere, 480/277 volt service switch with 900 ampere fuses be required to have ground-fault protection?

    Answer: _____

    _____

    _____

    _____

    Reference: _____

11. Are lighting fixtures operating at less than 15 volts between conductors required to be protected by a GFCI over hot tubs or spas?

    Answer: _____

    _____

    _____

    _____

    Reference: _____

12. Could you use EMT (electrical metallic tubing) in the earth below a concrete slab?

    Answer: _____

    _____

    _____

    _____

    Reference: _____

13. An industrial building will be served by several services of different voltage ratings. Is it required that each system be separately grounded?

    Answer: _____

    _____

    _____

    _____

    Reference: _____

170 Master Electrician's Review

14. In residential occupancies, locating equipment to comply with the code becomes a problem. Can a service panel or distribution panelboard be located over a counter or an appliance (washing machine, dryer) that extends away from the wall?

Answer: _____

_____

_____

_____

Reference: _____

15. A surface mounted fluorescent fixture with an integral receptacle is installed above the countertop in the kitchen of a dwelling and is not more than 5½ feet above the floor. Is this fixture required to be supplied by a 20-ampere circuit as are other kitchen receptacles? If within 6 feet of the sink, is the fixture receptacle required to be protected by GFCI?

Answer: _____

_____

_____

_____

Reference: _____

16. What are the requirements for grounding an agricultural building where livestock is housed?

Answer: _____

_____

_____

_____

Reference: _____

17. Can a fused disconnect switch or circuit breaker located on the meter pole of a farm installation be used as the service equipment for the residence?

Answer: _____

_____

_____

_____

Reference: _____

18. A type UF cable feeds a 120-volt yard light location on residential property. The circuit is protected by a 15-ampere overcurrent protection device. It is not GFCI protected. What is the minimum barrel depth permitted for the cable?

Answer: _____

_____

_____

_____

Reference: _____

19. Can a fixed storage-type water heater having a capacity of 120 gallons or less be cord and plug connected?

Answer: _____

_____

_____

_____

Reference: _____

20. A TV dish antenna is located in the side yard of a single-family dwelling. (a) What is the minimum depth required by the *NEC*® for the coaxial signal cable? (b) How are these antenna units required to be grounded?

Answers: _____

_____

_____

_____

Reference: _____

21. Does the Code permit low voltage control cables to be supported from the conduit containing the power circuit conductors feeding an air conditioning unit?

Answer: _____

_____

_____

_____

Reference: _____

22. Is it true that if the underground metallic water piping is not at least 10 feet long, the underground piping system is not adequate as a grounding electrode?

Answer: _____

_____

_____

_____

Reference: _____

172 Master Electrician's Review

23. Does the lighting outlet in the attic have to be switch controlled?

    Answer: _____

    _____

    _____

    _____

    Reference: _____

24. With regards to recessed lighting fixtures installed over showers, is it permissible to use a gasketed vapor-proof trim only to comply with the Code or must the entire fixture be rated for a wet location?

    Answer: _____

    _____

    _____

    _____

    Reference: _____

25. Are overhead outside branch circuit conductors of No. 10 copper to be used in an unsupported open span up to 50 feet? Is No. 10, listed copper multiconductor UF cable suitable for such an application?

    Answer: _____

    _____

    _____

    _____

    Reference: _____

PRACTICE EXAM 5

1. Is a nonmetallic cable tray permitted by the *NEC®*?

    Answer: _____

    _____

    _____

    _____

    Reference: _____

2. The Code requires clearance for lighting fixtures installed in clothes closets.
    a. What are the clearance requirements for surface mounted fixtures mounted on the wall above the door?
    b. What are clearance requirements for surface mounted fixtures mounted on the ceiling?
    c. What are the clearance requirements for recessed incandescent or fluorescent fixtures ceiling mounted?

    Answers: _____

    _____

    _____

    _____

    Reference: _____

Chapter Fourteen 173

3. Are No. 16 fixture wires counted for the number of conductors in outlet boxes to determine the box fill?

   Answer: _____
   _____
   _____
   _____

   Reference: _____

4. Are switchboards and control panels rated 1200 amperes or more, 600 volts, and over 6 feet wide required to have one entrance at each end of the room?

   Answer: _____
   _____
   _____
   _____

   Reference: _____

5. What size grounding electrode conductor is required where a grounding conductor is routed to a driven rod and from the rod to a concrete encased electrode?

   Answer: _____
   _____
   _____
   _____

   Reference: _____

6. Can handle locks on circuit breakers be locked so the power to loads such as emergency lighting, sump pumps, alarm warning circuits, and other types of equipment cannot be cut off by mistake? If so, what Sections of the 1993 *NEC*® would apply?

   Answer: _____
   _____
   _____
   _____

   Reference: _____

7. Can a service be mounted on a mobile home if the home is designed to be installed on a foundation?

   Answer: _____
   _____
   _____
   _____

   Reference: _____

174 Master Electrician's Review

8. Is an insulated equipment ground required for a pool panelboard feeder fed from the service equipment?

   Answer: _____

   _____

   _____

   _____

   Reference: _____

9. Would the Code allow the use of stainless steel ground rods?

   Answer: _____

   _____

   _____

   _____

   Reference: _____

10. The Code now allows flexible metal conduit for services. What about liquidtight flexible conduit, metallic or nonmetallic?

    Answer: _____

    _____

    _____

    _____

    Reference: _____

11. The grounding conductor can be connected to the grounding electrode by listed connectors, clamps, or other listed means and

    Answer: _____

    _____

    _____

    _____

    Reference: _____

12. Does the Code require a ground wire in all flexible raceways?

    Answer: _____

    _____

    _____

    _____

    Reference: _____

13. Is AC cable a permitted wiring method in places of assembly?

    Answer: _____
    _____
    _____
    _____

    Reference: _____

14. Does the oblisk notation in the footnotes to Tables 310-16 and 17 apply to motor circuits?

    Answer: _____
    _____
    _____
    _____

    Reference: _____

15. Can liquidtight flexible nonmetallic conduit be used on residential work?

    Answer: _____
    _____
    _____
    _____

    Reference: _____

16. Are there any Code requirements against installing a bare neutral wire inside the service conduit?

    Answer: _____
    _____
    _____
    _____

    Reference: _____

17. Can the branch circuit and Class II control wires for two three-phase motors be installed in a common raceway of the proper size?

    Answer: _____
    _____
    _____
    _____

    Reference: _____

18. The Code explicitly requires bonding around concentric knockouts on service equipment, but does the Code require that equipment downstream from the service equipment be bonded in the same manner?

    Answer: _____

    _____

    _____

    _____

    Reference: _____

19. Is nonmetallic sheathed cable (Romex) permitted to run through cold air returns if it is sleeved with thin-wall conduit or greenfield in short lengths?

    Answer: _____

    _____

    _____

    _____

    Reference: _____

20. Is it necessary to pigtail the neutral when connecting receptacles to a multi-wire circuit?

    Answer: _____

    _____

    _____

    _____

    Reference: _____

21. When a motor control circuit extends beyond the control enclosure, what is the maximum overcurrent protection acceptable for No. 12 AWG copper control wires?

    Answer: _____

    _____

    _____

    _____

    Reference: _____

22. Is an equipment bonding conductor required to bond a cover mounted receptacle to a surface mounted outlet box?

    Answer: _____

    _____

    _____

    _____

    Reference: _____

23. When computing branch circuit and feeder loads, what is the normal system voltages that shall be used?

   Answer: _____

   _____

   _____

   _____

   Reference: _____

24. Is it permissible to run four No. 12 AWG TW copper conductors in a cable or raceway where they encounter 100°F ambient temperatures to supply cord and plug connected loads protected at 20 amperes?

   Answer: _____

   _____

   _____

   _____

   Reference: _____

25. When flat conductor cable, type FCC, is installed under carpet what is the maximum size carpet squares permitted?

   Answer: _____

   _____

   _____

   _____

   Reference: _____

PRACTICE EXAM 6

1. Are bonding jumpers required on metal feeder and branch circuit raceways containing circuits of more than 250 volts to ground where oversize concentric or eccentric knockouts are encountered?

   Answer: _____

   _____

   _____

   _____

   Reference: _____

2. What is the maximum number of service disconnects allowed for a set of service entrance conductors installed in an apartment for the "93" Code?

   Answer: _____

   _____

   _____

   _____

   Reference: _____

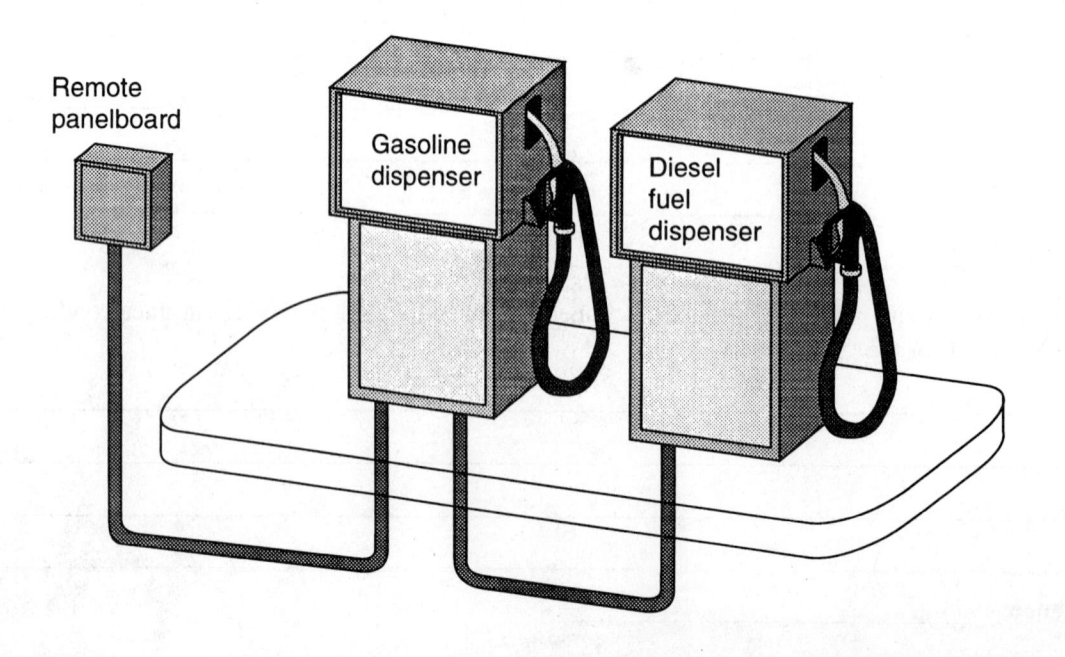

178   Master Electrician's Review

3. (a) Is a GFCI required on or within an outdoor portable sign? (b) Would a GFCI protected branch circuit supplying the sign be acceptable?

   Answers: _____
   _____
   _____
   _____

   Reference: _____

4. Is a lighting outlet required in the crawl space of a manufactured building if equipment requiring service is installed in the space?

   Answer: _____
   _____
   _____
   _____

   Reference: _____

5. Can an office partition assembly be cord and plug connected?

   Answer: _____
   _____
   _____
   _____

   Reference: _____

6. A diesel fuel dispenser is mounted adjacent to a gasoline dispenser on a service station island. A sealing fitting is installed in the conduit entering the gasoline dispenser. Is a sealing fitting required in the conduit entering the diesel fuel dispenser?

   Answer: _____

   _____

   _____

   _____

   Reference: _____

7. In residential type occupancies, does the wall space behind doors where the door opens against the wall count when spacing the receptacles in a room?

   Answer: _____

   _____

   _____

   _____

   Reference: _____

8. A four-family apartment building is supplied by an overhead drop to a 300-kcmil THW aluminum riser. Four electric meters are grouped and have 100-ampere breakers below the meters to supply a panel in each apartment. Number 3 THW copper feeders are run to each apartment unit. An individual water meter is provided for each apartment. Each water meter is supplied with nonmetallic pipe. The water system for each apartment is copper above grade. (a) What size grounding electrode conductor is required? (b) How would this conductor be attached to the metallic water piping in each apartment? (c) Is a supplementary driven electrode required?

   Answers: _____

   _____

   _____

   _____

   Reference: _____

9. Is a No. 14 copper conductor allowed in a kitchen of a dwelling unit if it serves a single fixed appliance such as a dishwasher?

   Answer: _____

   _____

   _____

   _____

   Reference: _____

180 Master Electrician's Review

10. Can UF type cable be used to supply a swimming pool motor?

    Answer: _____

    _____

    _____

    _____

    Reference: _____

11. Are No. 16 and No. 18 fixture wires counted for the number of conductors in outlet boxes for box fill?

    Answer: _____

    _____

    _____

    _____

    Reference: _____

12. Is a receptacle outlet required for a wet-bar and/or if an outlet is installed within 6 feet of the wet-bar sink, would it be required to be GFCI protected?

    Answer: _____

    _____

    _____

    _____

    Reference: _____

13. When calculating the demand for a 3 KW and a 6 KW range in a dwelling, would you use Table 220-19, Column B for the 3 KW and Column C for the 6 KW range or would you combine them and use either B or C as two ranges?

    Answer: _____

    _____

    _____

    _____

    Reference: _____

14. When sizing the branch circuit conductors to a 4.5-KW dryer, would you use 4.5 KW or 5 KW in making the calculations in as much as Exception 5 to 220-3(c) seems to only be a suggested method rather than a required method?

    Answer: _____

    _____

    _____

    _____

    Reference: _____

15. What is the maximum weight of a ceiling fan that can be supported by an outlet box?

    Answer: _____
    _____
    _____
    _____

    Reference: _____

16. Where does the *NEC*® list or calculate the number of conductors allowed in flexible metal conduit?

    Answer: _____
    _____
    _____
    _____

    Reference: _____

17. Are ceiling grid support wires that rigidly support a lay-in ceiling acceptable for the sole support of junction boxes above the lay-in ceiling?

    Answer: _____
    _____
    _____
    _____

    Reference: _____

18. Can electrical equipment approved for a Class I location be used in a Class II location?

    Answer: _____
    _____
    _____
    _____

    Reference: _____

19. Can THHN insulated conductors be used outside for service entrance conductors where exposed to the elements?

    Answer: _____
    _____
    _____
    _____

    Reference: _____

182 Master Electrician's Review

20. Who has the authority to determine acceptability of electrical equipment and materials?

    Answer: _____

    _____

    _____

    _____

    Reference: _____

21. Can more than one receptacle in a laundry room be supplied by the laundry branch circuit?

    Answer: _____

    _____

    _____

    _____

    Reference: _____

22. Many air conditioner units are being installed on roof tops of apartment buildings for different reasons. Is a receptacle adjacent to such equipment required for servicing this equipment?

    Answer: _____

    _____

    _____

    _____

    Reference: _____

23. Section 430-72 applies to motor control circuits that are tapped from the load side of motor branch circuit short-circuit and ground-fault protective devices. Since Column B of Table 430-72(b) covers control circuits that do not extend beyond the motor control equipment enclosure and Column C covers those that do extend beyond, what does Column A cover?

    Answer: _____

    _____

    _____

    _____

    Reference: _____

24. A 240-volt, single-phase circuit with a rated load of 1.8 KW draws 9 amperes of current. The power factor is _____ %.

    Answer: _____

    _____

    _____

    _____

    Reference: _____

25. Can you run multi-outlet assembly from one room to another room through a sheetrock wall?

   Answer: _____
   _____
   _____
   _____

   Reference: _____

PRACTICE EXAM 7

1. Illuminated exit signs are required by local building codes, and battery packs will be used. Do these signs have to be installed on a lighting circuit in the area of the sign or could they be on a separate branch circuit?

   Answer: _____
   _____
   _____
   _____

   Reference: _____

2. Can a metal cable tray be used as an equipment grounding conductor?

   Answer: _____
   _____
   _____
   _____

   Reference: _____

3. Can surface nonmetallic raceway be used outside in a wet location?

   Answer: _____
   _____
   _____
   _____

   Reference: _____

4. Switchboards and control panels rated 1200 amperes or more, 600 volts or less, and over 6 feet wide are required to have one entrance at each end of the room. What conditions would permit only one entrance to this room?

   Answer: _____
   _____
   _____
   _____

   Reference: _____

184 Master Electrician's Review

5. Is Type MC cable required to have a bushing installed like the bushings used on armored cable?

   Answer: _____
   _____
   _____
   _____

   Reference: _____

6. Can electrical nonmetallic tubing be surface mounted in a warehouse building?

   Answer: _____
   _____
   _____
   _____

   Reference: _____

7. How deep must a residential branch circuit be buried if it passes under the driveway?

   Answer: _____
   _____
   _____
   _____

   Reference: _____

8. Where more than one building or structure is on the same property and under single management, the Code requires each building to have a disconnecting means at each such building or structure of 600 volts or less. Under what conditions can the disconnecting means be located elsewhere on the premises?

   Answer: _____
   _____
   _____
   _____

   Reference: _____

9. Are cable trays permitted in a grain elevator?

   Answer: _____
   _____
   _____
   _____

   Reference: _____

10. If a fixture is approved for use as a raceway, can this raceway be used for conductors to supply a circuit or circuits beyond the fixture?

    Answer: _____

    _____

    _____

    _____

    Reference: _____

11. What Code section prohibits bare neutral feeders?

    Answer: _____

    _____

    _____

    _____

    Reference: _____

12. Would a large mercantile store be considered a place of assembly if it can hold more than 100 people?

    Answer: _____

    _____

    _____

    _____

    Reference: _____

13. Can the neutral be reduced by 70 percent for data processing equipment over 200 amperes?

    Answer: _____

    _____

    _____

    _____

    Reference: _____

14. What ampacity correction factor would you use for No. 12 THW copper conductors used in an ambient temperature of 78°F? (Not more than three conductors in raceway, in free air, 240-volt circuit.)

    Answer: _____

    _____

    _____

    _____

    Reference: _____

186   Master Electrician's Review

15. How are box volumes calculated when the box contains different size conductors and a combination of clamps, devices, and studs?

   Answer: _____

   _____

   _____

   _____

   Reference: _____

16. What is the minimum service drop clearance required over a residential yard where the drop conductors do not exceed 150 volts to ground?

   Answer: _____

   _____

   _____

   _____

   Reference: _____

17. An equipment room containing electrical, telephone, and air-handling equipment uses the space in the room for air-handling purposes. Is this room considered a plenum?

   Answer: _____

   _____

   _____

   _____

   Reference: _____

18. As applied to lighting in a clothes closet, is the entire 12 inch or width of shelf area required to be unobstructed to the floor or could part of this area be used for storage?

   Answer: _____

   _____

   _____

   _____

   Reference: _____

19. Would an outdoor lighting fixture be considered as grounded by the physical connection to a grounded metal support pole or would an equipment grounding conductor connection directly to the fixture be required?

   Answer: _____

   _____

   _____

   _____

   Reference: _____

20. Would entrance type lighting fixtures that are installed under roof overhangs, porch roofs, or canopies be required to be suitable for damp locations?

    Answer: _____

    _____

    _____

    _____

    Reference: _____

21. Table 430-147 through 430-150 list the horsepower and the full-load currents of motors. How could you determine the full-load current of a motor not listed, such as a 35-horsepower, 460-volt motor?

    Answer: _____

    _____

    _____

    _____

    Reference: _____

22. Where a single NM (Romex) cable is pulled into a raceway, what percent may the cross section of the conduit be filled?

    Answer: _____

    _____

    _____

    _____

    Reference: _____

23. Can you use Note 3 to 0–2000 Volt Ampacity Tables for a feeder to each apartment in a building regardless of whether the mains are in the apartment units or on the outside of the building?

    Answer: _____

    _____

    _____

    _____

    Reference: _____

24. Can swimming pool equipment such as a pump motor be calculated as a fixed appliance when applying 220-17 to a dwelling?

    Answer: _____

    _____

    _____

    _____

    Reference: _____

25. Is a GFCI receptacle required in a detached garage of a dwelling?

   Answer: _____
   _____
   _____
   _____

   Reference: _____

PRACTICE EXAM 8

1. Are all recessed lighting fixtures that are to be installed in a suspended ceiling required to be thermally protected?

   Answer: _____
   _____
   _____
   _____

   Reference: _____

2. How can you ground an agricultural building and comply with the Code?

   Answer: _____
   _____
   _____
   _____

   Reference: _____

3. Is Table 220-19 applicable to microwave ovens and convection cooking ovens?

   Answer: _____
   _____
   _____
   _____

   Reference: _____

4. Is an auto transformer recognized for use in a motor control circuit?

   Answer: _____

   _____

   _____

   _____

   Reference: _____

5. Can service equipment be mounted on a floating dwelling unit that is moved frequently?

   Answer: _____

   _____

   _____

   _____

   Reference: _____

6. Is it permissible to use neon lighting on or in a dwelling unit?

   Answer: _____

   _____

   _____

   _____

   Reference: _____

7. In a large industrial plant, does the Code allow a branch circuit over 50 amps to feed several outlets such as to supply power to portable equipment that moves from place to place?

   Answer: _____

   _____

   _____

   _____

   Reference: _____

8. Can nonmetallic raceways be used in health care facilities?

   Answer: _____

   _____

   _____

   _____

   Reference: _____

9. Can a conductor be protected by an overcurrent device sized for its ampacity even though the allowable load current is only 50% of the assigned ampacity after derating?

   Answer: _____
   _____
   _____
   _____

   Reference: _____

10. Does the Code permit single-pole circuit breakers in lighting panelboards to be used as a switch for controlling a lighting circuit as an off and on switch?

    Answer: _____
    _____
    _____
    _____

    Reference: _____

11. When determining the required size box, items such as devices, cable clamps, hickeys, where combinations of different size conductors are used, what size conductors do you deduct?

    Answer: _____
    _____
    _____
    _____

    Reference: _____

12. Where rigid metal conduit is used as service raceways, are double locknuts permitted for the service equipment required continuity?

    Answer: _____
    _____
    _____
    _____

    Reference: _____

13. A receptacle outlet is to be added to an existing installation, and NM cable is to be "fished in" the partition wall. Does the Code allow a nonmetallic box without cable clamps to be used for such an installation?

    Answer: _____
    _____
    _____
    _____

    Reference: _____

14. What size general use switch is required on an air conditioning compressor, sealed (hermetic-type) when the name plate states FLA of 100 amperes, three-phase, 230 volts, LRA 580?

    Answer: _____

    _____

    _____

    _____

    Reference: _____

15. A fluorescent fixture is supplied with 3/8 inch flexible conduit 5 feet long. Is an equipment grounding required to ground the fixture?

    Answer: _____

    _____

    _____

    _____

    Reference: _____

16. What is the full-load current of an electric furnace rated 240 volts, single phase, 10 KW when connected to a 208-volt circuit? (Show calculations.)

    Answer: _____

    _____

    _____

    _____

    Reference: _____

17. Can four 400 watt wet-niche lights be connected to one GFCI for a swimming pool?

    Answer: _____

    _____

    _____

    _____

    Reference: _____

18. How does one determine how many conductors conduit bodies (condulets) they are approved for?

    Answer: _____

    _____

    _____

    _____

    Reference: _____

19. How can a recessed fixture approved for direct contact with thermal insulation be identified?

    Answer: _____

    _____

    _____

    _____

    Reference: _____

20. Does the 10-foot tap rule apply only to conductors such as wires and not panel board bus?

    Answer: _____

    _____

    _____

    _____

    Reference: _____

21. How are working clearances measured between enclosed electrical equipment facing each other across an aisle?

    Answer: _____

    _____

    _____

    _____

    Reference: _____

22. Can either the disposal, dishwasher, or trash compactor be installed on a 20-ampere kitchen small appliance circuit?

    Answer: _____

    _____

    _____

    _____

    Reference: _____

23. What is the required horizontal distance between service drops and swimming pools?

    Answer: _____

    _____

    _____

    _____

    Reference: _____

Chapter Fourteen 193

24. Can a bare No. 4 copper grounding electrode conductor be buried below a concrete floor slab and, if so, is there any depth requirement?

    Answer: _____

    _____

    _____

    _____

    Reference: _____

25. Where the grounding electrode conductor connects to the water system on the house side of the water meter, is bonding required around valves with sweated connections? Also, what other equipment is required to be bonded around?

    Answer: _____

    _____

    _____

    _____

    Reference: _____

PRACTICE EXAM 9

1. Can a flood light and/or receptacle be on the GFCI circuit protecting an underwater light?

    Answer: _____

    _____

    _____

    _____

    Reference: _____

2. For a 600-ampere service where 350 MCM copper conductors are parallel in two conduits and only three-phase, three-wire is needed, but the transformer is Y-connected and grounded.
    (a) Must the grounded conductor be brought into the service?
    (b) Can the grounded conductor be brought in through a single raceway or is it required to be installed in each of the parallel raceways?
    (c) What is the maximum size of the grounded conductor?

    Answers: _____

    _____

    _____

    _____

    Reference: _____

Minimum size
21 x 24" x 6

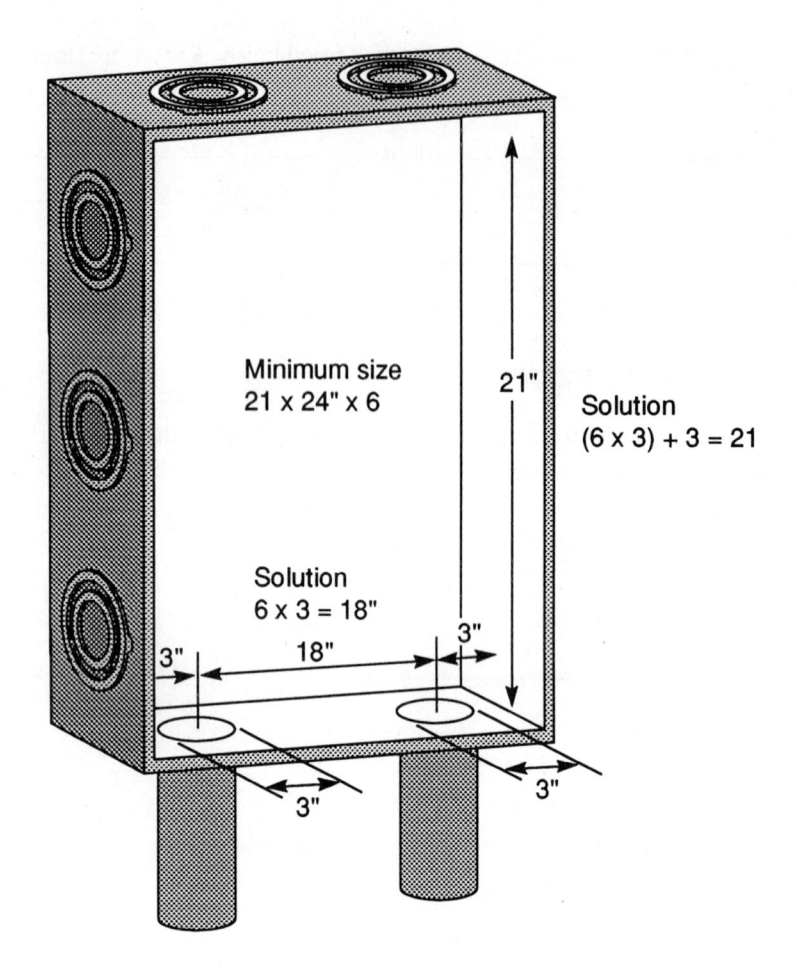

21"

Solution
(6 × 3) + 3 = 21

Solution
6 × 3 = 18"

3" 18" 3"

3" 3"

3. What are the minimum dimensions of a junction box illustrated above with only a U-pull of two 3-inch conduits in one side?

Answer: _____

_____

_____

_____

Reference: _____

4. With two 42 circuit panel boards bolted together, can the 42 circuits from one panel board feed through the gutter of the second panel board if the 40% fill in that gutter is not exceeded?

Answer: _____

_____

_____

_____

Reference: _____

5. Can suspended fluorescent fixtures be connected together with unsupported EMT between them if the length of the EMT is less than 3 feet?

   Answer: _____

   _____

   _____

   _____

   Reference: _____

6. (a) Are three phase generators permitted to be connected to a three-pole transfer switch with a solid neutral or is a four-pole transfer switch that breaks the neutral required? (b) If either is OK, must a grounding electrode be provided at the generator in both cases?

   Answers: _____

   _____

   _____

   _____

   Reference: _____

7. A 600-ampere underground service with four No. 4/0 conductors per phase paralleled is installed in two PVC conduits. Must the phase conductors and neutral be grouped in a single raceway if the conduits are PVC?

   Answer: _____

   _____

   _____

   _____

   Reference: _____

8. The raceway for a 240-volt motor circuit includes both the power circuit and control circuit conductors. The power circuit conductors use 600-volt insulation. Must the control circuit conductors also have 600-volt insulation or would 300-volt insulation be acceptable?

   Answer: _____

   _____

   _____

   _____

   Reference: _____

9. Does the Code either permit or prohibit a sprinkler head over a switchboard?

   Answer: _____

   _____

   _____

   _____

   Reference: _____

10. A 2000-ampere, three-phase, 480-volt ungrounded service is located 300 feet from the water line within a building. Can the grounding electrode conductor from this service be connected to the building steel in the vicinity of the service and then 300 feet away bond the steel to the water line such that the building steel is being used as the grounding electrode conductor?

    Answer: _____

    _____

    _____

    _____

    Reference: _____

11. Four No. 12 THNN conductors in one conduit supply a continuous lighting load. What is the maximum permissible circuit breaker size and permissible load allowed on each conductor?

    Answer: _____

    _____

    _____

    _____

    Reference: _____

12. A receptacle outlet is located in a soffit or roof overhang for connection of Christmas decorations or roof heating cable at a dwelling. When this receptacle is located out of reach from the grade, approximately 9 feet above grade, is this receptacle required to be GFCI protected?

    Answer: _____

    _____

    _____

    _____

    Reference: _____

13. Does the *NEC®* permit more than one NM cable under one device box cable clamp?

    Answer: _____
    _____
    _____
    _____

    Reference: _____

14. A 400-ampere service uses two 200-ampere panel boards, each with a main breaker. Additional load is added so that one more 100-ampere service disconnect is required. Must the service entrance conductors be increased in size? (The calculated load is not increased.)

    Answer: _____
    _____
    _____
    _____

    Reference: _____

15. Can an attachment plug be used as a disconnecting means for a 3-horsepower motor?

    Answer: _____
    _____
    _____
    _____

    Reference: _____

16. Where in the Code does it permit using No. 4 copper THW conductors for service conductors supplying a 100-ampere service?

    Answer: _____
    _____
    _____
    _____

    Reference: _____

17. In three-way and four-way switch circuitry in metal raceway, is the neutral required to be in the same raceway with the travelers or dummy's (switch legs)?

    Answer: _____
    _____
    _____
    _____

    Reference: _____

18. Single conductor UF cable comes underground from a submersible pump through a building wall to the pump controller. Are these conductors permitted to be run inside the building?

Answer: _____

_____

_____

_____

Reference: _____

19. In commercial garages, can regular receptacles and plates be used above 18 inches of the floor? What are the wiring restrictions in a garage that has a minimum of four air changes per hour?

Answer: _____

_____

_____

_____

Reference: _____

20. A single-family dwelling has two electric dryer receptacle outlets wired with two No. 10 black conductors to each one and one No. 10 white conductor neutral. Each outlet has a separate 30-ampere circuit breaker. Is one No. 10 neutral conductor permitted to be used with this installation?

Answer: _____

_____

_____

_____

Reference: _____

21. Can 3/8 inch flexible conduit be used on machinery and boilers for limit switches, flow switches, can switches, and solenoids?

Answer: _____

_____

_____

_____

Reference: _____

22. Article 100, Definitions, states that the definition of a building is a structure that stands alone or that is cut off from adjoining structures by fire walls with all openings therein protected by approved fire doors. Could this definition apply to a building with fire-rated floors and all elevator shafts and stairways protected by approved fire rating and fire doors? In other words, could each floor be considered a separate building?

    Answer: _____

    _____

    _____

    _____

    Reference: _____

23. Can unit equipment used for emergency lighting be directly connected on the branch circuit rather than using a plug-in receptacle?

    Answer: _____

    _____

    _____

    _____

    Reference: _____

24. Section 517-20 specifies GFCI receptacles for wet locations that by definition in Article 100 include outside areas exposed to the weather. Are receptacles required to be GFCI protected when outdoors at health care facilities?

    Answer: _____

    _____

    _____

    _____

    Reference: _____

25. Does a gas furnace in a dwelling require a disconnect switch on the furnace? May the furnace be directly connected onto a basement lighting circuit?

    Answer: _____

    _____

    _____

    _____

    Reference: _____

PRACTICE EXAM 10

1. Are open tube fluorescent fixtures acceptable in a basement garage under an apartment building?

   Answer: _____
   _____
   _____
   _____

   Reference: _____

2. Hot tubs are popular. Are there size and depth requirements to classify them as a pool or are they classified as a pool regardless of size for wiring of circulation pumps, heaters, and so forth?

   Answer: _____
   _____
   _____
   _____

   Reference: _____

3. Rigid metal conduit runs in commercial garage floor and extends up through 18-inch hazardous area with no couplings or fittings in that 18-inch area. Is a seal-off fitting required in this conduit?

   Answer: _____
   _____
   _____
   _____

   Reference: _____

4. Three single-phase, 480-volt transformers rated at 104 amperes each are connected in a wye bank for a three-phase system. What is the maximum overcurrent protection permitted for the primary when the secondary is also protected? Would the primary protection be different if connected in a delta bank?

   Answer: _____
   _____
   _____
   _____

   Reference: _____

5. How is the load for a section of multi-outlet assembly (plug-mold) determined?

   Answer: _____
   _____
   _____
   _____

   Reference: _____

6. A motor is connected to a branch circuit breaker in a panel that is out of sight of the motor location. The breaker is used as the controller and also the disconnecting means for the motor. If this breaker is of the "lock off" type, would this meet code requirements?

   Answer: _____
   _____
   _____
   _____

   Reference: _____

7. A No. 12 remote control circuit is protected by 80 ampere fuses in the motor disconnect (magnetic contact). Is this control circuit properly protected?

   Answer: _____
   _____
   _____
   _____

   Reference: _____

8. Bus duct is reduced in size from 1000 amperes to 400 ampere bus duct and runs 40 feet to where it terminates in a distribution panel. Is overcurrent protection required?

   Answer: _____
   _____
   _____
   _____

   Reference: _____

9. Can UF cable be used from the meter, down the pole, then underground 20 feet to the disconnect for a pasture pump?

   Answer: _____
   _____
   _____
   _____

   Reference: _____

10. On a 480/277 volt, three-phase, four-wire system where the neutral is not used, must it be grounded at the service panel?

    Answer: _____

    _____

    _____

    _____

    Reference: _____

11. If available on the premises, the Code requires the bonding of all grounding electrodes together to form the grounding electrode system. Are these requirements the same for a separately derived system?

    Answer: _____

    _____

    _____

    _____

    Reference: _____

12. There is no definition in the Code of a "tap." Define a tap and is it permissible to make a tap from a tap?

    Answers: _____

    _____

    _____

    _____

    Reference: _____

13. Can two tandem circuit breakers mounted adjacent to each other in a panel board with adjacent handles tied together be used in place of two double pole breakers to supply two 220-volt electric baseboard heaters?

    Answer: _____

    _____

    _____

    _____

    Reference: _____

14. What minimum size copper grounding electrode conductor can you use to ground a 400-ampere lighting service when two No. 3/0 copper conductors in parallel are used for the phase conductors?

    Answer: _____

    _____

    _____

    _____

    Reference: _____

15. Can *aluminum* rigid metal conduit be used in hazardous locations, such as in sewer plants?

    Answer: _____
    _____
    _____
    _____

    Reference: _____

16. A two lamp exit light is supplied from both a battery operated unit equipment and a normal source. Must the normal supply to the exit light be from a recognized emergency source such as ahead of the service disconnect?

    Answer: _____
    _____
    _____
    _____

    Reference: _____

17. A service is changed and moved to a different location. The old electric range run is SE cable that is too short to reach the new service so a metal junction box is installed and a short piece of new SE cable is installed to reach the new service. This is in a residential basement on a 7-foot ceiling. Does this metal junction box require grounding, and if so, does it require a separate grounding conductor to the service equipment (not the neutral of the three-wire SE cable)?

    Answer: _____
    _____
    _____
    _____

    Reference: _____

18. Can service conductors be run directly to a fire pump controller?

    Answer: _____
    _____
    _____
    _____

    Reference: _____

19. A trucking firm has added numerous times. They have 240-volt service to the old part of the building and 208-volt service to the new part. Common computer equipment is connected to both services with shielded TWINAX (signal cable) between devices. The two services are not tied to the same ground. This causes ground loop on signal shield. What is the best way per *NEC®* to achieve equal potential bonding on network?

    Answer: _____

    _____

    _____

    _____

    Reference: _____

20. A lighting fixture is installed over a hydromassage bathtub. Is GFCI protection required for the fixture?

    Answer: _____

    _____

    _____

    _____

    Reference: _____

21. Frequently the *NEC®* requires GFCI protection of individuals. Is there any requirement that limits the number of receptacles that can be protected by a single GFCI *circuit breaker or a feed through receptacle*?

    Answer: _____

    _____

    _____

    _____

    Reference: _____

22. Can an electrical discharge fixture such as a fluorescent strip be used as a branch circuit junction box or termination point for circuit conductors?

    Answer: _____

    _____

    _____

    _____

    Reference: _____

23. The Code allows ENT (electrical nonmetallic tubing) to be installed above suspended ceiling provided a thermal barrier of material has at least a 15-minute rating of fire-rated assemblies.

    Answer: _____

    _____

    _____

    _____

    Reference: _____

24. A 4 × 2⅛ inch square nonmetallic box is marked with the number of 14, 12, and 10 conductors that can be installed in it. Does this mean that No. 8 conductors are not permitted in the box?

    Answer: _____

    _____

    _____

    _____

    Reference: _____

25. Must lay-in fluorescent lighting fixtures be fastened or can they lay in grid T-bars of a suspended ceiling system?

    Answer: _____

    _____

    _____

    _____

    Reference: _____

PRACTICE EXAM 11

1. When UF cable is used for interior wiring as permitted are the conductors required to be rated at 90°C?

    Answer: _____

    _____

    _____

    _____

    Reference: _____

2. Are swimming pool underwater lighting fixtures operating at less than 15 volts between conductors required to be protected by GFCI device or circuit breaker?

   Answer: _____

   _____

   _____

   _____

   Reference: _____

3. Can handle locks on circuit breakers be locked so the power to loads such as emergency lighting, sump pumps, alarm warning circuits, and other types of equipment cannot be cut off by mistake? If so, what sections of the 1993 *NEC*® would apply?

   Answer: _____

   _____

   _____

   _____

   Reference: _____

4. Can (Romex) nonmetallic sheathed cable be run parallel to framing members through cold air returns?

   Answer: _____

   _____

   _____

   _____

   Reference: _____

5. Is a GFCI properly protected circuit supplying an outdoor portable sign an acceptable method for installation in accordance with the *NEC*®?

   Answer: _____

   _____

   _____

   _____

   Reference: _____

Grounding
electrode

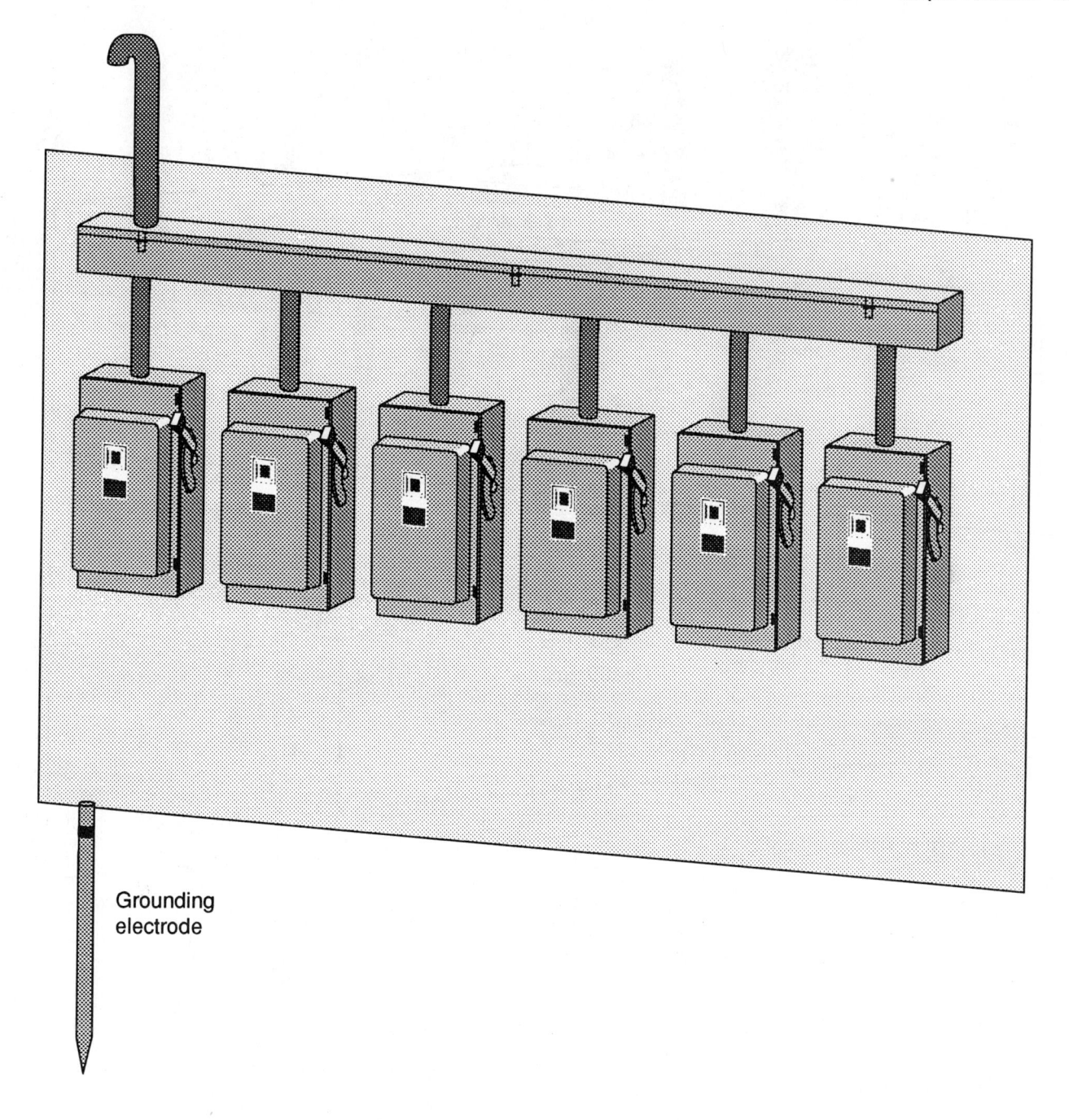

6. Refer to diagram. Where should the grounding electrode conductor terminate in a multiple disconnect service? (See above.)

   Answer: _____

   _____

   _____

   _____

   Reference: _____

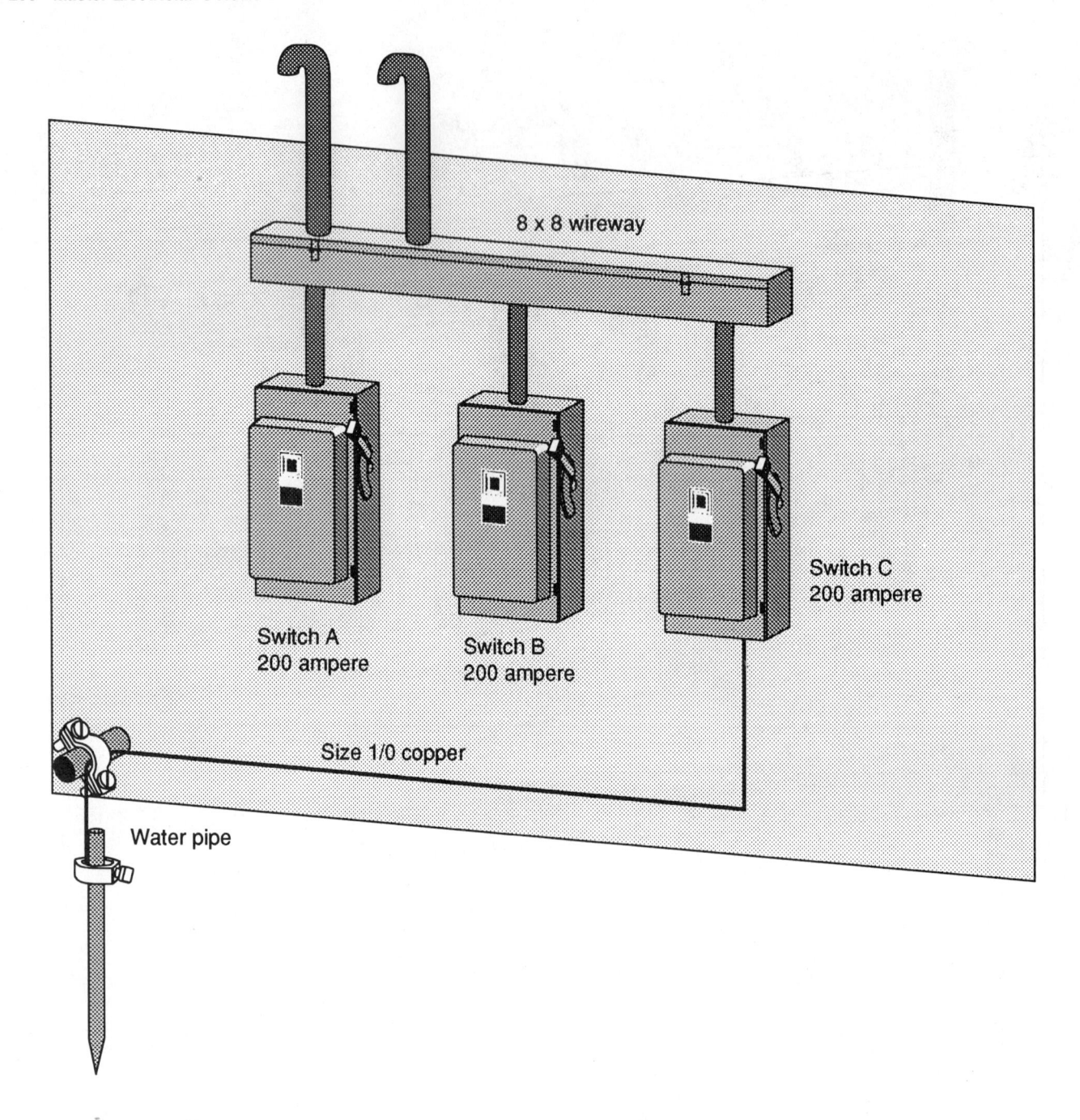

208 Master Electrician's Review

[Diagram shows an 8 x 8 wireway at top with two service conductors entering from above. Three fused disconnect switches hang below the wireway: Switch A 200 ampere, Switch B 200 ampere, and Switch C 200 ampere. A Size 1/0 copper grounding electrode conductor runs from Switch C to a water pipe ground connection with a ground rod.]

7. Refer to diagram. A 600-ampere service is built with parallel sets of 350 KCM CU THWN service conductors. Three 200-ampere fused service disconnects supplied with No. 4/0 THWN service taps within a 8" × 8" wireway. All service bonding is continuous from the CT cabinet through all service enclosures. Does the Code permit only one grounding electrode conductor connection at switch "C" if sized for No. 700 KCM service conductors? (See above.)

Answer: _____

_____

_____

_____

Reference: _____

8. With reference to the question above, how should the trough above the three service disconnects be sized? Is the trough to be sized as a wireway or a junction box?

   Answer: _____
   _____
   _____
   _____

   Reference: _____

9. An EMT circuit and steel box system was installed throughout an office/warehouse building. An additional equipment grounding conductor was pulled with all receptacle branch circuits and terminated at all receptacle grounding screws. Must the equipment grounding conductor also be attached to each metal box?

   Answer: _____
   _____
   _____
   _____

   Reference: _____

10. Is an insulated equipment grounding conductor required in nonmetallic sheathed cable that is supplying a swimming pool pump motor installed in the garage of a single-family dwelling?

    Answer: _____
    _____
    _____
    _____

    Reference: _____

11. A 225-ampere rated power panel has main lugs only and powers a calculated load of 180 amperes. It is supplied by a 500-KCM CU feeder with 400-ampere overcurrent protection. Does this installation comply with the Code?

    Answer: _____
    _____
    _____
    _____

    Reference: _____

210   Master Electrician's Review

12. If a small dental office uses limited amounts of inhalation anesthetics, does Section 517-32 require standby emergency power to supply emergency lighting and the anesthetic equipment?

    Answer: _____
    _____
    _____
    _____

    Reference: _____

13. Does the Code allow any derating of loads in a commercial laundry? The normal load of a washing machine is 4 amperes while the spin cycle draws 9 amperes. Because it is unlikely that all machines will be spinning at the same time, what load figure should be used for load calculations?

    Answer: _____
    _____
    _____
    _____

    Reference: _____

14. Is the examining room of a dental office considered to be general care patient care area? Is NM cable a permitted wiring method in this area?

    Answer: _____
    _____
    _____
    _____

    Reference: _____

15. Do the metal enclosures for individual channel type neon letters (either front-lit or back-lit) require grounding and under what conditions?

    Answer: _____
    _____
    _____
    _____

    Reference: _____

16. Is it permissible to ground the secondary of a separately derived system back to the neutral terminal of a service switchboard instead of going directly to the water line when there is no effectively grounded structural steel nearby?

    Answer: _____

    _____

    _____

    _____

    Reference: _____

17. When conductors are paralleled, does the ampacity double (assuming the same type, size, length)?

    Answer: _____

    _____

    _____

    _____

    Reference: _____

18. Suppose you use three 300 MCM conductors in a raceway in free air with 75°C temperature rating of 285 amperes for each conductor. Instead of this, can you use six No. 1/0 conductors with 75°C rating and two conductors in parallel per phase, all in the same conduit at 300 amperes ampacity, or must you derate because you have more than three conductors? (Assume that you are using Type THWN conductors in a 2½-inch conduit with device terminations rated for 75°C.

    Answer: _____

    _____

    _____

    _____

    Reference: _____

19. Are medium base HID lampholders permitted to be installed on a 277-volt circuit?

    Answer: _____

    _____

    _____

    _____

    Reference: _____

20. Is GFCI protection required for receptacles located on decks where the receptacles are located less than 6 feet 6 inches above grade and access to the deck is by steps or stepping onto a low deck? (The receptacle cannot be reached while standing on the ground.)

    Answer: _____

    _____

    _____

    _____

    Reference: _____

21. Are bathroom GFCI receptacles permitted to supply lighting outlets because tripping of the GFCI causes the lights to go out?

    Answer: _____

    _____

    _____

    _____

    Reference: _____

22. Is it acceptable to connect a garbage disposal and dishwasher on the same branch circuit?

    Answer: _____

    _____

    _____

    _____

    Reference: _____

23. Are receptacles required in a four-car garage at an apartment building if the garage is detached and provided with electric power and lights?

    Answer: _____

    _____

    _____

    _____

    Reference: _____

24. Does an outdoor entrance lighting fixture with integral photocell satisfy the Code without a wall switch control? What about a motion detector used to activate the lighting fixture?

    Answer: _____

    _____

    _____

    _____

    Reference: _____

25. Does the Code require feeder conductors to have an ampacity of 125% of the continuous load plus the noncontinuous load or is it only necessary for the feeder overcurrent device to have this rating? If only the feeder overcurrent device requires this rating, can the feeder conductors be sized only for total continuous plus noncontinuous load?

Answer: _____

_____

_____

_____

Reference: _____

PRACTICE EXAM 12

1. In calculating feeder loads for electric space heating, is it required to figure 125% of space heating loads or is the 125% required only for branch circuit loads?

   Answer: _____

   _____

   _____

   _____

   Reference: _____

2. For household range load calculations, it is acceptable to combine ovens and countertops and treat them as one appliance for feeder calculations?

   Answer: _____

   _____

   _____

   _____

   Reference: _____

3. Because the disconnecting means for separate buildings on one property are required to be suitable for use as service equipment, are the number of disconnects at each building's entrance limited to six?

   Answer: _____

   _____

   _____

   _____

   Reference: _____

214 Master Electrician's Review

4. The Code requires multiple services to be grounded to the same electrode. Must the grounding connections be made to same point on the electrode or can they be grounded to same electrode (water line or structural steel) at different locations?

   Answer: _____

   _____

   _____

   _____

   Reference: _____

5. Does Article 514 apply to underground wiring (such as to a sign) that is at least 30 feet from any hazardous location in a service station?

   Answer: _____

   _____

   _____

   _____

   Reference: _____

6. What is the required minimum clearance between a thermally protected Type IC recessed incandescent fixture and wood framing?

   Answer: _____

   _____

   _____

   _____

   Reference: _____

7. Does the Code permit the metal coaxial sheath of CATV cable to be grounded to a separate driven rod or is it required to be bonded to the building's power system ground or service equipment enclosure?

   Answer: _____

   _____

   _____

   _____

   Reference: _____

8. An automotive repair garage is classified as Class I, Division 2 up 18 inches above the floor. An attached sales and office area are not classified. Is Type NM-B cable permitted to be used in the sales and office area?

   Answer: _____

   _____

   _____

   _____

   Reference: _____

9. Do Type AC and MC cables installed on 277-volt circuits require bonding if connected into concentric knockouts?

   Answer: _____

   _____

   _____

   _____

   Reference: _____

10. A swimming pool is installed at a single-family dwelling. Can two-wire Type NM-B cable with a bare grounding conductor be used to connect the pump motor located in the basement?

    Answer: _____

    _____

    _____

    _____

    Reference: _____

11. Ceiling lighting consists of lay-in fluorescent fixtures. Is it acceptable to use ⅜ inch flexible metal conduit from fixture to fixture in lengths less than 6 feet, provided an equipment grounding conductor is installed in the conduit?

    Answer: _____

    _____

    _____

    _____

    Reference: _____

12. Can a 12-volt dry-niche lighting fixture mounted on the outside of a permanent aboveground swimming pool be supplied by flexible cord connected to a receptacle located 11 feet from the pool?

    Answer: _____

    _____

    _____

    _____

    Reference: _____

13. Can a timer switch for a spa be located within 5 feet of the spa? What if it is located on the unit and not on the wall?

    Answer: _____

    _____

    _____

    _____

    Reference: _____

14. Is there any requirement relative to the location of the wall switch for an outdoor entrance light? As an example, must the switch for an entrance light at a sunroom door be located at the entrance door from outside or can it be located 10 feet away at another door leading to the sunroom?

    Answer: _____

    _____

    _____

    _____

    Reference: _____

15. When calculating the demand for a 3-KW and a 6-KW range, would you use Table 220-19, Column B for the 3 KW and Column C for the 6 KW or would you combine them and use either B or C for 2 ranges?

    Answer: _____

    _____

    _____

    _____

    Reference: _____

16. Can more than one receptacle in a laundry room be supplied by the laundry branch circuit?

    Answer: _____

    _____

    _____

    _____

    Reference: _____

17. For a large bathroom with provisions for a washing machine, is GFCI protection required for an inaccessible receptacle located behind and dedicated to the washing machine?

    Answer: _____

    _____

    _____

    _____

    Reference: _____

18. Is there an acceptable method whereby control wiring can be run in the same conduit with power conductors to a central air conditioning unit?

    Answer: _____

    _____

    _____

    _____

    Reference: _____

19. Does the Code require grounding the metal sheath of CATV cable entering a dwelling?

    Answer: _____

    _____

    _____

    _____

    Reference: _____

20. The Code requires bonding a metallic conduit enclosing a grounding electrode conductor. How is the bonding jumper sized when the conduit enters a concentric knockout?

    Answer: _____

    _____

    _____

    _____

    Reference: _____

21. Does the Code require a GFCI protected receptacle when installed within 6 feet of a dwelling wet bar or laundry sink?

    Answer: _____

    _____

    _____

    _____

    Reference: _____

22. A 500-kcmil Type THW CU feeder protected with 400-ampere overcurrent device is routed through a junction box. What is the minimum size tap conductor required to supply a 30-ampere load using the 25-foot tap rule? What is the minimum size using the 10-foot tap rule?

    Answer: _____

    _____

    _____

    _____

    Reference: _____

23. A large recreation room in a dwelling has three sets of sliding doors opening onto a covered porch. Is a switch required at each door to control porch lighting?

    Answer: _____

    _____

    _____

    _____

    Reference: _____

218 Master Electrician's Review

24. Existing feeder conduits are being repulled using conductors with 90°C Type THHN insulation. What ampacity values are allowed for these conductors that are connected to a circuit breaker or panel main lug terminals?

    Answer: _____

    _____

    _____

    _____

    Reference: _____

25. Branch circuits serving patient care areas must be installed in metal raceways. Which areas in nursing homes and residential custodial care facilities are designated as patient care areas?

    Answer: _____

    _____

    _____

    _____

    Reference: _____

PRACTICE EXAMINATION

The following test was developed and administered by a nationally known seminar presenter and author at a meeting of Code authorities. Although this test is not representative of national electricians' tests, it does provide a good exercise in researching the *NEC®*. You will find the questions to be somewhat tricky in some cases. However, if you can pass this examination within one hour, you should be able to pass most national examinations. It will exercise your knowledge of the *NEC®*. Your ability to use the document as a reference and your ability to research variable difficult questions. (Courtesy Mike Holt Enterprises)

1. Equipment enclosed in a case or cabinet that is provided with a means of sealing or locking so that live parts cannot be made accessible without opening the enclosure is said to be
    A. guarded.
    B. protected.
    C. sealable equipment.
    D. lockable equipment.

    Answer: _____   Reference: _____

2. The letter(s) _____ indicate two insulated conductors laid parallel within an outer nonmetallic covering.
    A. D
    B. M
    C. T
    D. II

    Answer: _____   Reference: _____

3. Class 1 circuit conductors shall be protected against overcurrent
    A. in accordance with the values specified in Table 310-16 through 310-31 for No. 14 and larger.
    B. shall not exceed 7 amperes for No. 18.
    C. shall not exceed 10 amperes for No. 16.
    D. and derating factors do not apply.
    E. all of the above

    Answer: _____   Reference: _____

4. Enclosures for switches or circuit breakers shall not be used as
   A. junction boxes.
   B. auxiliary gutters.
   C. raceways.
   D. all of the above

   Answer: _____ Reference: _____

5. A plug fuse of the edison base has a maximum rating of _____ amperes.
   A. 20
   B. 30
   C. 40
   D. 50
   E. 60

   Answer: _____ Reference: _____

6. Power-limited fire-protective signaling circuit conductors shall be
   A. solid copper.
   B. bunch-tinned stranded copper.
   C. bonded stranded copper.
   D. any of the above

   Answer: _____ Reference: _____

7. When conduit or tubing nipples having a maximum length not to exceed _____ inches are installed between boxes and similar enclosures, the fill shall be permitted to 60 percent.
   A. 6
   B. 12
   C. 24
   D. 30

   Answer: _____ Reference: _____

8. Circuit breakers shall be
   A. capable of being opened by manual operation.
   B. capable of being closed by manual operation.
   C. trip free.
   D. all of the above
   E. A and B only

   Answer: _____ Reference: _____

9. For a circuit operating at less than 50 volts, standard lampholders shall have a rating not less than _____ watts shall be used.
   A. 300
   B. 550
   C. 660
   D. 770

   Answer: _____ Reference: _____

220 Master Electrician's Review

10. No. 18 TFF wire is rated _____ amperes.
    A. 14
    B. 10
    C. 8
    D. 6

    Answer: _____   Reference: _____

11. A No. 10 AWG solid copper wire has a cross-section area of _____ square inches.
    A. .008
    B. .101
    C. .022
    D. .160

    Answer: _____   Reference: _____

12. Conductors of light or power can occupy the same enclosure or raceway with conductors of power-limited fire-protective signaling circuits.
    A. True
    B. False

    Answer: _____   Reference: _____

13. Explanatory material in the *National Electrical Code®* is in the form of
    A. footnotes.
    B. fine print notes.
    C. obelisks and asterisk.
    D. red print.

    Answer: _____   Reference: _____

14. Straight runs of 1¼ inch rigid metal conduit using threaded couplings can be secured at not more than _____ foot intervals.
    A. 5
    B. 10
    C. 12
    D. 14

    Answer: _____   Reference: _____

15. An isolating switch is one that
    A. is not readily accessible to persons unless special means for access are used.
    B. is capable for interrupting the maximum operating overload current of a motor.
    C. is intended for use in general distribution and branch circuits.
    D. is intended for isolating an electrical circuit from the source of power.

    Answer: _____   Reference: _____

16. Where damage to remote control circuits of safety-control equipment would produce a hazard, all conductors of this Class 1 circuit shall be installed in _____ or otherwise suitably protected from physical damage.
    I. mineral-insulated or metal clad cable
    II. rigid metallic conduit
    III. rigid nonmetallic conduit
    A. I only
    B. II only
    C. III only
    D. I, II, or III

    Answer:_____   Reference:_____

17. _____ on equipment to be grounded shall be removed from contact surfaces to assure good electrical continuity.
    A. Conductive coatings
    B. Nonconductive coatings
    C. Manufacturers instructions
    D. all of the above

    Answer:_____   Reference:_____

18. A _____ conductor is one having one or more layers of nonconducting materials that are not recognized by this Code.
    A. noninsulating
    B. bare
    C. covered
    D. none of these

    Answer:_____   Reference:_____

19. To guard live parts, _____ accessible to qualified persons only.
    A. isolate in a room
    B. locate on a balcony
    C. enclose in a cabinet
    D. any of these

    Answer:_____   Reference:_____

20. Conductors larger than No. 4/0 are measured in
    A. inches.
    B. circular mils.
    C. square inches.
    D. AWG.

    Answer:_____   Reference:_____

21. Material identified by the superscript letter "x" includes text extracted from
    A. other NEC articles and sections.
    B. other NFPA documents.
    C. other IOWA publications.
    D. none of the above

    Answer:_____   Reference:_____

222 Master Electrician's Review

22. The voltage of a circuit is defined by the Code as the _____ root-mean-square (effective) difference of potential between any two conductors in the circuit.
    A. lowest
    B. greatest
    C. average
    D. nominal

    Answer: _____ Reference: _____

23. Doorbell wiring, rated as Class 2, in a residence _____ run in the same raceway with light and power conductors.
    A. is permitted with 600-volt insulation
    B. shall not be
    C. shall be
    D. is permitted if insulation is equal to highest installed

    Answer: _____ Reference: _____

24. Equipment or materials to which has been attached a symbol or other identifying mark acceptable to the authority having jurisdiction is known as
    A. listed.
    B. labeled.
    C. approved.
    C. rated.

    Answer: _____ Reference: _____

25. The grounded conductor of a branch circuit shall be identified by a _____ color.
    I. natural gray
    II. continuous green
    III. continuous white
    IV. green with yellow stripe
    A. I and III
    B. II and IV
    C. III only
    D. II only

    Answer: _____ Reference: _____

26. Circuits for lighting and power shall not be connected to any system containing
    A. hazardous material.
    B. trolley wires with ground returns.
    C. poor wiring methods.
    D. dangerous chemicals or gasses.

    Answer: _____ Reference: _____

27. The minimum clearance between an electric space heating cable and an outlet box shall not be less than _____ inches.
    A. 8
    B. 12
    C. 18
    D. 6

    Answer: _____ Reference: _____

28. The ampacity requirements for a disconnecting means of x-ray equipment shall be based on _____% of the input required for the momentary rating of the equipment if greater than the long-time rating.
    A. 125
    B. 100
    C. 50
    D. none of the above

    Answer: _____ Reference: _____

29. Open conductors installed outside shall be separated from open conductors of other circuits by not less than _____ inches.
    A. 4
    B. 6
    C. 8
    D. 10
    E. 12

    Answer: _____ Reference: _____

30. The minimum headroom of working spaces about motor control centers shall be
    A. 3½ feet.
    B. 5 feet.
    C. 6 feet, 6 inches.
    D. 6 feet, 3 inches.

    Answer: _____ Reference: _____

31. Where fixed multi-outlet assemblies are employed in locations where a number of appliances are likely to be used simultaneously, each foot or fraction thereof shall be considered as an outlet of not less than _____ VA.
    A. 180
    B. 200
    C. 60
    D. 35

    Answer: _____ Reference: _____

32. Soft-drawn or medium-drawn copper lead in conductors for television equipment antenna systems shall be permitted where the maximum span between points of support is less than _____ feet.
    A. 35
    B. 30
    C. 20
    D. 10

    Answer: _____ Reference: _____

33. For a one-family dwelling with an initial computed load of 10 KW or more or the initial installation has more than _____ two-wire branch circuits, the service disconnect shall not be less than 100 amperes.
    A. 6
    B. 5
    C. 3
    D. 2

    Answer: _____ Reference: _____

224 Master Electrician's Review

34. Which of the following machines shall be provided with speed limiting devices or other speed limiting means?
    I. series motors
    II. induction
    III. self excited DC motors
    A. I only
    B. II only
    C. III only
    D. I and III

    Answer: _____ Reference: _____

35. Where within _____ feet of any building or other structure, open wiring on insulators shall be insulated or covered.
    A. 4
    B. 3
    C. 6
    D. 10

    Answer: _____ Reference: _____

36. High-voltage conductors in tunnels shall be installed in
    A. rigid conduit.
    B. MC cable.
    C. other metal raceways.
    D. any of the above

    Answer: _____ Reference: _____

37. For garages and outbuildings on residential property, a _____ suitable for use on branch circuits shall be permitted as the disconnecting means.
    A. snap switch
    B. set of three-way or four-way snap switches
    C. set of two-way or four-way snap switches
    D. A or B
    E. none of the above

    Answer: _____ Reference: _____

38. Utilization equipment fastened in place connected to a branch circuit with other loads shall not exceed _____% of the branch circuit rating.
    A. 50
    B. 60
    C. 80
    D. 100

    Answer: _____ Reference: _____

39. The maximum amperage rating of a 4 inch × ½ inch busbar is _____ amperes.
    A. 500
    B. 1000
    C. 700
    D. 2000

    Answer: _____ Reference: _____

40. For temporary wiring over 600 volts, nominal, _____ shall be provided to prevent access of other than authorized and qualified personnel.
    I.   fending
    II. barriers
    III. signs
    A. I only
    B. II only
    C. III only
    D. I or II

    Answer: _____   Reference: _____

41. Each continuous duty motor _____ horsepower or less, not permanently installed, is nonautomatically started, and is within sight of the controller shall be permitted to be protected against overload by the branch circuit protective device.
    A. ⅛
    B. ½
    C. ¾
    D. 1

    Answer: _____   Reference: _____

42. The neutral (grounded) conductor in a mobile home shall be insulated from the equipment grounding system to which of the following location(s)?
    A. range and oven
    B. distribution panel
    C. clothes dryer
    D. all of the above

    Answer: _____   Reference: _____

43. The bottom of sign and outline lighting enclosures shall not be less than _____ feet above areas accessible to vehicles.
    A. 12
    B. 14
    C. 16
    D. 18

    Answer: _____   Reference: _____

44. The conductor used to ground the outer cover of a coaxial cable shall be
    I.   insulated.
    II. No. 14 AWG minimum.
    III. guarded from physical damage when necessary.
    A. I only
    B. II only
    C. III only
    D. I, II, III

    Answer: _____   Reference: _____

226 Master Electrician's Review

45. Where energized live parts are exposed, the minimum clear work space shall not be less than _____ feet high for over 600 volts.
    A. 3
    B. 5
    C. 3½
    D. 6¼
    E. 6½

    Answer: _____   Reference: _____

46. Exposed metal raceways' surfaces shall not exceed _____ mV differences at frequencies of 1000 Hertz or less measured across a 1000-ohm resistance between surfaces in a general patient care area of a health care facility.
    A. 100
    B. 250
    C. 500
    D. 1000

    Answer: _____   Reference: _____

47. Conductors No. _____ or larger supported on solid knobs shall be securely tied thereto by tie wires having an insulation equivalent to that of the open wire.
    A. 14
    B. 12
    C. 10
    D. 8
    E. 6

    Answer: _____   Reference: _____

48. A branch circuit that supplies a number of outlets for lighting and appliances is known as a
    A. general purpose branch circuit.
    B. a multi-purpose branch circuit.
    C. a utility branch circuit.
    D. none of the above

    Answer: _____   Reference: _____

49. The minimum size conductor that can be used for an overhead feeder from a residence to a remote garage is No.
    A. 10 cu.
    B. 12 cu.
    C. 6 al.
    D. 10 al.

    Answer: _____   Reference: _____

50. Transformers with a primary over 600 volts for outline lighting installations shall have secondary current ratings not more than
    A. 30 amperes.
    B. 20 amperes.
    C. 30 milliamperes.
    D. 60 amperes.

    Answer: _____   Reference: _____

FINAL EXAMINATION

Examination Instructions:

For a positive evaluation of your knowledge and preparation awareness, you must

1. locate yourself in a quiet atmosphere (room by yourself)
2. have with you at least two sharp No. 2 pencils, the 1993 *NEC®*, and a hand-held calculator
3. time yourself (three hours) with no interruptions
4. after the test is complete, grade yourself honestly and concentrate your studies on the Sections of the *NEC®* in which you missed the questions.

Caution: Do not just look up the correct answers, as the questions in this examination are only an exercise and not actual test questions. Therefore, it is important that you be able to quickly find answers from throughout the *NEC®*.

1. According to the *National Electrical Code®*, open conductors for communication equipment on a building shall be separated at least ____ feet from *lightning* conductors.
   A. 2
   B. 4
   C. 6
   D. 8

   Answer: _____   Reference: _____

2. A megohmmeter is an instrument used for
   A. polarizing a circuit.
   B. measuring high resistances.
   C. shunting a generation system.
   D. determining amperes.

   Answer: _____   Reference: _____

3. The total opposition to alternating current in a circuit that includes resistance, inductance, and capacitance is called
   A. reactance.
   B. resistance.
   C. reluctance.
   D. impedance.

   Answer: _____   Reference: _____

4. A 240-volt single-phase circuit has a resistive load of 8,500 watts. The net computed current to supply this load is ____ amps.
   A. 35
   B. 39
   C. 44
   D. 71

   Answer: _____   Reference: _____

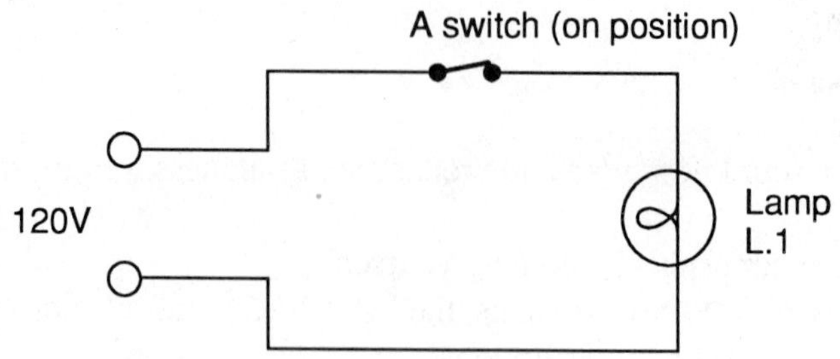

228 Master Electrician's Review

5. In the diagram, switch S1 is in the "on" position, but light L1 does **not** come on. Voltage across L1 is measured to be 120 volts. Voltage across S1 is measured to be 0 volts. The light does not come on because
   A. the light is open (burned out).
   B. the light and switch are shorted.
   C. the light is good but the switch does **not** make contact.
   D. there is a break in the wire of the circuit.

   Answer: _____   Reference: _____

6. In the diagram above, with a 9-amp current in the circuit, the power factor is ____ percent.
   A. 71
   B. 83
   C. 93
   D. 108

   Answer: _____   Reference: _____

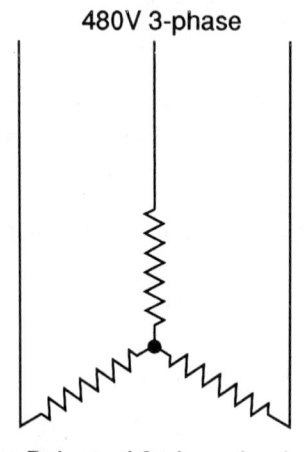

7. In the diagram, three *balanced resistive* 100 ampere loads are connected to a 480-volt, three-phase, three-wire circuit. The total power of this circuit is ____ kilowatts.
   A. 48
   B. 72
   C. 83
   D. 144

   Answer: _____   Reference: _____

8. A 100-horsepower induction motor is loaded to 30 horsepower. To improve the power factor of this motor, the amount of load should be
   A. increased.
   B. left unchanged because load has **no** effect.
   C. decreased.
   D. removed to give 0 load.

   Answer: _____   Reference: _____

9. An electrical installation requires a total of 200 feet of conductor with a maximum line resistance of 0.5 ohm. The minimum size conductor must have a resistance of no more than ____ ohms per 1,000 feet.
   A. 0.5
   B. 1.0
   C. 2.5
   D. 5.0

   Answer: _____   Reference: _____

10. To get the maximum total resistance using three resistors,
    A. all three resistors should be connected in series.
    B. all three resistors should be connected in parallel.
    C. two resistors should be connected in parallel then one connected in series.
    D. two resistors should be connected in series then one connected in parallel.

    Answer: _____   Reference: _____

11. Four resistance heaters are connected in parallel. Their resistances are heater 1, 20 ohms; heater 2, 30 ohms; heater 3, 60 ohms; and heater 4, 10 ohms. The total resistance of the parallel circuit is ____ ohms.
    A. 5
    B. 12
    C. 50
    D. 120

    Answer: _____   Reference: _____

12. For circuits supplying lighting units having ballasts, transformers, or autotransformers, the computed load shall be based on the ____ of the lighting units.
    A. total amp rating
    B. size of the conductors
    C. total wattage of the lamps
    D. voltage rating

    Answer: _____   Reference: _____

13. A building on a blueprint is 16 inches × 10 inches. If the drawing scale is ¼ inch = 1 foot, what is the area of the building in square feet?
    A. 160 square feet
    B. 640 square feet
    C. 2560 square feet
    D. 5120 square feet

    Answer: _____   Reference: _____

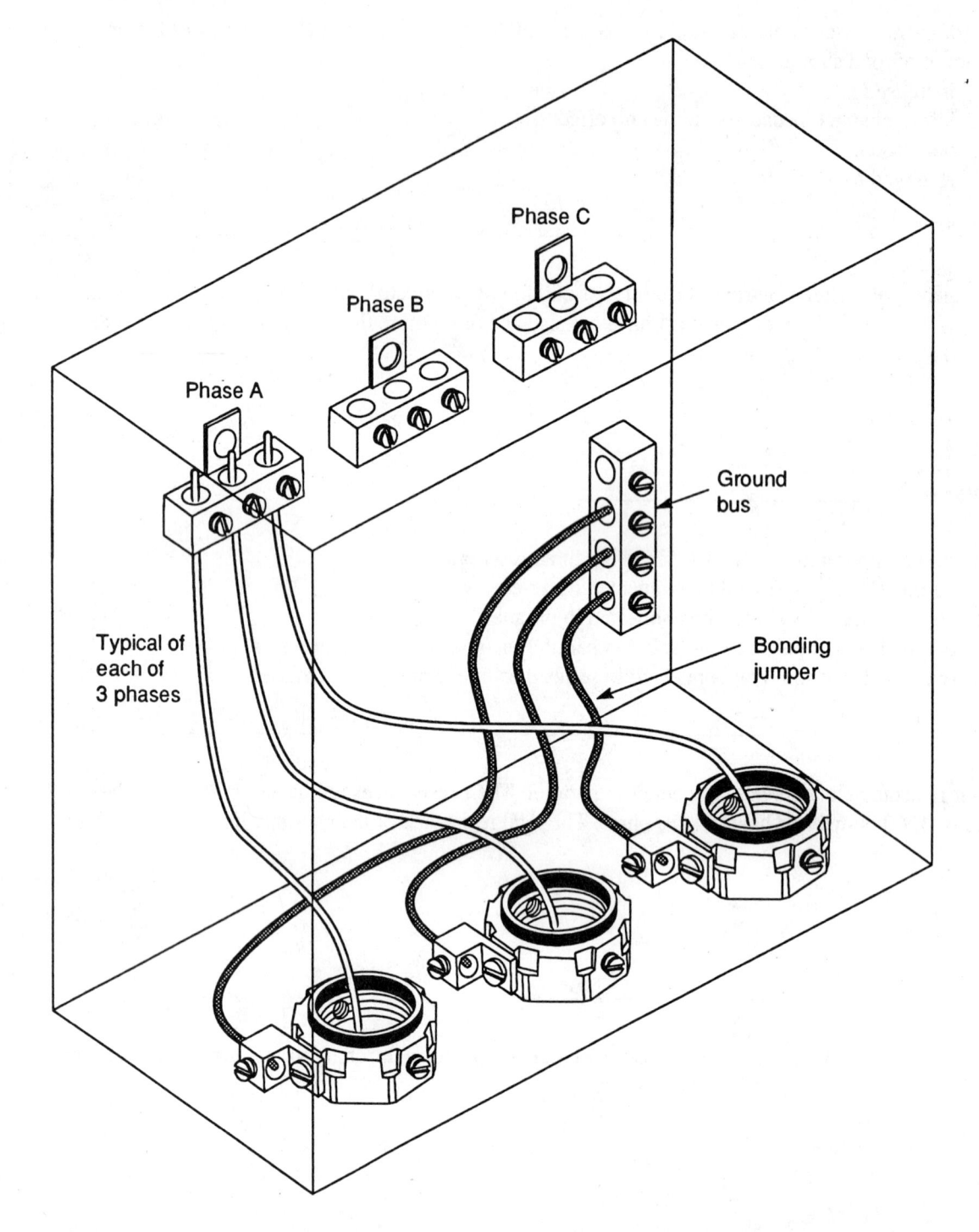

14. In the diagram, each of the supply side EMT conduits contain three 500-MCM copper THW service conductors in parallel. As shown, **a separate bonding jumper is installed from each conduit** to the grounded bus terminal.

   Each bonding jumper must be at least size _____ AWG copper.
   A. 1/0
   B. 2/0
   C. 3/0
   D. 4/0

   Answer: _____   Reference: _____

15. A group of conveyors is operated by individual motors in a manufacturing facility. The total ampacity required to run the group is 58 amps at 240 volts single phase. There is 250 feet of conduit between the main service panel and the subpanel. The voltage drop in the feeder conductors can be a maximum of 3%.

Each conductor in the feeder circuit to supply these conveyors must be at least size ____ AWG copper THW. Use resistance values in the *National Electrical Code®*, Chapter 9, Table 8.
A. 2
B. 3
C. 4
D. 6

Answer: _____   Reference: _____

16. A single-phase, three-wire service has two ungrounded conductors of size 2/0 AWG copper THWN. The neutral conductor is size 1 AWG copper THWN. The conduit size needed for the service entrance conductors is at least size ____ inch.
A. 1½
B. 2
C. 2½
D. 3

Answer: _____   Reference: _____

17. Two No. 1 AWG, one No. 1/0 AWG, and one No. 2/0 AWG copper conductors, Type THHN are in a 10-foot conduit. The size of IMC conduit must be at least ____ inch.
A. 1¼
B. 1½
C. 2
D. 2½

Answer: _____   Reference: _____

18. A one-family dwelling contains the following:

    1,200 square feet of floor space;
    Service: 120/240 volt, single-phase;
    Heat (separate control): two 500W, 240V baseboard heaters, three 1,000W, 240V, baseboard heaters, one 2,000W, 240V baseboard heaters;
    Range: one 11,000W, 240V;
    Dryer: one 6,000W, 240V;
    Water: one 4,000W, 240V water heater

Using the optional calculation method, each ungrounded conductor in the service for this dwelling has a total net computed load of ____ amps.
A. 146
B. 120
C. 99
D. 84

Answer: _____   Reference: _____

19. A metallic cold water piping system is available within the structure being supplied by an electrical service with size 1/0 THW copper service entrance conductors. The copper grounding electrode conductors run to the metal water pipe shall have a minimum size of:
    A. No. 8 AWG
    B. No. 6 AWG
    C. No. 4 AWG
    D. No. 2 AWG

    Answer: _____ Reference: _____

20. Ground fault protection is required to be installed on a 2000-amp, solidly grounded, wye service. What is the maximum setting of the ground fault protection?
    A. 800 amperes
    B. 1000 amperes
    C. 1200 amperes
    D. 1600 amperes

    Answer: _____ Reference: _____

21. Overhead service conductors shall be installed so that the minimum clearance from a window opening is
    A. two feet.
    B. three feet.
    C. six feet.
    D. eight feet.

    Answer: _____ Reference: _____

22. The 240-volt, single-phase feeder shown in the diagram supplies a branch circuit panel that has an improperly balanced load of 57 amperes on one ungrounded conductor and 45 amperes on the other ungrounded conductor. What is the load on the grounded (neutral) conductor?
    A. no load
    B. 12 amperes
    C. 57 amperes
    D. 102 amperes

    Answer: _____ Reference: _____

23. The maximum number of power or lighting conductors that can be installed in a raceway before the derating factors must be applied is
    A. 1.
    B. 3.
    C. 5.
    D. 7.

    Answer: _____   Reference: _____

24. A set of blueprints for an electrical installation calls for a ¾-inch conduit to be installed as a raceway to enclose a three-wire circuit consisting of No. 8 AWG, Type TW conductors. The load to be served is 36 amperes. After the conduit is installed, it is discovered that ten feet of the conduit pass through an area where the ambient temperature is 115° Fahrenheit (115°F). Which of the following actions is an acceptable action that can be taken to correct the condition?
    I. No. 8 THW conductors can be substituted for the No. 8 TW conductors
    II. No. 6 TW conductors can be substituted for the No. 8 TW conductors
    A. I only
    B. II only
    C. Both I and II
    D. Neither I nor II

    Answer: _____   Reference: _____

25. A circuit consisting of three No. 6, THW insulated copper conductors is run through an area of a building where the temperature is normally 120° Fahrenheit (120°F). Which of the following is the maximum allowable load current for each conductor?
    A. 48.75 amperes
    B. 56.25 amperes
    C. 60.00 amperes
    D. 65.50 amperes

    Answer: _____   Reference: _____

26. A 230-volt, single-phase, 100-ampere circuit is installed in a nonmetallic raceway. Which of the following conditions must apply to the equipment grounding conductor installed with the circuit?
    I. It must be counted when determining conductor fill of the raceway.
    II. It must be the same size as that of the circuit conductor.
    A. I only
    B. II only
    C. Both I and II
    D. Neither I nor II

    Answer: _____   Reference: _____

27. Three No. 4 THWN and four No. 1/0 THW conductors are to be installed in a single run of conduit. The minimum size of conduit permitted is
    A. 1 inch.
    B. 2 inches.
    C. 3 inches.
    D. 4 inches.

    Answer: _____   Reference: _____

28. Eight No. 4 THWN conductors are to be installed in a single run of conduit. The minimum size of conduit permitted is
    A. 1 inch.
    B. 1½ inches.
    C. 2 inches.
    D. 2½ inches.

    Answer: _____   Reference: _____

29. A municipality has adopted the *National Electrical Code®* without amendments. A conflict occurs about the interpretation of a section of the adopted electrical code. Which of the following is responsible for making the interpretation of the Code?
    A. The engineer overseeing the construction
    B. The electrical contractor performing the work
    C. The chief electrical inspector
    D. The International Association of Electrical Inspectors

    Answer: _____   Reference: _____

30. A service disconnecting means can be installed at which of the following locations?
    I. Outside a building at a readily accessible point nearest the point of entrance of the service-entrance conductors.
    II. Inside a building at a readily accessible point nearest the point of entrance of the service-entrance conductors.
    A. I only
    B. II only
    C. Either I or II
    D. Neither I nor II

    Answer: _____   Reference: _____

31. Figures 1 and 2 shown below are standard electrical symbols used on blueprints representing crossing conductors. Which symbol represents two conductors crossing but not connecting?

Conductors crossing and connecting

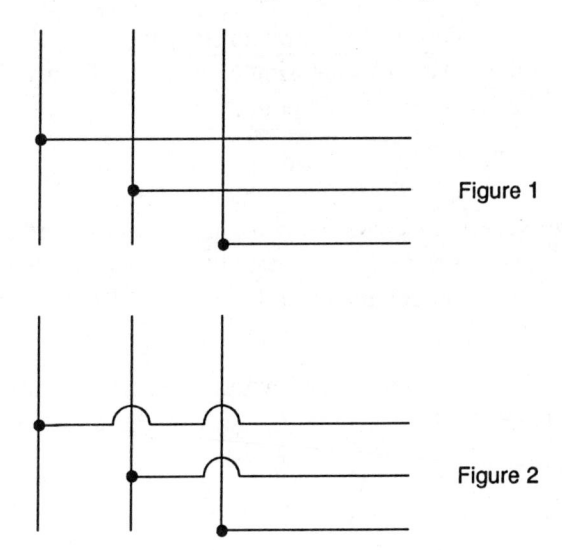

Figure 1

Figure 2

   A. Figure 1 only
   B. Figure 2 only
   C. Figures 1 and 2
   D. Neither Figure 1 nor Figure 2

   Answer: _____   Reference: _____

32. The following figure is a standard symbol used on blueprints. The symbol represents which of the following?
   A. An air circuit breaker
   B. A lightning arrestor
   C. A fuse
   D. A thermal element

   Answer: _____   Reference: _____

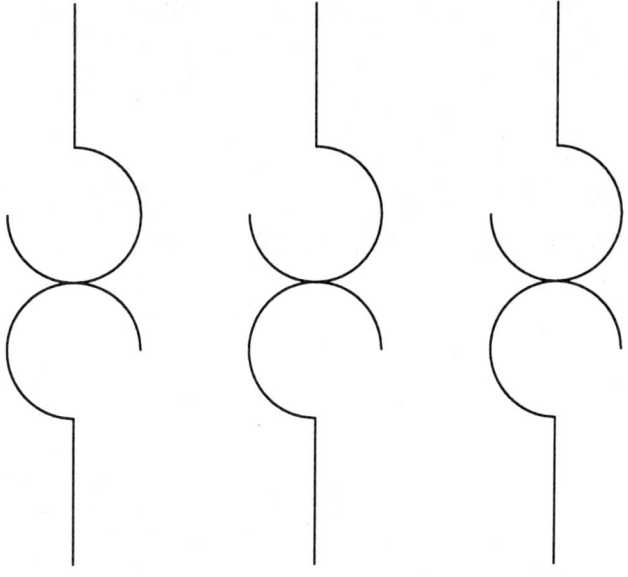

33. An electrical contractor installs the wiring in a new 1400 square-foot single-family dwelling unit before a permit is obtained from the city electrical inspector. The electrical inspector stops the contractor from continuing the job. The city where the structure is located has adopted an electrical code that contains the following provisions:

Permits for New Construction
1. One- and two-family dwelling: On all new single- and two-family dwelling construction, the electrical fee shall be as follows: $.02/square feet under roof.
2. All other

Penalty for Failure to Obtain Permit
In case it shall be discovered that any electrical work has been installed or put into use for which no permit has been issued, the violator shall pay a fee equal to three times the permit fee that shall have been paid for work done in violation thereof and no additional permits shall be granted until all fees have been paid.

Before the electrical contractor can complete the work, a permit fee must be paid. Which of the following is the minimum permit fee required?
A. $ 2.80
B. $28.00
C. $56.00
D. $84.00

Answer: _____   Reference: _____

34. A service disconnect is supplied by conductors having 38,500 amps RMS fault current available at the supply terminals of the disconnect. Which of the following statements is correct?
A. Only the overcurrent device has to have an interrupting rating at or above the 38,500 amps.
B. The overcurrent device and the panel must be rated for the maximum fault current available.
C. The overcurrent device must be rated at 200,000 amps RMS.
D. Neither the panel nor the overcurrent device has to be rated with fault current interrupting rating.

Answer: _____   Reference: _____

35. A motor is protected against short-circuit and ground-fault by an adjustable instantaneous trip circuit breaker that is part of a combination controller having motor overload and short-circuit and ground-fault protection in each conductor. The setting of the instantaneous trip breaker shall be permitted to exceed the motor full-load current by not more than
A. 250 percent.
B. 1300 percent.
C. 700 percent.
D. 1000 percent.

Answer: _____   Reference: _____

36. Thermal overload relays are used for the protection of polyphase induction motors. Their primary purpose is to protect the motor in case of
A. reversal of phases in the supply.
B. low line voltage.
C. short circuit between phases.
D. sustained overload.

Answer: _____   Reference: _____

37. A motor controller and motor branch circuit disconnecting means a 2300-volt motor shall have a continuous ampere rating of
    A. not less than the trip setting of the short-circuit protective device rating.
    B. not less than the trip setting of the overload protection device.
    C. not less than the trip setting of the fault-current protection device.
    D. not less than the locked-rotor rating of the motor.

    Answer: _____   Reference: _____

38. The disconnecting means for both a motor and the controller shall be
    A. permitted within the same enclosure as the controller.
    B. located separately from the controller enclosure.
    C. located at the service equipment.
    D. permitted within the watt-hour meter enclosure.

    Answer: _____   Reference: _____

39. Speed limiting devices shall be provided with which of the following?
    A. polyphase squirrel-cage motors
    B. synchronous motors
    C. compound motors
    D. series motors

    Answer: _____   Reference: _____

40. Which of the following appliances can be grounded to the grounded (neutral) conductor?
    A. electric water heater
    B. kitchen disposal
    C. dishwasher
    D. electric dryer

    Answer: _____   Reference: _____

41. A metal lighting fixture shall be grounded if located
    A. 10 feet vertically or 6 feet horizontally from a kitchen sink.
    B. 8 feet vertically or 5 feet horizontally from a kitchen sink.
    C. 6 feet vertically or 3 feet horizontally from a kitchen sink.
    D. 8 feet vertically or 3 feet horizontally from a kitchen sink.

    Answer: _____   Reference: _____

42. When used, a driven ground rod shall be installed so the soil will be in contact with a length of the rod not less than
    A. 4 feet.
    B. 6 feet.
    C. 8 feet.
    D. 10 feet.

    Answer: _____   Reference: _____

43. Portable and stationary electrically heated appliances having metal frames are connected to an electrical system operating at 277 volts to ground. Which of the following statements about the metal frames of each appliance is (are) correct?
   A. It must be grounded.
   B. It must be grounded only if supplied by a portable cord.
   C. It can be permanently insulated from ground by special permission.
   D. It is connected by the manufacturer to the grounded conductor thermal.

   Answer: _____   Reference: _____

44. Which of the following most accurately describes the condition of a motor known as "locked-rotor?"
   A. When the electrician places a lock on the motor controller to keep the motor from being energized.
   B. Mechanical brakes used on the motor shaft to stop the motor during shutdown.
   C. An electronic control device used to lock the speed of the motor at that specified by the manufacturer.
   D. When the circuits of a motor are energized but the rotor is not turning.

   Answer: _____   Reference: _____

45. The following diagram represents overhead conductors between two buildings on an industrial site. The voltage is 240/480 volts AC. The conductors pass over a driveway leading to a loading dock at one of the buildings. What is the minimum vertical clearance permitted between the overhead conductors and the driveway?

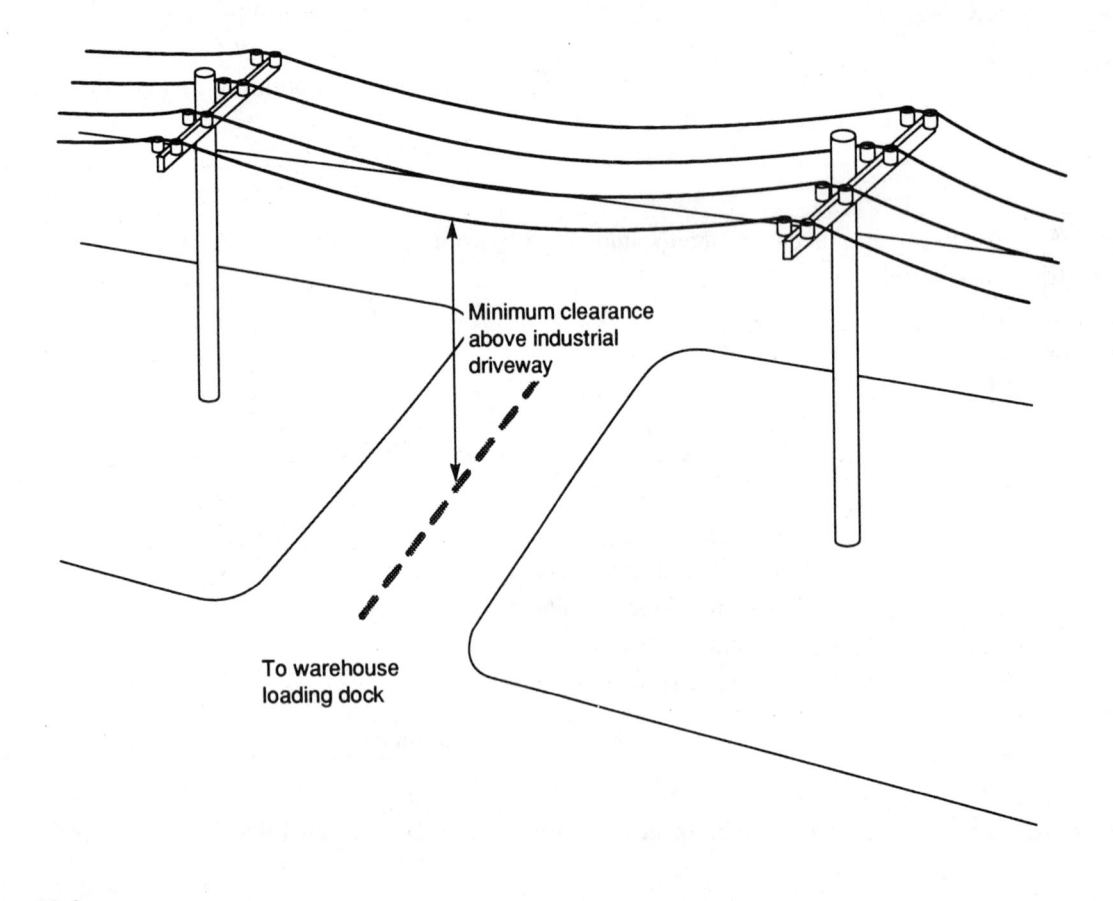

   A. 10 feet
   B. 12 feet
   C. 15 feet
   D. 18 feet

   Answer: _____   Reference: _____

46. When a ground fault protection for equipment is installed within service equipment, it shall be performance tested
    A. at the factory before shipment.
    B. before being installed on site.
    C. when first installed on site.
    D. after the electrical system has been used for one week.

    Answer:_____   Reference:_____

47. Three continuous duty motors with full-load current ratings of 5.6 amperes, 4.5 amperes, and 4.5 amperes, respectively, are to be installed on a single-branch circuit. The circuit conductors shall have a minimum ampacity of
    A. 16 amperes.
    B. 20 amperes.
    C. 24 amperes.
    D. 30 amperes.

    Answer:_____   Reference:_____

48. Which of the following statements about the number of electrical services to a building is (are) correct?
    I. A service supplied by a wind-powered generator can be installed on a building in addition to the service supplied by the local electrical utility.
    II. A service supplied by a solar photovoltaic system can be installed on a building in addition to the service supplied by the local electrical utility.
    A. I only
    B. II only
    C. Both I and II
    D. Neither I nor II

    Answer:_____   Reference:_____

49. Which of the following statements about the termination and bonding of conductors at service equipment is (are) true?
    I. The grounded service conductor and the equipment grounding conductor shall be bonded together within the service equipment.
    II. The grounding electrode conductor and the system grounded conductor shall be bonded together within the service equipment.
    A. I only
    B. II only
    C. Both I and II
    D. Neither I nor II

    Answer:_____   Reference:_____

50. When a universal series AC motor is caused to run without being connected to a load, the motor will
    A. run at a constant speed for an indefinite time.
    B. vary in running speed from about 80% to 125% of the normal rated speed.
    C. decrease in running speed until a locked-rotor condition occurs.
    D. increase in running speed to a dangerous level that can damage the motor.

    Answer:_____   Reference:_____

51. The following diagram represents an electrical service with a fused switch as the main disconnect. What is the maximum height the center of the grip of the operating handle of the switch is permitted to be located above the ground?
A. 5 feet

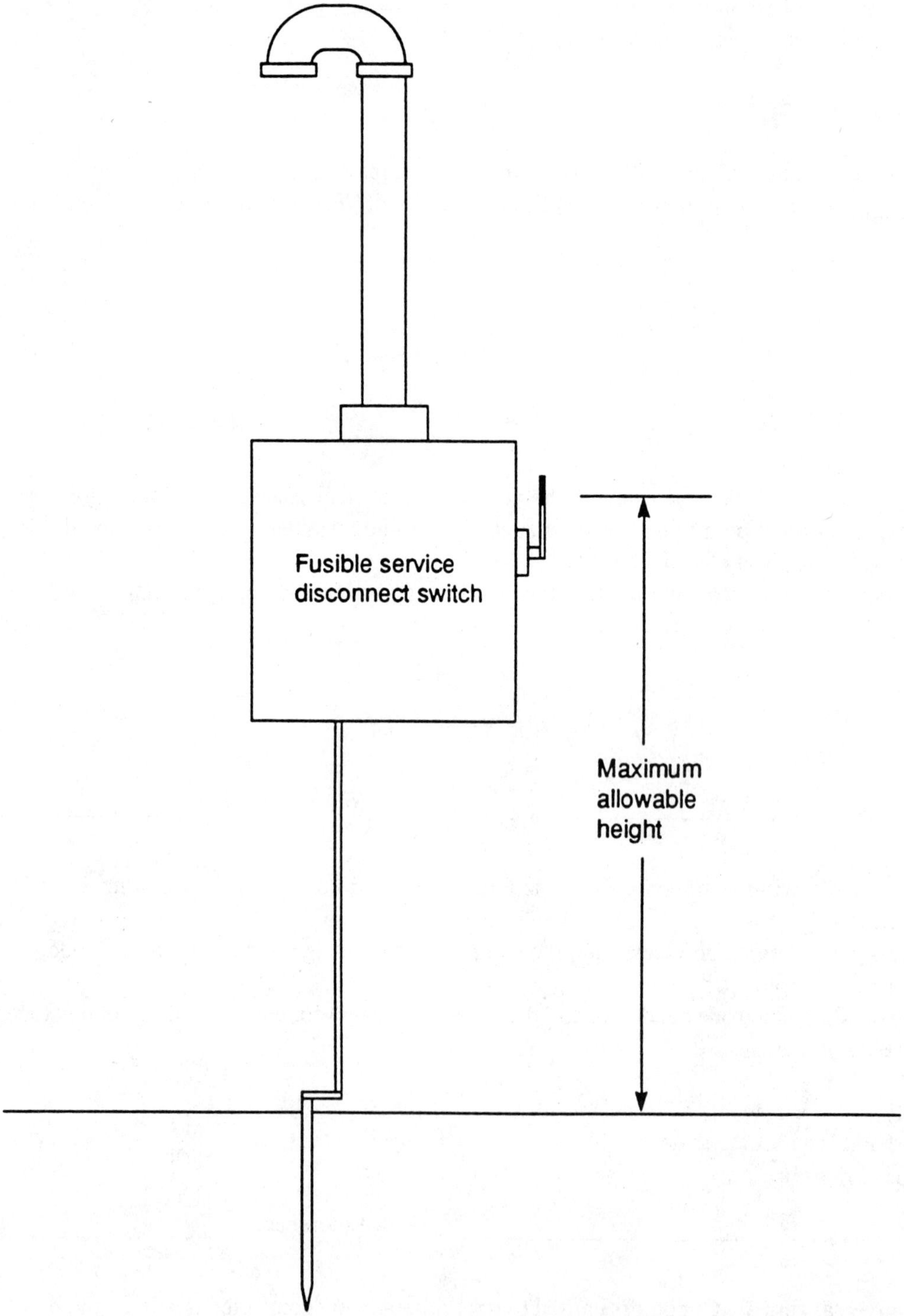

B. 6½ feet
C. 7 feet
D. ½ foot

Answer: ____________________ Reference: ____________________

52. When fuses are used for motor overload protection for a three-wire, three-phase motor, the fuse shall protect
    A. each ungrounded conductors.
    B. only two ungrounded conductors.
    C. only one ungrounded conductor.
    D. the equipment grounding conductor.

    Answer: _____ Reference: _____

53. How is the size of an electrical conductor supplying a circuit determined?
    A. voltage
    B. amperage
    C. the length
    D. all of the above

    Answer: _____ Reference: _____

54. Ohm's Law is
    A. measurement of the I^2R losses.
    B. the relationship between voltage, current, and power.
    C. an equation for determining power.
    D. a relationship between voltage, current, and resistance.

    Answer: _____ Reference: _____

55. Electrical pressure is the measure in
    A. volts.
    B. amperes.
    C. coulombs.
    D. watts.

    Answer: _____ Reference: _____

56. The resistance of a 1500 watt, 120-volt resistance heater element is
    A. 14.4 ohms.
    B. 9.6 ohms.
    C. 11.2 ohms.
    D. 12.5 ohms.

    Answer: _____ Reference: _____

57. Total ampacity of stranded or solid conductors is the same if they have the same
    A. diameter.
    B. circumference.
    C. cross-sectional area.
    D. insulation.

    Answer: _____ Reference: _____

58. Two 6-volt batteries connected in parallel will give
    A. longer service than one battery.
    B. 12 volts.
    C. higher currents.
    D. lower currents.

    Answer: _____ Reference: _____

242 Master Electrician's Review

59. A branch circuit that supplies a number of outlets for lighting and appliances is called a _____ branch circuit.
    A. general purpose
    B. utility
    C. multi-purpose
    D. none of the above

    Answer: _____   Reference: _____

60. An electric heater will produce less heat on low voltage because
    A. its total watt output decreases.
    B. the current will decrease.
    C. the resistance does not change.
    D. all of the above.

    Answer: _____   Reference: _____

61. A 10-ohm resistance carrying 10 amperes of current uses _____ watts of power.
    A. 100
    B. 200
    C. 500
    D. 1,000

    Answer: _____   Reference: _____

62. A _____ stores energy in much the same manner as a spring stores mechanical energy.
    A. coil
    B. capacitor
    C. resistor
    D. none of the above

    Answer: _____   Reference: _____

63. More heat is created when current flows through which of the following?
    A. a 10-ohm capacitor
    B. a 10-ohm inductance coil
    C. a 10-ohm resistor
    D. all of the above are equal

    Answer: _____   Reference: _____

64. _____ means that it is constructed or protected so exposure to the weather will not interfere with its successful operation.
    A. weatherproof
    B. weather tight
    C. weather resistant
    D. all of the above

    Answer: _____   Reference: _____

65. The conductor clearance from windows can be satisfactory if
    A. not less than 24 inches from the window.
    B. run above the top level of the window.
    C. run below the bottom level of the window.
    D. any of the above.

    Answer: _____   Reference: _____

66. The *NEC®* states that electrical equipment shall be installed
    A. not exceeding the provisions of the Code.
    B. not less than the Code permits.
    C. according to the Code and local code amendments.
    D. none of the above

    Answer: _____ Reference: _____

67. The equivalent resistance of three resistors of 8 ohms, 8 ohms, and 4 ohms that are connected in parallel is
    A. 2 ohms.
    B. 4 ohms.
    C. 6 ohms.
    D. 8 ohms.

    Answer: _____ Reference: _____

68. Because fuses are rated by both amperage and voltage, a fuse will operate correctly on
    A. AC only.
    B. AC or DC.
    C. DC only.
    D. any voltage.

    Answer: _____ Reference: _____

69. A 10-horsepower, 480-volt, three-phase motor operating at 80% efficiency has an input current of _____ amperes. (HP = 746 watts)
    A. 11.5
    B. 21.5
    C. 33.3
    D. 42

    Answer: _____ Reference: _____

70. A branch circuit supplying a detached garage requires a total of 380 feet of circuit conductor with a maximum conductor resistance of 0.8 ohms. The minimum size conductor must have a resistance of no more than _____ ohms per 1000 feet.
    A. 1.6
    B. 2.1
    C. 3.4
    D. 4.8

    Answer: _____ Reference: _____

71. An electric resistance heater is rated for 2400 watts at 240 volts. What power is consumed when the heater is operated at 120 volts?
    A. 600 watts
    B. 1200 watts
    C. 2400 watts
    D. 4800 watts

    Answer: _____ Reference: _____

72. A 230-volt, single-phase circuit has 10 kilowatts of load and 50 amps of current. The power factor is _____ %.
    A. 78
    B. 87
    C. 94
    D. 115

    Answer: _____ Reference: _____

73. A 1½ horsepower single-phase motor that has an efficiency of 80%, operates at 230 volts, and has an input current of _____ amps. (1 hp = 746 watts.)
    A. 1.3
    B. 3.2
    C. 4.9
    D. 5.7

    Answer: _____ Reference: _____

74. A 30-horsepower wound-rotor induction motor with no code letter is to be installed with 460-volt, three-phase, alternating current. Disregarding all the exceptions, the nontime delay fuse for short-circuit protection of the motor branch circuit must be rated at a maximum _____ amps.
    A. 40
    B. 50
    C. 60
    D. 80

    Answer: _____ Reference: _____

75. The following 480-volt, three-phase, three-wire, intermittent use equipment is in a commercial kitchen: two 5,000 watt water heaters; four 3,000 watt fryers; and two 6,000 watt ovens. Each ungrounded conductor in the feeder circuit for this kitchen equipment must be sized to carry a minimum computed load of _____ amps.
    A. 15
    B. 27
    C. 41
    D. 46

    Answer: _____ Reference: _____

Appendix 1: Symbols

(Courtesy of American Iron and Steel Institute)

Electrical Wiring Symbols
Selected from American National Standard Graphic for
Electrical Wiring and Layout Diagrams Used in Architecture and Building Construction
ANSI Y32.9-1972

1. Lighting Outlets

Ceiling   *Wall*

1.1 Surface or pendant incandescent, mercury-vapor, or similar lamp fixture

1.2 Recessed incandescent, mercury-vapor, or similar lamp fixture

1.3 Surface of pendant individual fluorescent fixture

1.4 Recessed individual fluorescent fixture

1.5 Surface or pendant continuous-row fluorescent fixture

1.6 Recessed continuous-row fluorescent fixture

1.8 Surface or pendant exit light

1.9 Recessed exit light

1.10 Blanked outlet

1.11 Junction box

1.12 Outlet controlled by low-voltage switching when relay is installed in outlet box

2. Receptacle Outlets

Grounded   *Ungrounded*

2.1 Single receptacle outlet

2.2 Duplex receptacle outlet

2.3 Triplex receptacle outlet

2.4 Quadrex receptacle outlet

2.5 Duplex receptacle outlet – split wired

2.6 Triplex receptacle outlet – split wired

2.7 Single special-purpose receptacle outlet – split wired

NOTE 2.7A: Use numeral or letter as a subscript alongside the symbol, keyed to explanation in the drawing list of symbols, to indicate type of receptacle or use

2.8 Duplex special-purpose receptacle outlet
See note 2.7A

246 Master Electrician's Review

2.9 Range outlet (typical)
See note 2.7A

2.10 Special-purpose connection or provision for connection
Use subscript letters to indicate function (SW – dishwasher; CD – clothes dryer, etc).

2.12 Clock hanger receptacle

2.13 Fan hanger receptacle

2.14 Floor single receptacle outlet

2.15 Floor duplex receptacle outlet

2.16 Floor special-purpose outlet
See note 2.7A

2.17 Floor telephone outlet – public

2.18 Floor telephone outlet – private

2.19 Underfloor duct and junction box for triple, double, or single duct system (as indicated by the number of parallel lines)

2.20 Cellular floor header duct

3. Switch Outlets

3.1 Single-pole switch

S

3.2 Double-pole switch

S2

3.3 Three-way switch

S3

3.4 Four-way switch

S4

3.5 Key-operated switch

SK

3.6 Switch and pilot lamp

SP

3.7 Switch for low-voltage switching system

SL

3.8 Master switch for low-voltage switching system

SLM

3.9 Switch and single receptacle

S

3.10 Switch and double receptacle

S

3.11 Door switch

SD

3.12 Time switch

ST

3.13 Circuit breaker switch

SCB

Appendix 1   247

3.14 Momentary contact switch or pushbutton for other than signaling system

SMC

3.15 Ceiling pull switch

Ⓢ

5.13 Radio outlet

[R]

5.14 Television outlet

[TV]

6. Panelboards, switchboards, and related equipment

6.1 Flush-mounted panel board and cabinet
NOTE 6.1A: Identify by notation or schedule

6.2 Surface-mounted panel board and cabinet
See note 6.1A

6.3 Switchboard, power control center, unit substations (should be drawn to scale)
See note 6.1A

6.4 Flush-mounted terminal cabinet
See note 6.1A

NOTE 6.4A: In small-scale drawings the TC may be indicated alongside the symbol

[TC]

6.5 Surface-mounted terminal cabinet
See note 6.1A and 6.4A

[TC]

6.6 Pull box
Identify in relation to wiring system section and size

6.7 Motor or other power controller
See note 6.1A

[MC]

6.8 Externally operated disconnection switch
See note 6.1A

6.9 Combination controller and disconnection m
See note 6.1A

7. Bus Ducts and Wireways

7.1 Trolley duct
See note 6.1A

| T | | T | | T |

7.2 Busway (service, feeder, or plug-in)
See note 6.1A

| B | | B | | B |

7.3 Cable through, ladder, or channel
See note 6.1A

| BP | | BP | | BP |

7.4 Wireway
See note 6.1A

| W | | W | | W |

9. Circuiting

Wiring method indentification by notation on drawing or in specifications.

9.1 Wiring concealed in ceiling or wall

NOTE 9.1A: Use heavy weight line to identify service and feed runs.

9.2 Wiring concealed in floor
See note 9.1A

Appendix 2: Basic Electrical Formulas

Basic Electrical Formulas
(Courtesy of American Iron and Steel Institute)

DC Circuit Characteristics

Ohm's Law:

$$E = IR = \frac{E}{R}R = \frac{E}{I}$$

E = voltage impressed on circuit (volts)
I = current flowing in circuit (amperes)
R = circuit resistance (ohms)

Resistances in Series:

$R_t = R_1 + R_2 + R_3 + \ldots$

R_T = total resistance (ohms)
R_1, R_2 etc. = individual resistances (ohms)

Resistances in Parallel:

$$R_t = \frac{1}{\frac{1}{R_1}+\frac{1}{R_2}+\frac{1}{R_3}+\ldots}$$

Formulas for the conversion of electrical and mechanical power:

$HP = \frac{watts}{746}$ (watts × .00134)

$HP = \frac{kilowatts}{.746}$ (kilowatts × 1.34)

Kilowatts = HP × .746
Watts = HP × 746
HP = Horsepower

In direct-current circuits, electrical power is equal to the product of the voltage and current:

$$P = EI = I^2R = \frac{E^2}{R}$$

P = power (watts)
E = voltage (volts)
I = current (amperes)
R = resistance (ohms)

Solving the basic formula for I, E, and R gives

$$I = \frac{P}{E} = \sqrt{\frac{P}{R}}; \quad E = \frac{P}{I} = \sqrt{RP}; \quad R = \frac{E^2}{P} = \frac{P}{I^2}$$

Energy

Energy is the capacity for doing work. Electrical energy is expressed in kilowatt-hours (kWhr), one kilowatt-hour representing the energy expended by a power source of 1 kW over a period of 1 hour.

Efficiency

Efficiency of a machine, motor or other device is the ratio of the energy output (useful energy delivered by the machine) to the energy input (energy delivered to the machine), usually expressed as a percentage:

$$\text{Efficiency} = \frac{\text{output}}{\text{input}} \times 100\%$$

or $\text{Output} = \text{Input} \times \frac{\text{efficiency}}{100\%}$

Torque

Torque may be described as a force tending to cause a body to rotate. It is expressed in pound-feet or pounds of force acting at a certain radius:

Torque (pound-feet) = force tending to produce rotation (pounds) × distance from center of rotation to point at which force is applied (feet).

Relations between torque and horsepower:

$$\text{Torque} = \frac{33{,}000 \times HP}{6.28 \times rpm}$$

$$HP = \frac{6.28 \times rpm \text{ } time \text{ torque}}{33{,}000}$$

rpm = speed of rotating part (revolutions per minute)

AC Circuit Characteristics

The instantaneous values of an alternating current or voltage vary from zero to maximum value each half cycle. In the practical formula that follows, the "effective value" of current and voltage is used, defined as follows:

Effective value = 0.707 × maximum instantaneous value

Inductances in Series and Parallel:

The resulting circuit inductance of several inductances in series or parallel is determined exactly as the sum of resistances in series or parallel as described under dc circuit characteristics.

Impedance:

Impedance is the total opposition to the flow of alternating current. It is a function of resistance, capacitive reactance and inductive reactance. The following formulae relate these circuit properties:

$$X_L = 2\pi Hz L \quad X_C = \frac{1}{2\pi Hz C} \quad Z = \sqrt{R^2 + (X_L - X_C)^2}$$

X_L = inductive reactance (ohms)
X_C = capacitive reactance (ohms)
Z = impedance (ohms)
Hz - (Hertz) cycles per second
C = capacitance (farads)

L - inductance (henrys)
R = resistance (ohms)
π = 3.14

In circuits where one or more of the properties L, C, or R is absent, the impedance formula is simplified as follows:

Resistance only: Inductance only: Capacitance only:
$Z = R$ $Z = X_L$ $Z = X_C$
Resistance and   Resistance and   Inductance and
Inductance only: Capacitance only: Capacitance only:
$Z = \sqrt{R^2 + X_L^2}$ $Z = \sqrt{R^2 + X_C^2}$ $Z = \sqrt{X_L - X_C^2}$

Ohm's law for AC circuits:
$E = I \times Z$ $I = \dfrac{E}{Z}$ $Z = \dfrac{E}{I}$

Capacitances in Parallel:
$C_t = C_1 + C_2 + C_3 + \ldots$
C_t = total capacitance (farads)
$C_1 C_2 C_3 \ldots$ = individual capacitances (farads)

Capacitances in series:
$C_t = \dfrac{1}{\dfrac{1}{C_1} + \dfrac{1}{C_2} + \dfrac{1}{C_3} + \ldots}$

Phase Angle

An alternating current through an inductance lags the voltage across the inductance by an angle computed as follows:

Tangent of angle of lag = $\dfrac{X_L}{R}$

An alternating current through a capacitance leads the voltage across the capacitance by an angle computed as follows:

Tangent of angle of lead = $\dfrac{X_C}{R}$

The resultant angle by which a current leads or lags the voltage in an entire circuit is called the phase angle and is computed as follows:

Cosine of phase angle = $\dfrac{\text{R of circuit}}{\text{Z of circuit}}$

Power Factor

Power factor of a circuit or system is the ratio of actual power (watts) to apparent power (volt-amperes), and is equal to the cosine of the phase angle of the circuit:

$PF = \dfrac{\text{actual power}}{\text{apparent power}} = \dfrac{\text{watts}}{\text{volts} \times \text{amperes}} = \dfrac{kW}{kVA} = \dfrac{R}{Z}$

KW = kilowatts
kVA = kilowatt-amperes = volt-amperes ÷ 1,000
PF = power factor (expressed as decimal or percent)

Single-Phase Circuits

$kVA = \dfrac{EI}{1,000} = \dfrac{kW}{PF}$ $kW = kVA \times PF$

$I = \dfrac{P}{E \times PF}$ $E = \dfrac{P}{I \times PF}$ $PF = \dfrac{P}{E \times I}$

$P = E \times I \times PF$
P = power (watts)

Two-Phase Circuits

$I = \dfrac{P}{2 \times E \times PF}$ $E = \dfrac{P}{2 \times I \times PF}$ $PF = \dfrac{P}{E \times I}$

$kVA = \dfrac{2 \times E \times I}{1000} = \dfrac{kW}{PF}$ $kW = kVA \times PF$

$P = 2 \times E \times I \times PF$
E = phase voltage (volts)

Three-Phase Circuits, Balanced Star or Wye

$I_N = 0$ $I = I_P$ $E = \sqrt{3} E_P = 1.73 E_P$

$E_P = \dfrac{E}{\sqrt{3}} = \dfrac{E}{1.73} = 0.577 E$

I_N = current in neutral (amperes)
I = line current per phase (amperes)
I_P = current in each phase winding (amperes)
E = voltage, phase to phase (volts)
E_P = voltage, phase to neutral (volts)

Three-Phase Circuits, Balanced Delta

$I = 1.732 \times I_P$ $I_P = \dfrac{1}{\sqrt{3}} = 0.577 \times I$

$E = E_P$

Power:

Balanced 3-Wire, 3-Phase Circuit, Delta or Wye

For unit power factor (PF = 1.0):

$P = 1.732 \times E \times I$

$I = \dfrac{P}{\sqrt{3} E} = 0.577 \dfrac{P}{E}$ $E = \dfrac{P}{\sqrt{3} I} = 0.577 \dfrac{P}{I}$

P = total power (watts)

For any load:

$P = 1.732 \times E \times I \times PF$ $VA = 1.732 \times E \times I$

$E = \dfrac{P}{PF \times 1.73 \times I} = 0.577 \times \dfrac{P}{PF \times I}$

$I = \dfrac{P}{PF \times 1.73 \times E} = 0.577 \times \dfrac{P}{I} \times E$

$PF = \dfrac{P}{1.73 \times I \times E} = \dfrac{0.577 \times P}{I \times E}$

VA = apparent power (volt-amperes)
P = actual power (watts)
E = line voltage (volts)
I = line current (amperes)

Power Loss:

Any AC or DC Circuit

$P = I^2 R$ $I = \sqrt{\dfrac{P}{R}}$ $R = \dfrac{P}{I^2}$

P = power heat loss in circuit (watts)
I = effective current in conductor (amperes)
R = conductor resistance (ohms)

Load Calculations

Branch Circuits—Lighting & Appliance 2-Wire:

$I = \dfrac{\text{total connected load (watts)}}{\text{line voltage (volts)}}$

I = current load on conductor (amperes)

3-Wire:

Apply same formula as for 2–wire branch circuit, considering each line to neutral separately. Use line-to-neutral voltage; result gives current in line conductors.

USEFUL FORMULAS

TO FIND	SINGLE PHASE	THREE PHASE	DIRECT CURRENT
AMPERES when kVA is known	$\dfrac{kVA \times 1000}{E}$	$\dfrac{kVA \times 1000}{E \times 1.73}$	not applicable
AMPERES when horsepower is known	$\dfrac{hp \times 746}{E \times \%\,eff \times pf}$	$\dfrac{hp \times 746}{E \times 1.73 \times \%\,eff \times pf}$	$\dfrac{hp \times 746}{E \times \%\,eff}$
AMPERES when kilowatts are known	$\dfrac{kW \times 1000}{E \times pf}$	$\dfrac{kW \times 1000}{E \times 1.73 \times pf}$	$\dfrac{kW \times 1000}{E}$
KILOWATTS	$\dfrac{I \times E \times pf}{1000}$	$\dfrac{I \times E \times 1.73 \times pf}{1000}$	$\dfrac{I \times E}{1000}$
KILOVOLT-AMPERES	$\dfrac{I \times E}{1000}$	$\dfrac{I \times E \times 1.73}{1000}$	not applicable
HORSEPOWER	$\dfrac{I \times E \times \%\,eff \times pf}{746}$	$\dfrac{I \times E \times 1.73 \times \%\,eff \times pf}{746}$	$\dfrac{I \times E \times \%\,eff}{746}$
WATTS	$E \times I \times pf$	$E \times I \times 1.73 \times pf$	$E \times I$

I = amperes   E = volts   kW = kilowatts   kVA = kilovolt-amperes
hp = horsepower   % eff = percent efficiency   pf = power factor

EQUATIONS BASED ON OHM'S LAW

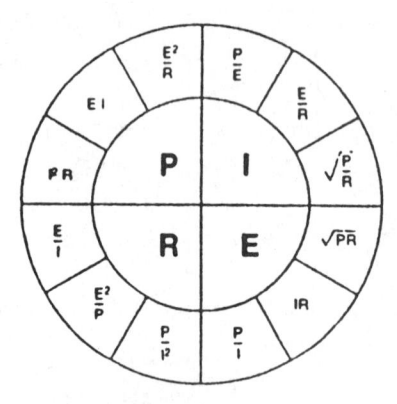

Inner wheel: P, I, R, E

Outer segments:
- For P: $\dfrac{E^2}{R}$, $\dfrac{P}{E}$: $E \cdot I$, $\dfrac{E^2}{R}$
- For I: $\dfrac{P}{E}$, $\dfrac{E}{R}$, $\sqrt{\dfrac{P}{R}}$
- For E: $\sqrt{PR}$, IR, $\dfrac{P}{I}$
- For R: $\dfrac{E}{I}$, $\dfrac{E^2}{P}$, $\dfrac{P}{I^2}$

P = POWER IN WATTS
I = CURRENT IN AMPERES
R = RESISTANCE IN OHMS
E = ELECTROMOTIVE FORCE IN VOLTS

Appendix 3: 1993 Highlighted Code Changes

The following is a list of some (but not all) changes that were accepted in the 1993 Code. They are noted in the code book by a vertical line in the margin or a bullet where a deletion was made. These changes have gone through the consensus process and were enacted at the May 1992 annual NFPA meeting held in New Orleans. The 1993 Code book was then published incorporating all adopted additions and deletions to that document.

Article 90—Introduction

Section 90-2a(2) and (3)—The word "equipment" was added to the scope (1), which states that the code covers installations of conductors *and equipment* that connect to the supply of electricity and (3) applies to installations of outside conductors *and equipment* on the premises.

Section 90-2b(2)—It should be noted this section outlines those things not covered by the Code. Self-propelled mobile surface mining machinery and the attendant electrical trailing cable were exempted from this code and are no longer covered by the Code.

Section 90-5—This section was appropriately moved from Article 110 to Section 90-5, which covers the mandatory rules and explanatory material characterized by the use of the word "shall" for mandatory, and explanatory material is in the form of a fine print note (FPN). The remaining sections were then renumbered correspondingly.

Article 100—Definitions

The definition of a *multi-wire branch circuit* was revised to clarify that a multi-wire branch circuit consists of two or more ungrounded conductors of different potential. The circuit is connected to a neutral or the grounded conductor of the system. Revisions were made to the *definition of circuit breaker*, adding a new fine print note that states automatic opening means can be integral, direct acting with the circuit breaker, or remote from the circuit breaker, and the *definition of an adjustable circuit breaker* was clarified to note the circuit breaker can trip at various values of current, time, or both within the predetermined range. A change in the *definition of conduit body* clarifies that boxes such as FS and FD or larger cast or sheet metal boxes are not classified as conduit bodies. Under *Explosion Proof Apparatus*, a new fine print was added for further information to reference UL 1203. A new definition was added for *grounded effectively* and covers intentionally grounded to earth through a ground connection or connections of sufficiently low impedance having sufficient current-carrying capacity to prevent the buildup of voltages that may result in undue hazards to connected equipment or to persons. The definition of *grounded conductor equipment* was reworded. The definition of *grounding electrode conductor* was modified slightly and now clarifies the conductor used to connect the grounding electrode to the equipment grounding conductor to the grounded conductor or both of the circuit at the service equipment or at the source of the separately derived system. *Non-incendive circuit* definition was added, along with a fine print note referencing ANSI/ISA-S12.12 for non-incendive circuits. This new definition is needed because of non-incendive equipment introduced in hazardous locations and is defined as a circuit in which any arc or thermal effect produced under intended operating conditions of the equipment or due to open shorting or grounding of field wire is not capable under specified test conditions of igniting the flammable gas vapor or dust/air mixture. The definition of *premise wiring systems* has been clarified so now both permanently and temporary installed wiring extending from the service point to the utility conductors or source of a separately derived system to the outlet does not include wiring internal to appliances, fixtures, motors, controllers, motor control centers, and similar equipment. A new definition has been added for a *separately derived system*, that clarifies that the premise wiring system, whose power is derived from a generator or transformer or converter winding that has no direct electrical connection including a solidly grounded circuit conductor to supply conductors originating in another system, is a separately derived system. *Service point* is a new definition that clarifies that the service point applies to all services regardless of voltage and is the point of connection between the facilities of the serving utility and premise wiring. Similar language was formerly found in Part H of Article 230, which covers services over 600 volts. This definition being relocated to Article 100 clarifies that it applies to all voltages. A similar change was made in Section 230-10 for the *service conductors*.

Article 110—Requirements for Electrical Installations

In *Section 110-7*, the words "as required or" were added to the last sentence for clarity. In *Section 110-9*, "nominal circuit" was added before voltage to clarify the wording. *Section 110-10* was revised slightly for clarity, and a new fine print note was added explaining that for circuit breakers (see Section 240-83(c)). In *Section 110-11*—fine print note 2 was added, stating that some cleaning and lubricating compounds can cause severe deterioration to many plastic materials used for insulating in structural applications and equipment. This is needed because of the deteriorating effect of some cleaning lubricants on nonmetallic components. In *Section 110-12*, (c) was added, an important change, to state that the internal parts of electrical equipment including bus bars, wiring terminals, insulators, and other surfaces shall not be damaged or contaminated by foreign materials such as paint, plaster, cleaners, or abrasives. This new change in 110-12 makes it mandatory that installers cover electrical components during the construction phases so painters, plasterers, and other mechanical disciplines do not get these contaminants on the electrical components, as they do have a deteriorating effect. A new section, *Section 110-14(c)*, was added. Although these requirements are not new because they have appeared for many years in the UL Underwriters Laboratories Construction Materials Guidebook, they were added to the *National Electrical Code®* in this section and relate to the temperature limitations associated with electrical connections. This change is covered in this text. Read it carefully; it is an important new section in the *National Electrical Code®*. In *Section 110-16(a)*, the second paragraph, the workspace clearances and locations were clarified by stating that the workspace shall be clear and extend from the floor or platform to the height required by this section. In *Section 110-16(d)*, under illumination, a new fine print note was added to clarify that additional lighting is not intended where the workspace is illuminated by an adjacent light source; for example, if a piece of electrical equipment such as a panelboard is located in a hallway or in another room with adequate light, that special lighting would not be necessary over the electrical equipment. In *Section 110-16(f)*, the minimum headroom working clearances was raised from 6.25 feet to 6.5 feet, and a new exception was added to exempt existing dwelling units that do not exceed 200 amperes. In *Section 110-21*, under the marking requirements, the words "indicating voltage" were added to the text. In *Section 110-22*, new language was added to the last sentence, clarifying that the marking of equipment for series rated systems should be readily available and state "Caution–Series Rated System _____ A available identified replacement component required." In *Section 110-31*, electrical installations, clarity was added to the last paragraph related to the barrier wall, fence, or screen required to enclose outdoor electrical installations and deter the access of persons not qualified, clarifying that it not be less than seven feet in height. A new fine print note was added to that same section, FPN 3, stating that a fence made of a combination of six feet or more of fence fabric and one foot or more extension using three or more strands of barbed wire, is equivalent to a 7-foot fence.

Article 200—Use and Identification of Grounded Conductors

In Article 200, there were only minor changes that clarified in *Section 200-6(d)* that one system must comply with 200-6(a) or (b), which requires a white or natural gray outer finish along its entire length and that the second system would be white with an identified colored stripe, not green. Proposals submitted to this section would have permitted white for one system, gray for the second system, and the white with the stripe for the third system. However, the panel clarified that they intended the white or natural gray be used as one system and that they were not readily distinguishable in existing installations or after the building has been in use for some time it would be difficult to distinguish one from the other; therefore, they clarified that the white or natural gray would be used for one system and the white with the stripe for the second system.

Article 210—Branch Circuits

A minor change was made to *Section 210-2*, adding central heating equipment, other than fixed electrical space heating equipment for the specific purpose of branch circuits and theaters, audience areas of motion picture and television studios, and similar locations. Three minor changes were made to *Section 210-3*, branch circuit rating, for clarity. In *Section 210-4*, multi-wire branch circuits, a new fine print note was added to state a three-phase, four-wire power system used to supply power to computer systems or other similar electronic loads may necessitate a power system design allow for the possibility of high harmonic neutral currents. This is one of the few sections of the code that now brings the term harmonics into the *National Electrical Code®*. In *Section 210-6(c)(2)*, information was added to cover listed incandescent lighting fixtures with medium-based screwshell lampholders where supplied as 120 volts or less from the output of a step-down auto transformer that is an integral part of the fixture, and the outer screwshell terminal is electrically connected to the grounded conductor of the branch circuit. This permits these auto transformer equipped fixtures to be supplied from 277-volt circuits. *Section 210-7(d)* has been revised for the third consecutive Code cycle. This change permits ground fault circuit interrupter protected receptacles as replacements. It now permits either a circuit breaker or a receptacle to protect downstream circuits in an existing dwelling when no ground conductor is present with grounding type receptacles. However, they must be marked GFCI Protected. No grounding conductor is to be connected to the green terminal. Note: It has been brought to the attention of many of those circuit riders traveling around the country that many of these receptacles are installed incorrectly, because for the GFCI receptacle to work properly, it is mandatory that the line conductors be connected correctly. If they are not connected correctly, they will not work properly. This section now requires when replacing receptacles in an existing structure required to be GFCI protected by the present Code, they must be replaced with GFCI type. *Section 210-8(a)(2)*, Exception 2 has been revised to require a single receptacle for one appliance or duplex receptacle for two appliances located within a designated space that in normal use is not easily moved. This change permits non-GFCI protected receptacles in garages for one or two dedicated appliances. If you have only one, then a single receptacle is necessary. If there are two in the same location, then a duplex is permitted. *Section 210-8(a)(3)* is clarified by adding the word "direct" for direct grade access to clarify the intent of where the outdoor receptacles must be installed. In *Section 210-8(a)(4)*, unfinished basements are now defined. This section provides the requirements for 125-volt, single-phase receptacles requiring GFCI protection in crawlspaces at or below grade level in unfinished basements, and the new wording states that for the purposes of this section unfinished basements are defined as portions or areas of a basement not intended as habitable rooms or limited to storage areas, work areas, and the like. *Section 210-8(a)(5)* has now been revised to require ground-fault circuit protection for personnel within six feet of a wet bar sink. *Section 210-8(b)* has been revised and now requires that all 125-volt, 15- and 20-amp receptacles installed in bathrooms in commercial, industrial, and other nondwelling locations shall have ground-fault circuit interrupter protection for personnel. In the substantiation for this change it was noted that the *NEC®* does not require a receptacle in these commercial and industrial bathrooms. However, if they are installed, they must be GFCI protected. The panel agreed. *Section 210-8(b)(2)* has been added to state that all 125-volt, single-phase, 15- and 20-amp receptacles installed on roofs shall have ground-fault circuit interrupter protection. *Section 210-9* has been revised; both the text of this section and Exceptions 1 and 2 now permit auto transformers (typically known as buck-boost transformers) to be used in new installations. Formerly, these were only permitted to efficiently use existing equipment whose voltages did not correspond with the new installation. *Section 210-24* has been revised, stating that branch circuits in dwelling units shall supply only loads within the dwelling unit or loads associated only with that dwelling unit. Branch circuits required for lighting, central alarms, signal communications, or other needs for public or common areas shall not be supplied from the dwelling unit panelboard. This change clarifies that in multiple occupancy buildings the common area branch circuits, such as hallway lighting and outdoor lighting, may not be supplied from one of the occupancy panelboards. A house panelboard would now be necessary. In *Section 210-52(a)*, the words "or railings" have been added to clarify that the wall space also includes those areas where railings may be installed. The purpose of the general provisions in this section are to ensure that regardless of how the furniture is arranged in habitable rooms a receptacle will be available within six feet of any furniture or equipment requiring electricity. This change should eliminate at least one more area where extension cords are being used. In *Section 210-52(b)*, small appliances, two new exceptions, 4 and 5, have been added to exempt the receptacle served by a circuit supplying only motor loads and receptacles installed to provide power for electrical ignition systems or clock timers for gas fires, ranges, ovens, or counter-mounted cooking units, and now clarifies that the small appliance loads are not required to supply the motor circuit. *Section 210-52(c)*, countertops, has been revised again and now clarifies that for countertops in kitchen and dining areas of dwelling units a receptacle outlet shall be installed in each counterspace 12 inches or wider and that no point along the wall line is more than 24 inches from a receptacle outlet. It further clarifies that for each island or peninsula countertop, the long dimension of 24 inches or greater and a short dimension of 12 inches or greater, receptacle outlets to serve island and peninsula countertops shall be installed above or within 12 inches below, and the receptacle shall be installed so no point measured along the centerline of the long dimension is more than 24 inches from the receptacle outlet in that space. It also stipulates that receptacles shall not be mounted face-up in countertops. A fine print note was added to clarify this language. However, it is safe to assume that for peninsula and island countertops a receptacle is required each 24 inches of linear space, and the width can be generally ignored. The receptacle can be installed on either side, above, or within 12 inches below. *Section 210-52(e)* now requires that outdoor outlets be installed in front and back of each dwelling. In *Section 210-52(h)*, the words "dwelling units" have been added for hallways of 10 feet or more in length and are required to have one receptacle outlet. *Section 210-63* has been revised, requiring a 125-volt, single-phase, 15- to 20-ampere receptacle outlet for servicing heating and air conditioning equipment in rooftops, attics, and crawl spaces, and shall be within 25 feet of that equipment. This change was made to correlate with the present requirements found in the building and mechanical code. In *Section 210-70*, a fine print note was added in 210-70(a) to state that a vehicle door in a garage is not considered an outdoor entrance. *Section 210-70(c)* has been revised so that one wall switch to control lighting outlet shall be installed at or near equipment requiring servicing, such as heating, air conditioning, and refrigeration equipment in attics and under floor spaces. The wall switch must be located at the point of entry to the attic or under floor space.

Article 215—Feeders

No significant changes were made to this article.

Article 220—Branch Circuits, Feeder and Service Calculations

In *Section 220-1* under the scope, the word "service" has been added to clarify that this article covers branch circuit, feeder and service loads. *Section 220-3(c)(5)* has been added for track lighting (see Section 410-102). *Section 220-22* has been revised, alerting that three-phase systems and grounded conductor of a three-wire circuit consisting of two-phase wires and a neutral of a four-wire, three-phase wye system should not have a reduced neutral. Fine print notes alert the use of Examples 1A, 1B, 2B, 4A, and 5A in Chapter 9. Fine print note 2 relates to the three-phase, four-wire power systems that supply power to computer systems or other similar electronic loads that may necessitate that power system design to allow for the possibility of high harmonic and neutral currents.

Article 225—Outside Branch Circuits and Feeders

The scope has been revised in this article to cover outside branch circuits, feeder runs between buildings, structures or poles on premises, and electrical equipment or wiring for the supply of utilization equipment located on or attached outside of buildings, structures, or poles. *Section 225-8* has been added. This language formerly appeared in Article 230, Section 84, and are the requirements for the disconnecting means required for more than one building or other structure on the same property under single management. It clarifies that those conductors leaving the service area of the primary building feeding adjacent or additional buildings on the same premises are feeder conductors and must meet the requirements in this section. It clarifies that the disconnecting means fuses shall be in accordance with Section 240-40 and that a disconnecting means is required and shall be located and installed in accordance with the requirements of Section 230-71 and 72. *Section 225-8(c)* states that the disconnecting means shall be suitable for use as service equipment, and *Section 225-9* requires that the overcurrent protection shall be in accordance with Section 210-20 for branch circuits and Article 240 for feeders. The fine print note advises users to see Section 240-24 for accessibility. *Section 225-18* has been revised for clearance requirements for overhead spans of open conductors and open multi-conductor cable. *Section 225-19* has a similar clarity change. A new Exception 4 has been added to this section, stating the requirements for maintaining vertical clearance three feet from the edge of the roof shall not apply to the final conductor span where the conductors are attached to the side of the building. This was necessary because the service mast on the side of the building often extends up near the roof edge, and where the conductors leave that building, they often pass within those parameters. However, no danger has been cited. In *Section 225-19(d)*, a new paragraph has been added stating that overhead branch circuit and feeder conductors shall not be installed through which materials may be moved, such as openings in farm and commercial buildings, and shall not be installed where they will obstruct entrance to these building openings.

Article 230—Services

Section 230-9 has been revised to clarify that the clearance requirements from windows apply only to those windows designed to be opened. This section has been further revised, clarifying that overhead service conductors should not be installed beneath openings through which materials may be moved, such as openings in farm and commercial buildings, and shall not be installed where they will obstruct entrance to these building openings. *Section 230-10* has been added for service conductors; the conductors from the service point to the service disconnecting means shall be considered service conductors. This section was formerly found in 230-201. *Section 230-23(b)* has been revised so that conductors shall not be smaller than No. 8 copper or No. 6 aluminum or copper-clad aluminum. *Section 230-23(c)* states grounded conductors shall not be less than the size required by Section 250-23(d). The clearance requirements in Section 230-24, both 10(a) and (b), have been revised, clarifying the clearances above roofs and vertical clearance from ground. *Section 230-24(a)*, above roofs, and Exception 4, which states the requirement for maintaining vertical clearance of three feet from the edge of the roof, do not apply to the final conductor span or the service drop as attached to the side of the building. In *Section 230-24(b)*, both the 12- and 15-foot clearance requirements have been revised for clarity. In *Section 230-28*, a new fine print note states it is the intent of this section to allow only power service drop conductors to be attached to the service mast. This fine print note was added because other entities, such as cable television and telephone, have been attaching to this service mast. The panel clarified it was not their intent to permit this practice. *Section 230-43* covers the wiring methods for service entrance conductors and now has a list of fourteen permitted wiring methods including liquidtight flexible metal conduit in lengths not over 6 feet long between raceways or between raceway and service equipment with equipment bonding jumper routed with the flexible metal conduit. *Section 230-66* has been added, stating that service equipment shall be marked identified as suitable for use in service equipment. This section of the code has been in the code for many years. It was inadvertently deleted out of the 1990 Code and has now been reinstated, and service rated equipment is required. *Section 230-90(a)*, Exception 3, has been rewritten, clarifying that no more than six circuit breakers or six sets of fuses shall be permitted as the overcurrent device to provide overload protection. The sum of the circuit breakers or fuses shall be permitted to exceed the ampacity of the service conductor, providing the calculated load does not exceed the ampacity of the service conductors. This is not a change in the code; it is merely a clarification, as this section has been misinterpreted in the past by some code authorities. The calculated load must be derived in Article 220. The overcurrent protection is permitted as stated. *Section 230-95*, ground-fault protection of equipment, includes two new exceptions. Exception 1 clarifies that the section does not apply to service disconnecting means for continuous industrial process where nonorderly shutdown would introduce additional or increased hazards. Exception 2 clarifies that the section does not apply to fire pumps.

Article 240—Overcurrent Protection

Several minor changes have been made to *Section 240-2*, clarifying that the additional Code sections cover overcurrent protection for busways, health care facilities, phase converters, and theaters, audience areas of motion picture, television studios and similar locations. *Section 240-3*, the protection of conductors, has been extensively revised; however, there are no major changes. This section formerly contained a long list of exceptions. It has been rewritten by the panel, changing each exception to a positive statement, and now states unless otherwise permitted in A through M below, each of those alpha suffix were originally in the form of exceptions. To clarify where these formerly appeared, you may want to do a comparison between the 1990 Code and the 1993 Code. The new (M) permissiveness is for fire protective signaling systems or circuit conductors. That is the only substantive change in this section. In *Section 240-8*, an exception has been added for circuit breakers or fuses factory-assembled in parallel and listed as a unit, which would generally not be acceptable except for this new exception. *Section 240-20(b)* has been extensively revised, clarifying where circuit breaker handle-ties are required serving line-to-line connected loads. Individual circuit breakers are permitted in some locations. *Section 240-21* should be studied to clarify the application. Like Section 240-3, this section contained a list of eleven exceptions and has been rewritten with positive statements listed in alpha order. There are no major changes to this section; however, (M), covering outside feeder taps, is new, permitting unlimited lengths for outside feeder taps. However, Sections 240-3 and 450-3 must be complied with. In *Section 240-24*, new section (E) now prohibits branch circuit overcurrent devices being located within dwelling units or guest rooms of hotels and motels bathrooms. This has been a controversial section prohibited by many code authorities throughout the country. This change now clarifies that bathrooms should not contain branch circuit overcurrent devices. In *Section 240-60*, a new exception has been added for 300-volt type cartridge fuses and fuseholders, stating that in single-phase line-to-line circuits supplied from three-phase, four-wire solid-to-ground systems where the line to neutral voltage does not exceed 300 volts. The highlight of this section is that because of the vector angle on three-phase solidly grounded systems, the voltage may exceed over 300 volts in any one phase under interrupting conditions. In *Section 240-83(e)*, voltage marking requirements, a new fine print note has been added, clarifying the requirements for straight voltage rating circuit breakers protecting three-phase corner-grounded delta systems and that slash rating circuit breakers, such as 120/240, 480 wye/277 volt can be only applied to a circuit in which the nominal voltage to ground for any conductor does not exceed the lower of the two values.

Article 250—Grounding

In *Section 250-1*, fine print notes have been rewritten, clarifying why systems and circuit conductors are grounded. Two fine print notes clearly explain the rationale for applying Article 250 correctly. In *Section 250-5*, a new fine print note has been added to give an example as a corner-grounded delta system as permitted to be grounded. In *Section 250-6(a)*, (1) has been added for portable generators that supply only equipment mounted on the generator or cord-and-plug-connected equipment through receptacles mounted on the generator or both. *Section 250-21(b)* has been revised, clarifying that the requirements of 250-51 must always be met, regardless of the objectional currents. *Section 250-24*, which covers the grounding requirements for two or more buildings or structures supplied from a common service, Sections A through D have undergone clarification editing, and now reads clearer because of these minor changes. However, there has been no major change to this section. In *Section 250-27(b)*, a new paragraph has been added for high-impedance grounded neutral systems to state that the neutral conductor shall have an ampacity of not less than the maximum current rating of the grounding impedance and in no case shall this conductor be smaller than No. 8 copper or No. 6 aluminum or copper-clad aluminum. In *Section 250-32* and *Section 250-33*, exceptions have been added that permit metal elbows installed in an underground installation of rigid nonmetallic conduit and isolated from possible contact by the minimum cover requirements of 18 inches to any part of the elbow not to be grounded. These two sections cover services and feeders and branch circuit installations. This now legitimizes a common practice used in the field since PVC was first introduced in the industry; long runs of PVC have been traditionally difficult to pull wire in because of burn-through. The conductors or the pulling rope or cable would burn through the elbow, requiring the digging up of the raceway system and replacing the elbow. Therefore, many engineers, electricians and installers have traditionally installed metal elbows at these hard-to-pull locations. The new exceptions to Section 250-32 and Section 250-33 now permit this installation practice. However, where the elbow is turned up out of the ground, the Code clearly requires that this metal portion be bonded. In *Section 250-43*, a new subsection (L) has been added for metal well casings and requires these metal well casings where a submersible pump is used shall be bonded to the pump circuit equipment grounding conductor, regardless of voltage. In *Section 250-45(c)*, for equipment required to be grounded in residential occupancies, the list has been expanded to cover many hand-held and stationary tools and motor-operated appliances often found in residential occupancies. These are required to be grounded unless they are a double insulated listed tool. A similar change has been made in *Section 250-45(d)* for other than residential occupancies. In *Section 250-53(b)*, main bonding jumper requirements, a new exception has been added to clarify that where the service disconnecting means is located in an assembly listed for use in service equipment, that one grounded conductor shall be required to be run to the assembly, and it shall be bonded to the assembly enclosure, therefore clarifying that a connection to each disconnecting means within a single enclosure is not required. A new Exception 3 has been added to *Section 250-57* to correlate with the requirements in Section 250-50(a) and (b) for existing installations, permitting that the equipment grounding conductor be run separate from the circuit conductors. This change is made for correlation only. *Section 250-71(a)(3)* covering the bonding requirements for service equipment has been revised to clarify that where a metallic raceway or armor enclosing a grounding electrode conductor is used, bonding shall apply to each end and all intervening raceways, boxes, and enclosures between the service equipment and the grounding electrode. In *Section 250-71(b)*, (3) has been added for intersystem bonding and for the purposes of providing an accessible means for intersystem bonding, the disconnecting means at a separate building or structure as permitted in Section 250-24 and the disconnecting means at a mobile home as permitted in Section 550-23(a) Exception 1 shall be considered the service equipment. *Section 250-72(b)* has been rewritten: the connections using threaded couplings or threaded boxes on enclosures shall be made of wrench tight where rigid metal conduit and intermediate metal conduit is involved. In *Section 250-74*, Exception 4, the wording has been rewritten, and the last sentence clarifies that the grounding terminal shall terminate within the same building or structure at an equipment grounding conductor terminal of an separately derived system or service. The fine print note also has been clarified, adding the word "system" and "outlet box," thereby clarifying these requirements for the connection. In *Section 250-75*, bonding other enclosures, the exception added in the 1990 Code to permit isolation of a raceway system containing circuits for data processing equipment for the reduction of noise and electromagnetic interference should be accomplished by the use of a listed nonmetallic raceway fitting located at the point of attachment of the raceway to the equipment enclosure. Although this is a permitted practice, it should be noted that where this is done, an equipment grounding conductor must be run, and all of the isolated equipment still must be grounded back to the system ground. This equipment grounding conductor is not permitted to ground any other equipment. In *Section 250-81*, requirements for grounding electrode system have been revised so interior metal water piping located more than 5 feet from the point of entrance of a building cannot be used as a conductor to connect electrodes and the grounding electrode conductor. Exception will permit this practice in industrial and commercial buildings where condition of maintenance and supervision ensure that only qualified persons will service the installation and the entire length of the interior metal water piping is exposed. This requirement was justified by stating that sections of piping have been replaced, interrupting the continuity of the system without notifying the electrician or the electrical maintenance person. Therefore, all connections must now be made accessible within the first 5 feet of that interior metal water piping. Exceptions are noted. In *Section 250-81(a)*, an exception has been added to correlate with this change in Section 250-81, Exception 2. *Section 250-81(c)* has been rewritten to clarify that epoxy-coated rebars that are now being used in some locations could not be used as a concrete encased electrode. *Section 250-83(c)(2)* has been revised to clarify that stainless steel grounding electrode rods less than $\frac{5}{8}$ inch in diameter or their equivalent shall be listed and shall be not less than $\frac{1}{2}$ inch. In *Section 250-91(a)*, a new Exception 3 has been added to permit irreversible compression type connectors listed for the purpose of splicing the grounding electrode conductor. In *Section 250-91(b)*, Exceptions have been rewritten, clarifying that listed liquidtight flexible metal conduit is permitted as grounding means under the designed conditions. This change was brought about because of the large amounts of nonlisted liquidtight flexible metal conduit that are in the market today. *Section 250-95* has been rewritten, clarifying that in-parallel or multiple raceway or cable runs, the equipment grounding conductor shall be run in parallel and shall be sized based on the ampere rating of the overcurrent device protecting the circuit conductors. This change clarifies that users may not reduce the equipment grounding conductor in parallel runs. The impact of this change would require that those using cable such as MC cable would need to specify and special order that equipment, because when ordered in a standard configuration the enclosed equipment grounding conductor is sized for 250-95 based on the conductors enclosed. When paralleled, that equipment grounding conductor would have to be significantly larger in most cases.

Article 280 — Surge Arresters
No significant changes were made.

Article 300 — Wiring Methods
In *Section 300-3(b)*, two exceptions have been added. Exception 1—The conductors of a single conductor Type MI cable with nonmagnetic sheath installed in accordance with Section 330-16 shall be permitted to be run in separate cables. Exception 2 permits column type panelboards using auxiliary gutter and pull boxes with neutral terminations. These changes are necessary because this section requires that all conductors of the same circuit where used shall be contained in the same raceway, cable tray, trench, cord, etc. *Table 300-5* has been revised; Column 3 heading has been revised as "Non-Metallic Listed Raceways for Direct Burial Without Concrete Encasement or Other Approved Raceways." This change was not in the early editions of the *NEC®* and may not appear in your code book. Also added was the verbiage under one- and two-family dwelling driveways and added outdoor parking areas and used only for dwelling related purposes. This was brought about because somebody submitted a proposal stating that dwelling driveways are often used for storage and other activities, not only for

driving onto the property right-of-way. In *Section 300-6(c)*, a new exception has been added that permits nonmetallic raceways, box, and fittings to be installed without airspace in indoor wet locations, on concrete masonry tile, and other similar surfaces. The rationale is that the nonmetallic raceways are not affected by standing moisture. *Section 300-11* has been rewritten in the last three code cycles and relates to the securing and supporting of raceway cables, boxes, and cabinets. It now states that only branch circuit wiring associated with equipment that is located within, supported by, or secured to a nonfire rated ceiling assembly shall be permitted to be supported by the ceiling support wires. Therefore, branch circuit wiring passing through an area must be supported through the structural surface. All feeders must be supported to the structural surface, and all wiring in fire-rated ceiling spaces must be supported to the structural ceiling. These conductors must provide secure support. Many code authorities do not permit this method of support under any conditions. In *Section 300-15*, which lists the requirements for boxes and fittings, a new (d) has been added, stating that equipment and integral junction box wiring compartment as part of a listed equipment shall be permitted at an outlet in lieu of a box. In *Section 300-21*, spread of fire and products of combustion, a fine print note has been added stating that directories of electrical construction materials published by qualified testing laboratories contain many of the listing installation restrictions necessary to maintain the fire-resistive rating of assemblies. *Section 300-23* has been added requiring that panels designed to allow excess, including suspended ceiling panels, shall be so arranged and secured as to allow the removal of the panels in excess of the equipment. This change was brought about because of the many computer and telephone cables installed across these ceiling panels without being properly supported, thereby rendering these panels inaccessible. This requirement clarifies that all cables and conductors must be raised up off of these ceiling panels so the area above the suspended ceiling continues to be accessible.

Article 305—Temporary Wiring

Section 305-4(i) clarifies that cables entering enclosures containing devices requiring termination shall be secured to the box with fittings designed for the purpose. *Section 305-6(a)* clarifies that a receptacle installed as part of the permanent wiring but used for temporary electrical power by construction personnel requires GFCI protection. This change was brought about because many installers were installing the permanent service (primarily in residences), before having the building completed, thereby not interpreting the code as requiring GFCI protection for the construction crews. This change now clarifies that regardless where the power is coming from during the construction, GFCI protection is required for the construction personnel.

Article 310—Conductors for General Wiring

Section 310-4, Exception 3 has been added so conductor sizes smaller than 1/0 shall be permitted to be run in parallel for frequencies of 360 Hz and higher where conditions a, b, and c of Exception 2 are met. This change permits high-frequency conductors to be run in parallel, thereby limiting the skin effect losses in these high frequency circuits. In *Section 310-7*, direct burial conductors, two fine print notes have been added, referencing Section 300-5 for installation requirements for conductors rated 600 volts or less and Section 710-4(b) for installation requirements for conductors rated over 600 volts. *Section 310-11*, (d) has been added, permitting optional markings of the conductor types listed in Table 310-13 to indicate special characteristics of the cable materials. A fine print note has been added to list an example of Type LS for limited smoke and markings such as Sunlight Resistant. In *Section 310-15*, ampacities of conductors have been rewritten to clarify the requirements for allowable ampacity and the application of the Tables 310-16 through 19 and the Neher-McGrath methods permitted under engineering supervision. (Examples of Neher-McGrath calculations are in Appendix B.) The Notes to the Ampacity Table 0–2000 Volts, Note 3 has been revised again. It is rewritten in each code cycle and continues to be controversial. It now clarifies that three-wire, single-phase service entrance conductors, service lateral conductors, and feeder conductors that supply the total load to the dwelling unit and installed in raceway or cable with or without equipment grounding conductor are covered by this note. Note 8, the Ampacity Adjustment Factors, has been revised, deleting one column, and now requires the allowable ampacities be reduced as shown in the table in Note 8. Diversity is no longer a consideration. However, Exception 5 to this note has been added, stating that for loading conditions, adjustment factors and ampacities shall be permitted to be calculated under Section 310-15(b) and a fine print note has been added to this exception, stating See Appendix B, Table 310-11 for adjustment factors for more than three current-carrying conductors in a raceway or cable with load diversity. Note 10 has been rewritten, clarifying that the neutral conductor that carries only the unbalanced current from other conductors of the same circuit need not be counted when applying Note 8. Note 10(c) has been revised to add loads of nonlinear loads and electronic computer, data processing loads, the neutral shall be considered as a current-carrying conductor and must be counted.

Article 318—Cable Trays

Section 318-3, which is a list of wiring methods permitted to be installed in cable tray systems, has been revised and now includes 18 different methods that may be used. It should be clarified that a cable tray system is not a raceway but is a support system for supporting approved wiring methods. This section lists those methods approved for installation. *Section 318-3(b)* has been revised for ventilated cable trays, 4 inches and 6 inches ventilated channel cable trays. *Section 318-6*, installation requirements for cable trays, has a new fine print note that states it is the intent to permit discontinuous segments and termination of cable tray installations where the system provides for support of cables in accordance with the corresponding articles and where adequate bonding is provided in the cable tray system design. *Section 318-6*, (j) has been added to permit incidental support and permits the support of raceways in accordance with Articles 345, 346, 347, and 348 to be incidentally supported where they cross under or over cable tray systems. *Section 318-9(a) through (e)* has been revised for clarity and covers the allowable cables in any mixture in ladder, ventilated, trough cable, and solid bottom cable trays for both power and control and signal cables. These numbers have been revised so cables will not exceed the maximum allowable cable fill permitted in Table 318-9. *Section 318-10* has been similarly revised for single conductor cables rated 2000 or less in cable trays. In *Section 318-11*, multiple conductor cables rated 2000 volts or less in cable trays, the text has been revised and Exception 2 has undergone extensive revision and clarifies that the ampacity shall not exceed the allowable ambient temperature corrected ampacities of the multi-conductor cables with not more than three insulated conductors in free air in accordance with Section 310-15(b).

Article 320—Open Wire on Insulators

There were no changes to this article.

Article 321—Messenger Supported Wiring

There were no significant changes to this article.

Article 324—Concealed Knob-and-Tube Wiring

There were no changes to this article.

Article 325—Integrated Gas Spacer Cable

There were no changes to this article.

Article 326—Medium Voltage Cable, Type MV

There were no significant changes to this article.

Article 328—Flat Conductor Cable, Type FCC

There were no significant changes to this article.

Article 330—Mineral-Insulated, Metal-Sheathed Cable, Type MI

Section 330-13 has been revised, clarifying the bending diameters of cable; five times the external diameter of the metal sheath cable of not more than $\frac{3}{4}$ inch in external diameter and ten times the external diameter of metal sheath cable greater than $\frac{3}{4}$ inch, but not more than 1 inch in external diameter. *Section 330-16* has been clarified that for single conductor cables, all phase conductors and, where used, the neutral conductor shall be grouped to minimize induced voltage on the sheath. Where they enter ferrous enclosures, the installation must also comply with Section 300-20 to prevent heating from induction. This usually is accomplished by providing a fiber entry so that metal does not separate the phase conductors. Caution also must be observed where running through metal roof trusses throughout a building; inductive heating can occur. These concerns are stated in Section 300-20.

Article 331—Electrical Nonmetallic Tubing

In *Section 331-3*, uses of permitted, electrical nonmetallic tubing and fittings are permitted in exposed and concealed work on any building not exceeding three floors above grade. However, where exposed it must not be subject to physical damage. The change here is that it is no longer required to be installed behind a 15-minute finish barrier, wall, floor, or ceiling in buildings not exceeding three floors. This also includes suspended ceilings. In *Section 331-11*, supports have been revised to clarify that where support has been provided by framing members additional support is not required. *Section 331-5* now permits electrical nonmetallic tubing in 2 inches electrical trade size. *Section 331-15* now permits the tubing to be marked with a suffix LS for limited smoke-producing characteristics.

Article 333—Armored Cable, Type AC

Although there are no significant changes to this article, uses permitted have been rewritten, *Section 333-3*, and it is permitted by the code where not subject to physical damage for branch circuits and feeders, both exposed and concealed and in cable trays, when identified for such use. It is permitted in dry locations imbedded in plaster finish, on brick or other masonry, except in damp or wet locations. It is permissible to run or fish the cable in air voids of masonry, block and tile walls where such walls are not exposed to excessive moisture or dampness, or are for low-grade lighting. *Section 333-10* has been added to clarify that AC cable must comply with Section 300-4 where installed through or parallel to studs, joists, rafters, or similar wood or metal members. *Part C of Article 333* has been extensively rewritten covering Sections 333-19 through 333-22, clarifying the construction specifications for Type AC cable, and permitting the manufacturer of cables that are flame retardant and have limited smoke characteristics to be identified with a suffix LS.

Article 334—Metal Clad Cable, Type MC

In *Section 334-10*, Exception 1 and 2 have been added, where installed as branch circuits in dwelling units, Type MC cable shall be secured within 12 inches from every outlet box, junction box, cabinet, and fitting, and Exception 2 where it is fished. This Exception 1 was added since MC cable is only required generally to be supported at intervals not exceeding 6 feet. In recent years MC cable has been increasingly manufactured and marketed as a branch circuit wiring method in small size multi-conductor No. 12 and No. 14. This new exception now requires that it be installed similar to other cable systems in these locations. *Section 334-10(f)* has been added to clarify that Type MC cable must comply with Section 300-4 where installed through or parallel to joist studs, rafters, similar wood, or metal members. *Section 334-10(g)* where installed in accessible attics, it must comply with Section 333-12.

Article 336—Nonmetallic Sheath Cable, Types NM and NMC

Section 336-11 has been added, clarifying that Type NM or NMC cable shall comply with Section 300-4 where installed through or parallel to joists, studs, rafters or similar wood or metal members. *Section 336-15* covering the support clarifies that two conductor cables shall not be stapled on edge, and where they are run through holes in wooden joists, rafters or studs, shall be considered to be supported. The fine print note was added: see Section 370-17(c) for support where nonmetallic boxes are used.

Article 337—Shielded Nonmetallic Sheath Cable, Type SNM

There were no significant changes to this article.

Article 338—Service Entrance Cable, Type SE and USE

An Exception has been added to *Section 338-1(b)* stating Type USE cable shall be permitted for service laterals above ground outside of buildings. Fine print notes have been added to this Exception, stating see Section 230-41, Exception Item B for direct buried uninsulated service entrance cables. Fine print note, see Section 300-5(d) for protection from physical damage.

Article 339—Underground Feeder and Branch Circuit Cables, Type UF

No changes were made to this article.

Article 340—Power and Control Tray Cable, Type TC

Section 340-3, the construction specifications have been rewritten to conform with the manufacturer's product guidelines. Type TC cable is available in sizes 18 through 1000 kcmil copper and sizes 12 through 1000 kcmil aluminum or copper clad aluminum. Insulated conductor sizes of No. 18 and No. 16 shall be in accordance with Section 225-16. The uses permitted in *Section 340-4* has been expanded to cover nonpower limited fire-protective signal circuits if the conductors comply with the requirements of Section 760-16. *Section 340-70*, the ampacity requirements, reference Section 402-5 for conductors smaller than No. 14 and Section 318-11.

Article 342—Nonmetallic Extensions

There were no changes to this article.

Article 343—New Article—Pre-assembled Cable and Nonmetallic Conduit

This is a new article to cover a polyethylene underground wiring method preassembled with conductors in the conduit. It is a listed factory assembly of conductors and cables and is to be installed from reels, direct burial. It is not permitted to be exposed or inside buildings. It cannot extend in a building. It is not permitted in hazardous classified locations. Its sizes are limited from $\frac{1}{2}$ inch to 2 inch. This is a specialty type product with special installation requirements. Alert: It is not to be installed inside buildings. It has no fire retardant characteristics.

Article 345—Intermediate Metal Conduit

Section 345-12, Supports, states it is now permitted that the support fastening be increased to a distance of 5 feet where structural members do not readily permit fastening within the 3 foot requirements. The word "fitting" has been removed from this section and replaced with "other conduit termination," to clarify that a coupling is not a fitting, and therefore supports should not be referenced from the coupling.

Article 346—Rigid Metal Conduit

Section 346-12 has been revised similarly to that in Section 345-12, permitting the increased distance up to 5 feet where structural members do not permit. The word "fitting" in this article has also been removed and replaced with the term "other conduit termination."

Article 347—Rigid Nonmetallic Conduit

Section 347-1 now permits the rigid nonmetallic conduit listed for the purpose to be installed underground in continuous lengths from a reel. *Section 347-3*, Uses not Permitted, has been revised with a list of exceptions so it is not permitted in a hazardous location except as covered in Sections 503-3(a), 504-20, 514-8, and 515-5, and in Class 1, Division 2 locations as permitted in the Exception to Section 501-4(b). In *Section 347-3(b)*, an Exception has been added, clarifying that this wiring method is permitted to support nonmetallic conduit bodies no larger than the largest trade size of an entering raceway, provided that the conduit bodies do not contain devices or support fixtures. *Section 347-8* covering the support requirements has been revised, stating the conduit shall be securely fastened. The word conduit termination has been added, and rigid nonmetallic conduit shall be fastened so movement from thermal expansion and contraction will be permitted. It is no longer permitted to solidly support this wiring method where expansion and contraction will occur. In *Section 347-9*, a new Exception has been added to clarify that where the computed length change due to thermal expansion or contraction exceeds .25 inches, an expansion fitting is required. *Section 347-7* has been revised, clarifying that it may be surface marked to indicate any special characteristics of the material. An example would be for limited smoke characteristics in which it may be marked with an LS designation.

Article 348—Electrical Metallic Tubing

A new Exception has been added to Section 348-12, permitting unbroken lengths be increased to 5 feet where structural members do not permit readily fastening within 3 inches. The word fitting has been removed in this section, also, and been replaced by "tubing terminations."

Article 349—Flexible Metallic Tubing

There were no significant changes to this article.

Article 350—Flexible Metal Conduit

Section 350-2 where subject to physical damage has been listed as a use not permitted.

Article 351—Liquidtight Flexible Metal Conduit and Liquidtight Flexible Nonmetallic Conduit

Section 351-4 flexible metal conduit is now permitted to be used in exposed or concealed locations for direct burial where listed and marked for the purpose and is not permitted where any combination of ambient and conductor temperature will produce an operating temperature in excess of that for which the material is approved. *Section 351-7*, covering fittings, has been revised so angle connectors shall not be used for concealed raceway installations. *Part B* covering liquidtight flexible nonmetallic conduit has been revised. *Section 351-23(b)*, not permitted uses, limiting this product to six feet lengths, has added a new exception that will permit longer length where essential for the required degree of flexibility. This section also limits this product to 600-volt nominal circuitry. An exception has been added for electric signs over 600 volts, now permitting the product to be used in this application. *Section 351-4*, which lists the size limitations at size $\frac{1}{2}$ through 4 inch inclusive, Exceptions 2 and 3 have been added permitting $\frac{3}{8}$ not in excess of 6 feet as part of a listed assembly for tap connections to lighting fixtures as required in Section 410-67(c) or for utilization equipment; Exception 3, $\frac{3}{8}$ trade size for electric sign conductors and insulators in accordance with Section 600-31(a). A similar change has been made in *Section 351-26*, stating angle connectors shall not be used in concealed installations. *Section 351-29 and Section 351-30* have been added, requiring that splices and taps only be made in junction boxes, outlet boxes, device boxes, and conduit bodies and that bends shall be limited to an equivalent of 4 quarter bends, 360° total between pole points.

Article 352—Surface Metal Raceways and Surface Nonmetallic Raceways

Section 352-9 requires that surface metal raceway enclosures making a transition from another wiring method shall have a means for connecting the equipment grounding conductor. *Section 352-21* permits the marking of special characteristics on surface nonmetallic raceways that have limited smoke producing characteristics to be identified with a suffix, LS. *Section 352-28* permits extension of unbroken lengths of surface nonmetallic raceways through drywall, dry partitions, and dry floors.

Article 353—Multi Outlet Assembly

There were no changes to this article.

Article 354—Under Floor Raceways

There were no changes to this article.

Article 356—Cellular Metal Floor Raceways

There were no significant changes to this article.

Article 358—Cellular Concrete Floor Raceways

There were no changes to this article.

Article 362—Metallic Wireways and Nonmetallic Wireways

The title of this article has been changed to include nonmetallic wireways, a new wiring method introduced to the 1993 Code. *Part A* now covers metal wireways. *Part B* covers nonmetallic wireways. The new Part B has been included and covers Sections 362-14 beginning with the definition through 362-27 including the marking. This is a limited use product, nonmetallic wireway. It is only permitted for exposed work. It is not permitted where subject to physical damage. It is not permitted outside where exposed to sunlight, unless marked for the purpose, and is not permitted where subject to ambient temperatures other than those for which the nonmetallic wireway is listed. This product would generally find its use indoors, such as dairies and food processing locations where the temperature is controlled. Expansion characteristics for PVC is significant and must be allowed for with this product if installed where temperature change occurs and will create an expansion or contraction of .25 inches or more. Read these sections carefully before selecting this product for installation.

Article 363—Flat Cable Assemblies, Type FC

There were no changes to this article.

Article 364—Busways

Section 364-8, branches from busways, has been revised so that a tension take-up device is required for cord and length of cord from the busway plug-in device does not exceed six feet. In *Section 364-11*, the text has been revised, requiring overcurrent protection shall be required where busways are reduced in ampacity. An exception has been added, permitting the reductions in industrial establishments only, provided the length of busway having the smaller ampacity does not exceed 50 feet and has an ampacity at least equal to $\frac{1}{3}$ the rating and setting of the overcurrent device next back on the line. *Section 364-14* requires busways used as a branch circuit must be designed so loads can be connected to any point and be limited to lengths which will not be overloaded in normal use.

Article 370—Outlet, Device, Pull and Junction Boxes, Conduit Bodies and Fittings

This article has been renumbered so that beginning with Part B, the installation requirements, 10 numbers have been added to each section. Therefore, what was formerly Section 370-5 in the 1990 Code now becomes Section 370-15 in the 1993 Code. *Section 370-5*, short radius conduit bends, has been added stating that conduit bodies such as capped elbows, service entrance elbows, and enclosing conductors No. 6 or smaller that are only intended to enable the installation of the raceway and contain conductors shall not contain splices, taps, or devices, and shall be of sufficient size for all conductors enclosed in the fitting. *Section 370-16(a)* has been revised, clarifying the box size calculations. This section now clarifies that for combination conductor sizes entering the box, the maximum number of conductors permitted shall be computed using the volume for the conductor listed in Table 370-16(b), and the deductions provided in this section for fixtures, stud cable clamps, or hickeys shall be based on the largest conductor entering the box. For each yoke or strap containing one of more device or equipment, a deduction shall be based on the largest conductor connected to the device or equipment supported by the yoke or strap. For equipment grounding conductors, the conductors shall be based on the largest equipment grounding conductor entering the box. The maximum number of conductors listed in Table 370-16(a) shall not be exceeded. An exception has been added so that where the equipment grounding conductor or not over four fixture wires smaller than No. 14 or both enter the box from a fixture canopy and terminate within that box, these conductors need not be calculated, or they can be omitted from the calculations. These new requirements clarify the calculation of a box. Examples for this are found in this text. *Section 370-16(c)* has been revised, stating that conduit bodies shall not contain splices, taps, or devices unless they are durably and legibly marked by the manufacturer of the cubic inch capacity. The maximum number of conductors permitted shall be computed using the provisions of Section 370-16(b). In *Section 370-17(c)*, the Exception has been revised, clarifying that multiple cable entries are permitted in a single knock-out opening. *Section 370-23(d) and (e)* have been rewritten. These sections cover raceway supported enclosures with devices or fixtures and without devices or fixtures. The language has been clarified as to the acceptability of a raceway supported enclosure under these conditions. *Section 370-40(b)* has been rewritten to include additional metallic boxes, such as brass, bronze, or zinc, and conduit bodies shall not be less than $3/32$ of an inch thick. Other cast metal boxes shall have a wall thickness of a minimum of $1/8$ inch. Two exceptions have been added permitting listed boxes and conduit bodies shown to have equivalent strength and characteristics made of thinner or other metals, and Exception 2 permitting the walls of listed short radius conduit bodies as covered in Section 370-5 to be made of a thinner metal. *Section 370-41* has been rewritten to cover the thickness of the metal covers.

Article 373—Cabinets, Cut-Out Boxes, and Meter Socket Enclosures

The title has been rewritten to include Meter Socket Enclosures. This reference was included into the scope of this document in the 1987 Code; however, now it has been expanded into the title of the document. No significant changes were made to this article.

Article 374—Auxiliary Gutters

No significant changes were made to this article.

Article 380—Switches

Section 380-4, wet locations, has been revised so that switches shall not be installed within wet locations in tub or shower spaces, unless installed as part of a listed assembly. *Section 380-12* has been revised so that metal face plates or strap switches shall be effectively grounded where used with wiring method that includes or provides an equipment ground. *Section 380-17* has been added to clarify that a fused switch shall not have fuses connected in parallel except as permitted in Section 240-8, Exception.

Article 384—Switchboards and Panelboards

Section 384-3(d) states the terminals of switchboards and panelboards shall be so located that it will not be necessary to reach across or beyond an ungrounded line bus in order to make connections. This is an existing requirement in UL 891. A fine print note 4 was added to *Section 384-4*, Installation Requirements, to state that it is not the intent that any of the provisions of this rule or the exceptions thereto allow any equipment to be located in the working space described in Section 110-16. *Section 384-16* covering overcurrent protection, (c) has a new Exception added, permitting an assembly including the overcurrent device listed for continuous operation at 100% of its rating. Before this change, overcurrent devices located in a panel were limited to 80% of the rating where in normal operation, the load will continue for three hours or more.

Article 400—Flexible Cords and Cables

Although there were a significant number of changes in this article, most related to product types and uses generally acceptable. Reference has been made in this section to nonlinear loads and electric discharge lighting, and the type cables acceptable for use in show windows and showcases, the equipment grounding conductor identification requirements, and attachment plug requirements. *Section 400-31*, construction requirements for flexible cords, the conductor shall be of a No. 8 copper or larger and it shall employ flexible stranding for portable cables over 600 volts nominal. The equipment grounding conductor shall be provided. The total area shall not be less than that of the size of equipment grounding conductor required in Section 250-95.

Article 402—Fixture Wires

No significant changes were made to this article.

Article 410—Light Fixtures, Lighting Fixtures, Lampholders, Lamps, and Receptacles

In *Section 410-8(d)(1),(2), (3), and (4)*, the word "space" was added at the end of each sentence to clarify that the storage space is not necessarily the entire room. *Section 410-15(b)(1) Exception* permits the omission of the hand hold required on a metal pole 20 feet or less in height above grade if the pole is provided with a hinged base and the grounding terminal is accessible. Both parts of the hinged pole must be bonded in accordance with Section 250-75. In *Section 410-16(h)*, fine print notes were added referencing Sections 225-26 and Section 300-5(d). *Section 410-30(c)* the words "listed fixture" or "listed fixture assembly" have been added and a new exception permitting a listed fixture or listed fixture assembly incorporating a cord and canopy shall not be required to terminate at the outer end in an attachment plug or busway plug. This change was a controversial change and now permits cord-wired fixtures, which are normally found as chain-hung fixtures in shops, etc.; however, the fixture must be listed and the cord assembly attached at the time of shipment. Field wiring of cord-connected fixtures still require that a cord cap or plug be used for terminating the fixture. A new section, *Section 410-56(i)* has been added that requires receptacles and raised covers shall not be secured solely by a single screw. An exception has been added for devices or assemblies listed and identified for such use. Exceptions 1, 2, and 3 have been added to *Section 410-73(e)* covering thermal protection requirements, and now permit appliances and fixtures using straight tubular lamps with simple reactance ballasts, ballasts for use in exit fixtures and so identified, an egress lighting energized only during emergency to not be protected by an integral thermal protection. A fine print note has been added to this section, clarifying that the thermal protection requirement may be accomplished by means other than the thermal protector. *Section 410-80(b)* covering special provisions for high discharge lighting of more than 1,000 volts, dwelling occupancy equipment is limited to open circuit voltage exceeding 1,000 volts shall not be installed in or on dwelling occupancies. Live parts, the terminal of the electric discharge lamps shall be considered as live where any lamp terminal is connected to a circuit of over 300 volts. A statement has been added to *Section 410-101(c)(8)* that limits lighting track shall not be installed less than 5 feet above the finished floor, except where protected from physical damage, or operating at less than 30 volts RMS open circuit voltage.

Article 422—Appliances

Section 422-8(d)(1) has a change concerning electrical operated kitchen waste disposers intended for dwellings and permits cord-and-plug connected with a flexible cord identified for the purpose. A similar change has been made in (2) for built-in dishwashers and trash compactors. Two new exceptions to (3) include Exception 1, for high-pressure spray washer protected by a system of double insulation, if the unit is provided with a permanent warning indicating that it shall be connected to a GFCI circuit interrupter receptacle, and Exception 2 for high-pressure spray washers rated for a three-phase supply. A fine print note has been added to *Section 422-23* that states for polarity of Edison-based lampholders see Section 410-42(a). A change to *Section 422-29* now requires that cord-and-plug connected pipe heating assemblies intended to prevent freezing of piping must be listed.

Article 424—Fixed Electric Space Heating Equipment

A change has been made to *Section 424-9*, the general requirements for installation that permit permanently installed electrical heaters to be equipped with a factory installed outlets or outlets provided as separate listed assemblies in lieu of electrical outlets required by Section 210-50(b) and shall not be connected to the heater circuits. In *Section 424-22(c)*, two fine print notes have been added. Fine print note 1 for supplementary overcurrent protection, see Section 240-10. Fine print note 2 for disconnecting means for cartridge fuses in circuits of any voltage, see Section 240-40.

Article 426—Fixed Outdoor Electrical Deicing and Snow Melting Equipment

A minor change has been made to the title of this article. *Section 426-27*, Grounding, noncurrent-carrying metal parts of equipment likely to become energized are required to be bonded together and grounded in a manner specified in Article 250. *Section 426-40* requires the current through the electrical insulated conductor inside the ferromagnetic envelope be permitted to exceed the ampacity values shown in Article 310, provided it is identified as suitable for this use. *Section 426-53* for equipment protection requiring ground fault protection on equipment supplying fixed outdoor electrical deicing and snow melting equipment has been added. *Section 426-54* has been added, requiring that cord-and-plug connected deicing and snow melting equipment be listed.

Article 427—Fixed Electrical Heating Equipment for Pipelines and Vessels

No significant changes were made to this article.

Article 430—Motors, Motor Circuits and Controllers

In *Section 430-1*, the scope now has been expanded to cover motor control centers. A new Exception 1 has been added stating the installation requirements for motor control centers are covered in Section 384-4; Exception 2 that air conditioning and refrigeration equipment are covered in Article 440; and a fine print note noting that the diagram is for information only. The diagram has been redrawn to now cover motor control centers. In *Section 430-22*, conductor requirements for a single motor, changes have been made in Section 430-22(a), general requirements for a multispeed motor. The selection of the branch circuit conductors shall be based on the highest full load current rating shown on the nameplate and the selection of the branch circuit conductors between the controller and motor shall be based on the current rating of the windings that the conductors energize. In *Section 430-24*, conductors for several motors, numerous changes have been made. However, none are significant. New language in this section clarifies that the full-load current rating of all of the motors plus 25% plus the ampere rating of other loads determined in accordance with Article 220 and other applicable sections. In *Section 430-25*, multimotors and combination load equipment, a new last sentence has been added where the equipment is not factory wired and individual nameplates are visible in accordance with Section 430-7(d)(2), the conductor ampacity shall be determined in accordance with Section 430-24. *Section 430-26* now permits the authority having jurisdiction to grant permission for feeder conductors having an ampacity less than specified in 424, provided the conductors have sufficient ampacity for the maximum load determined. In *Section 430-28(1)*, a new statement has been added for field installation be protected by an overcurrent device on the line side of the tap connector, the rating setting shall not exceed 1,000% of the tap conductor ampacity, and a new (g) to the Exception, stating that the tap shall not be made less than 30 feet from the floor. In *Section 430-34*, selection of overcurrent relays, a new fine print note has been added. A class 20 or 30 overload relay will provide longer motor acceleration than a class 10 or 20, respectively. Use of a higher class overload relay may preclude the need for selection of a higher trip current. The significance of these class overloads are at six times their rated current a Class 10 will open in 10 seconds; at six times its rated current a Class 20 will open in 20 seconds; and at six times its rated current a Class 30 will open in 30 seconds. Normal overloads are generally applied as Class 20. In *Section 430-52(a)*, Exception 1 has been revised now requiring that where the values for branch circuits, short circuit, and ground-fault protective device as determined in Table 431-52 do not correspond to the standard size and ratings of fuses or circuit breakers, as found in Section 240-6, the next lower standard size shall be used. If this next lower standard size is not adequate to carry the load, the next higher standard size or rating shall be permitted. The exception as related to instantaneous circuit breakers in Section 430-52(a) has been revised stating that in no case shall the instantaneous circuit breaker exceed 1300% of the full-load motor current. It further states that settings above 700% shall only be permitted where the need has been demonstrated by engineering evaluation. In *Section 430-62(a)*, new language has been added to the specific load requirements, requiring the branch circuit, short circuit, and ground-fault protective device for any motor or group of them shall be based on the maximum permitted value for the specific type protective device, as shown in Table 430-152 or Section 440-22(a) for hermetic refrigerant motor compressors, plus the sum of full load currents for other motors of the group. An exception has been added under Part G, Motor Controllers. *Section 430-83* has been revised, requiring that controllers have a horsepower rating not lower than the horsepower rating of the motor. *Section 430-83(d)* has been added, alerting the user that straight voltage controllers, such as 240 or 480, are permitted to be applied in a circuit in which the nominal voltage between any two conductors does not exceed the controller's voltage rating. Controllers with slant rating, such as 120/240 volt or 480/277 volt, shall only be applied to a circuit in which the nominal voltage to ground of any conductor does not exceed the lower of the two values. A new *Part H* has been added to Article 430, to cover motor control centers. Starting with Section 430-92, the general requirements, and continuing through Section 430-98, Section 430-109 has been rewritten, requiring the type disconnecting means to be a listed device, a motor circuit switch rated in horsepower, a circuit breaker, or a molded case switch, there are seven exceptions. *Exception 7* has been added, an instantaneous trip circuit breaker that is part of a listed combination for the controller is permitted as the disconnecting means. *Section 430-145(b)* has been revised to permit liquidtight flexible nonmetallic conduit and rigid nonmetallic conduit from the junction box to the motor to enclose leads to the motor, provided an equipment grounding conductor is connected both to the motor and the junction box.

Article 440—Air Conditioning and Refrigerating Equipment

In *Section 440-2*, the definition of branch circuit selection current and rated load current have been extensively revised.

Article 445—Generators

No changes were made to this article.

Article 450—Transformers and Transformer Vaults

Section 450-3, the overcurrent protection requirements, have been revised, now permitting the secondary overcurrent device to consist of not more than six circuit breakers or six sets of fuses grouped in one location. Where the multiple overcurrent devices are used, the total of all devices is not permitted to exceed the allowed value of a single overcurrent device, thus making this change impractical in many applications because it may not permit proper loading of the transformer. Fine print notes have been added to this section: Fine print note 1 stating to see Section 240-3, 240-21 and 240-100 for the overcurrent protection of conductors; fine print note 2 alerting us that nonlinear loads caused by loads such as electric discharge lighting, electronic computer, and data processing equipment, or similar equipment, can increase the heat in a transformer and the overcurrent device may not operate properly. In *Section 450-3(a)(1)*, electronic actuated fuses that may be set to open at a specific current are required to be set in accordance with the settings for circuit breakers. In *Section 450-3(b)*, transformers 600 volts nominal or less, the overcurrent protection of the transformers rated at 600 volts nominal or less must comply with (1) and (2) below. A *new section (d)* has been added to permit fire pump installations. Where the transformer is dedicated for supplying a fire pump installation, secondary overcurrent protection is not required. Primary overcurrent protection must be provided in accordance with Section 450-3(a) or (b), and the setting must be sufficient to carry the equivalent of the transformer secondary locked rotor current of the fire pump motor and all associated fire pump equipment indefinitely. *Section 450-11* has been revised, requiring clearance for transformers ventilated openings.

Article 455—Phase Converters

This is a new article to the 1993 Code and covers phase converters. Much of this information was formerly found in Article 430 of the 1990 *NEC®*. This new article covers both static and rotary phase converters, the installation and uses permitted.

Article 460—Capacitors

Section 460-2, enclosing and guarding requirements, (a) covering capacitors containing more than three gallons or flammable liquid has been revised to clarify that the limit shall apply to single unit in an installation of capacitors.

Article 470—Resistors and Reactors for Rheostats

See Section 430-82. There were no changes to this article.

Article 480—Storage Batteries

There were no changes to this article.

Article 500—Hazardous (Classified) Locations

In *Section 500-2*, fine print note 2 has been revised, updating the references to additional standards, both NFPA and the American Petroleum Institute. *Section 500-3* has been rewritten, reinstating the classes and group classifications for Class 1 and Class 2 locations.

Article 501—Class I Locations

In *Section 501-3(a)*, a fine print note has been added to see purged and pressurized enclosures for electrical equipment, NFPA 496-1989 (ANSI). *Section 501-3(b)(1)* the exception, a new statement has been added, stating in classified circuits that under normal conditions do not release sufficient energy to ignite the specific ignitable atmospheric mixture. In *Section 501-4*, wiring methods, (d) for Class 1, Division 2 locations, a new statement has been added permitting extra hard usage cords, an additional conductor for grounding shall be included in the flexible cord. A fine print note has been added to see Section 501-16(b) for the grounding requirements where flexible conduit is used. An exception for wiring nonincendive circuits shall be permitted using any of the methods suitable for the wiring in ordinary locations. In *Section 501-5*, a fine print note 2 has been added, advising gas and vapor leakage and propagation of flames may occur through interstices between the strands of standard stranded conductors larger than No. 2, special conductor constructions, compact strands, or sealing of the individual strands are means of reducing the leakage and preventing the propagation of flames. In *Section 501-5(b)*, a new Exception 2 has been added for conduit systems terminating at outdoor unclassified locations where a wiring method transition is made to cable tray, cable bus, ventilated busway type MI cable, or open wiring is now required to be sealed where passing from the Class 1, Division 2 location into the outdoor and classified areas. The conduits are not permitted to terminate at an enclosure containing an ignition source. In *Section 501-5(d)*, an Exception has been added for multiconductor cables with gas/vapor tight continuous sheath capable of transmitting gas and vapors through the cable core are permitted to be considered as a single conductor by sealing the cable in the conduit within 18 inches of the enclosure and cable end within the enclosure by an approved means to prevent entrance of gases or vapors and propagation of flames into the cable core or by other approved methods. In *Section 501-8(b)* for Class 1, Division 2 motor and generators, a new fine print note has been added, stating it is important to consider the risk of ignition due to currents arcing across discontinuities and overheating in parts of multisection enclosures of large motors and generators. Such motors and generators may require equipotential bonding jumpers across joints in the enclosures and from enclosure to ground and clean air purging immediately before and during the start-up periods. In *Section 501-10(b)(1)*, an Exception has been added where electric resistance heat tracing approved for Class 1, Division 2 locations. In *Section 501-11*, flexible cords in Class 1, Division 1 and 2 locations, new language has been added, covering the extension of a flexible cord within a suitable raceway between a wet pit and the power source permitted, and electric mixers intended to travel into and out of these open mixing tanks and vats are to be considered portable utilization equipment, thereby permitting the flexible cord connections. In *Section 501-14(b)*, Exception, a new alpha C location, as related to nonincendive circuits that under normal conditions do not release sufficient energy to ignite a specific ignitable atmospheric mixture. In *Section 501-16* covering the grounding of Class 1, Division 1 and 2 locations, an Exception has been added where the specific bonding means to the point of grounding of the building disconnecting means as specified in Section 250-24(a), (b), and (c), provided that branch circuit overcurrent protection is located on the load side of the disconnecting means. A fine print note has been added referencing Section 250-78 for additional bonding requirements in hazardous locations. In *Section 501-16(b)*, an Exception to this section covering a Class 1, Division 2 areas, the bonding jumper shall be permitted to be deleted where all of the following conditions are met: an approved liquidtight flexible metal conduit 6 feet or less in length with approved fittings, overcurrent protection in the circuit is limited to 10 amperes or less, and the load is not power utilization load. *Section 501-17* covering surge protection has been added for Class 1, Division 1 locations, and Class 1, Divisions 2 locations. Surge arrestors, including their installation and connection, must comply with Article 280 and installed in enclosure approved for the class and division. Surge protector capacitors shall be the type designed for the specific duty. A new section, *Section 501-18*, for multi-wire branch circuits has been added, requiring in a Class 1, Division 1 location a separate grounded conductor shall be installed in each single-phase branch circuit that is part of a multi-wire branch circuit. The exception permits the disconnect device for the circuit, all ungrounded conductors of the multi-wire circuit simultaneously (common trip type).

Article 502—Class 2 Locations

Section 502-1, the general requirements, the fine print note has been revised to provide information on dust ignition proof enclosures, Type 9 ANSI/NEMA 250-1985 and explosion proof dust ignition proof electrical equipment for hazardous location, ANSI/UL 1203/1988. The last sentence has been revised to clarify that where Group E dusts are present in hazardous locations, there are only Class 2, Division 1 locations. There are no Division 2s for Group E type dusts. *Section 502-4(2)* verifies that where flexible cords are used, they must comply with Section 502-12. *Section 502-4(b)*, Class 2, Division 2 wiring methods, has been revised, adding two types of cable tray, both the ladder and ventilated trough cable tray, to the permitted wiring methods. The Exception has been added to this section for wiring with nonincendive circuits that are permitted using any of the methods suitable for ordinary wiring locations. Justification for this change can be found in the definition for nonincendive in Article 100. This same exception has also been added to *Section 502-14(b)* for Class 2 Division wiring methods for signalling alarm, remote control and communication systems, leaders, instruments, and relays. In *Section 502-16(a)* covering the bonding requirements for Class 2, Division 1 and 2 locations, a new Exception has been added, permitting specified bonding means only required to the point of grounding of a building disconnecting means as specified in Section 250-24(a), (b) and (c), provided the branch circuit protection is located on the load side of that disconnecting means. A fine print note has been added, clarifying that Section 250-78, bonding requirements, must also be met. In *Section 502-16(b)*, a new Exception has been added, permitting the deletion of the bonding jumper in Class 2, Division 2 locations where liquidtight flexible metal conduit 6 feet or less in length with approved fittings is used, and the overcurrent protection in the circuit is limited to 10 amperes or less, and the load is not power utilization equipment. *Section 502-18* has been added for multi-wire branch circuits, requiring that in Class 1, Division 1 locations a separate grounded conductor must be installed in each single-phase branch circuit that is part of a multi-wire branch circuit. An Exception has been added to remove this requirement where the circuit opens all of the ungrounded conductors of the multi-wire branch circuit simultaneously (common trip type).

Article 503—Class 3 Locations

In *Section 503-16(a)*, an Exception has been added to the bonding requirements to permit specified bonding only required to the point of grounding of the building disconnecting means as specified in Section 250-24(a), (b) and (c), provided the branch circuit overcurrent protection is located on the load side of the disconnecting means. In *Section 503-16(b)*, an Exception has been added to permit the deletion of the bonding jumper in Class 3, Division 1 and 2 areas, provided approved liquidtight flexible metal conduit 6 feet in length with approved fittings is used, and the overcurrent in the circuit is limited to 10 amperes or less and the load is not a power utilization load.

Article 504—Intrinsically Safe Systems

In *Section 504-1*, the scope, has been clarified so it is now apparent that the article covers the installation of intrinsically safe apparatus, wiring, and systems for Class 1, 2 and 3 locations. The new definition has been added to *Section 504-2* for different intrinsically safe circuits that are intrinsically safe circuits in which possible interconnections have not been evaluated or approved as intrinsically safe. *Section 504-4* now clarifies that all intrinsically safe apparatus and associated apparatus must be approved. A new fine print note has been added to *Section 504-30(a)(3)*. This fine print note 2 states that physical barriers, such as grounded metal partitions or approved insulated partitions or approved restricted access wiring ducts separated from other such ducts by at least ¾ of an inch, can be used to help ensure the required separation of wiring. *Section 504-50*, which covers the grounding requirements, has been modified, clarifying in (a) that all intrinsically safe apparatus, associated apparatus, cable shields, enclosures, and raceways, if of metal shall be grounded, because any of these components including the cable shields can become charged, creating hazardous conditions and affecting the integrity of the intrinsically safe circuit. (c) has been added to clarify that where shielded conductors or cables are used, they must be grounded, except when the shield is part of the intrinsically safe circuit. Modifications have been made to *Section 504-60*, bonding. This section has been clarified so it is now apparent that in hazardous locations all intrinsically safe apparatus must be bonded in accordance with Section 250-78 and in nonhazardous locations where metal raceways are used with intrinsically safe system wiring, in hazardous locations associated apparatus must be bonded in accordance with Sections 501-16(a), 502-16(a) or 503-16(a) as applicable. *Section 504-70*, covering the sealing requirements, has been clarified so conduits and cables required to be sealed by Sections 501-5 and 502-5 must be sealed to minimize the passage of gases, vapors, or dust. A fine print note has been added to note that it is not the intent of this section to require an explosion-proof shield; however, for one to ensure sealing to prevent the migration of hazardous gases, vapors, or dust, it is unclear which sealing method could accomplish this other than listed seals manufactured especially for this purpose. An Exception has been added to *Section 504-80* covering the identification requirements for intrinsically safe systems to state that where circuits are run underground, they only need be identified where they become accessible after emergence from the ground. (C) covering color coding permits color coding to identify intrinsically safe conductors, whereas colored light blue and where no other conductors colored light blue are used.

Article 510—Hazardous Classified Locations, Specific

No changes were made to this article.

Article 511—Commercial Garages, Repair and Storage

In *Section 511-3(b)*, an Exception has been added, stating that lubrication in service rooms without dispensing shall be classified in accordance with Table 514-2. This exception clarifies that quick lubrication facilities such as "Jiffy Lubes" are no longer classified as a commercial repair or storage garage but are covered by Article 514. In *Section 511-6*, fixed wiring above Class 1 locations, four new wiring methods have been added to this section to conform to the wiring methods permitted in Section 501-4(b) for Class 1 Division 2 locations. The new wiring methods added are Type PL PC cable, flexible metal conduit, liquidtight flexible metal conduit, and liquidtight flexible nonmetallic conduit.

Article 513—Aircraft Hangars

In *Section 513-4*, wiring not within Class 1 locations, a change has been made to (a) clarifying that fixed wiring in an aircraft hangar but not within a Class 1 location as defined in Section 513-2 must be installed in metal raceways or Type MI, TC, SNM, or Type MC cable. The exception permits wiring in unclassified locations to be types recognized in Chapter 3. *Section 513-4(b)* has been modified to clarify that a separate equipment grounding conductor is required in all permitted cord types.

Article 514—Gasoline Dispensing and Service Stations

In *Section 514-1*, the definition, a subtle but important change has been made to the definition of a gasoline dispensing and service station, to clarify that where volatile flammable liquids or liquified flammable gases are transferred to fuel tanks, including auxiliary fuel tanks of self-propelled vehicles or approved containers, this change now impacts all locations where LPG (liquified petroleum gases) portable tanks are filled and are clearly to be covered by Article 514. Presently, many hardware stores, sporting goods stores, campgrounds, etc. fill LPG portable tanks. These must be wired in accordance with Article 514. *Section 514-5*, circuit disconnects, has been written to clarify that each circuit leading to or through dispensing equipment, including remote pumping systems, must be routed with a switch or acceptable means to disconnect simultaneously from the source of supply all conductors of the circuit, including the grounded conductor, if any. (B) covering attended service stations, as defined in NFPA 30A, requires that an emergency control acceptable by the authority having jurisdiction must be located not more than 100 feet from the dispensers and emergency disconnect control shall be located not more than 100 feet from the dispensers. (C) covering unattended service stations as defined in NFPA 30A now requires that emergency disconnecting controls must be installed at a location acceptable by the authority having jurisdiction not more than 20 feet but less than 100 feet from the dispensers, and additional emergency disconnecting controls must be installed on each group of dispensers or outdoor equipment used to control the dispensers. These emergency disconnecting controls shall shut off all power to the dispensing equipment at the station and must be the manual reset type in a manner approved by the authority having jurisdiction. *Table 514-2*, a new Class 1, Division 2 division has been added under lubrication for a service room without dispensing, stating that the entire unventilated area within any pit below grade area or subfloor area is a Class 1, Division 2 location. Additional Division 2 area in this same section is now described as an area up to 18 inches above any such unventilated pit below grade work area or subfloor work area and extending a distance of 3 feet horizontally from the edge of any such pit below grade work area or subfloor work area. Nonclassified area is now any pit below grade work area or subfloor work area that is ventilated in accordance with 5-1.3. NFPA 30 5-1.3 requires one CFM per square foot floor area. Interlocking is required; therefore, if there is no less than one CFM air change per square foot of floor area and the ventilation system is interlocked, the area may be permitted by the (AHJ) authority having jurisdiction to be classified less hazardous.

Article 515—Bulk Storage Plants

Section 515-2 has now been clarified so a Class 1 location is an area that does not extend beyond the floor, wall, roof, or other solid partition that has no communicating openings. A fine print note has been added to this section, stating that the Table 515-2 is based on the premise that the installation meets all applicable requirements for flammable and combustible liquids code, NFPA 30-1990 in Chapter 5 in all respects, and should this not be the case, the authority having jurisdiction has the authority to classify this area.

Article 516—Spray Application, Dipping and Coating Processes

Section 516-2, the classification of locations, has now been clarified that for dipping and coating operations, all space within a 5-foot radial distance from the vapor extending from the surfaces to the floor and pits within 25 feet horizontal from the vapor source, provided pits extend beyond 25 feet from the vapor space, the vapor stop shall be provided and the entire pit must be classified Class 1, Division 1. An Exception has been added for dip tanks to clarify that the space shall not be required as hazardous where the vapor source is 5 square feet or less and where the contents of the open tank, trough or container do not exceed 5 gallons. In addition, the vapor concentration during operation shut-down periods must not exceed 25% of the lower flammable limit outside the Class 1 location specified in Section 516-2(a)(4) above. (C) to this section has been modified for enclosed coating and dipping operations, and the space adjacent to the enclosed coating and dipping operations are considered nonclassified, except that space within 3 feet in all directions from any opening in the enclosure, which must be classified as Class 1, Division 2. In *Section 516-4*, fixed electrostatic equipment, a major rewrite to this section has been made. If one needs to apply this section for fixed electrostatic equipment, close scrutiny should be made to this new rewritten section that covers the power control equipment, the electrostatic equipment, high voltage leads, support of goods, automatic controls, the grounding requirements, isolation requirements, signs required in this area, insulators, and other than nonincendive equipment (spray equipment that cannot be classed as nonincendive).

Article 517—Health Care Facilities

Several changes have been made to this article to conform with the health care standard NFPA 99. *Section 517-2*, the general requirements, now clarify that Part B and C not only apply to single-function buildings but also are intended to be individually applied to their respective forms of occupancy within a multi-function building, such as doctors' examining rooms located within a limited care facility would be required to meet Section 517-10. Definitions have been revised or added to *Section 517-3*. Health care facilities are buildings or portions of buildings that contain, but are not limited to, occupancies such as hospitals, nursing homes, limited care, supervisory care clinics, medical and dental offices, and ambulatory care, whether permanent or moveable. A limited care facility is a building or part thereof used on a 24-hour basis for housing four or more persons who are incapable of self-preservation because of age, physical limitation due to accident or illness, or mental limitations such as mental retardation, mental disability, mental illness, or chemical dependency. A patient bed location is a location of an inpatient sleeping bed or the bed or procedure table used in a critical patient care area. The patient care area is any portion of a health care facility wherein patients are intended to be examined or treated. A fine print note clarifies that business offices, corridors, lounges, day rooms, dining rooms, or similar areas typically are not classified as patient care areas. Patient care areas of a hospital have now been clarified that the areas of a hospital in which patient care is administered are classified as general care areas or critical care areas, either of which may be classified as wet location. The governing body of the facility shall designate these areas in accordance with the type of patient care anticipated, with the following definitions of the area classifications: general care areas, critical care areas, or wet locations. Wet locations have been clarified that these are not routine housekeeping procedures or incidental spillage of liquids but are areas that include standing fluids on the floor or drenching of the work area. The reference to grounding point has now been clarified as the ground bus of a panelboard or isolation power system supply panel supplying the patient care area. *Section 517-13(b)*, covering the wiring methods permitted for grounding of receptacles and fixed electrical equipment in patient care areas, shall be a metal raceway system or cable armor or sheath assembly that qualifies as an equipment grounding path in accordance with Section 250-91(b). It further clarifies that Type MC or MI cable shall have an outer metal sheath armor identified as an acceptable grounding return path. There are two Type MC cables, both the smooth-tubed sheath and the corrugated solid-tube sheath, which meet this criteria in accordance with the UL requirements. However, UL clearly states that these types in combination with an insulated ground wire provide a ground path when the insulated ground wire is present. To meet this section of the Code, it is this author's opinion as taken from the UL construction materials guide, that Type MC cable would include an insulated green wire to be used in combination with the sheath and an insulated green wire with a yellow stripe to provide the redundancy required by this section. *Section 517-14*, panelboard rebonding, has been modified to clarify that where two panels serve the same location, a conductor shall be continuous from panel to panel, but permitted to be broken to terminate on the ground bus of each panel. In *Section 517-17*, a fine print note has been added, noting that additional levels of ground fault protection required by Section 517-17(a) are not intended on the load side of an essential electrical system transfer switch or between the on-site generating unit as described in Section 517-35(b) and the essential electrical system transfer switch or on the electrical systems that are not solidly grounded wye systems. *Section 517-18*, general care areas, (a) the patient bed location branch circuits has been revised and (b) the patient bed location receptacles, now requiring listed hospital grade or so identified receptacles, effective January 1, 1993, now clarifying that the standard green dot hospital grade receptacles are required in these locations. *Section 517-19*, covering critical care areas, Section 517-19(a) now clarifies that emergency system receptacles must be identified to indicate the panelboard and circuit number applying. In *Section 517-20*, a section has been added for wet locations stating that branch circuits supplying only listed and fixed therapeutic and diagnostic equipment are permitted to be supplied from normal grounded service or single-phase, three-phase, single or three-phase system providing the wiring for the grounded and isolated circuits of the same raceway, and all conductive surfaces of the equipment are grounded. A new Exception has been added to *Section 517-30(c)(3)* for the mechanical protection of emergency system wiring, now permitting that where enclosed in not less than 2 inches of concrete, Schedule 40 PVC is permitted except for branch circuits serving patient care areas, which require a metal raceway system or cable meeting the requirements of Section 250-91(b) and an insulated equipment ground for redundancy. *Section 517-41(b)*, covering essential electrical system transfer switch, has been rewritten to clarify that one transfer switch is permitted to serve more than one branch or system facility, provided the facility has a maximum demand on the essential electrical system of 150 kVA as shown in the diagram 517-41(3). *Section 517-61(c)* has been modified to now permit the wire serving other than hazardous classified locations be installed in a metal raceway system or cable assembly or cable armor or sheath assembly must qualify as the equipment grounding return path in accordance with Section 250-91(b). Type MC and Type MI cable must have an outer metal armor or sheath as identified as an acceptable grounding return path. See comments for this wiring method in the explanation of Section 517-13(b). *Section 517-62* now clarifies that in any anesthetizing area that all metal raceways and metal sheath cables and noncurrent-carrying conductive portions of the fixed electrical equipment shall be grounded.

Article 518—Places of Assembly

Section 518-5, supply, requires that portable switchboards and portable power distribution equipment shall be supplied only from listed power outlets of sufficient voltage and ampere rating. Such power outlets must be protected by overcurrent devices. Such overcurrent devices and power outlets shall not be accessible to the general public, and the provisions for connecting the equipment grounding conductor shall be provided. The neutral of the feeder supplying solid-state, three-phase, four-wire dimmer system shall be considered as a current-carrying conductor.

Article 520—Feeders, Audience Areas of Motion Picture and Television Studios, and Similar Locations

The title to this article has been modified, as well as the scope, in *Section 520-1*, to clarify that this article covers all buildings or parts of buildings or structures designed or used for presentation, dramatic and musical, motion picture projection, or similar purposes and as specified audience seating areas within motion picture or television studios. *Section 520-4*, the wiring methods have been modified, adding an Exception 3 to permit nonmetallic sheath cable, Type AC cable, electrical and nonmetallic tubing and rigid nonmetallic conduit in those buildings or portions thereof that are not required to be fire rated construction by the applicable building code. *Section 520-8*, covering branch circuits, now states that a branch circuit of any size supplying one or more receptacles shall be permitted to supply stage set lighting, the voltage rating of the receptacle shall not be less than the surface voltage, receptacle ampere ratings and branch circuit conductor ampacity shall not be less than the branch circuit overcurrent device ampere rating. Table 210-21(b)(2) does not apply to this article. In *Section 520-23*, new wording has been added to clarify that the reduction of lighting branch circuits including branch circuits supplying stage and auditorium receptacles used for cord-and-plug stage equipment. *Section 520-27*, covering stage and support feeders, (c) covering separate feeders to single primary stage switchboard dimmer banks, two paragraphs have been added indicating for the purpose of computing supply capacity to switchboards, it shall be permissible to consider the maximum load that the switchboard is to control in a given installation, provided that all feeders supplying the switchboard shall be protected by an overcurrent device with a rating not greater than the ampacity of the feeders and that the opening of the overcurrent device does not have any effect on the egress or emergency lighting systems. *Section 520-44*, (b) covering cables for border lights and *Table 520-44* covering the allowable ampacity for extra hard usage cord, has been added. An Exception for listed multi-conductor extra hard usage type cords not in direct contact with equipment containing heat producing elements is now permitted to have an ampacity determined by this table. The maximum load current of any conductor shall not exceed the values of this table. In *Section 520-45,* covering receptacles, a new last sentence has been added for conductors supplying receptacles must be in accordance with Articles 210 and 400. *Section 520-48* for curtain machines, requiring the curtain machines be listed. *Section 520-53(h)*, covering the supply conductors for construction and feeders for portable switchboards for use on stages, this section has been extensively revised, noting that a reduced size grounded (neutral) equipment grounding conductors are identified in accordance with Sections 200-6, 250-57(b) and 320-10, and the grounding conductors are permitted to be identified by a marking at least the first 6 inches from both ends of each length conductor with white or natural grey. Equipment grounding conductors can be identified in the same manner, with a green or green with yellow stripe at both ends. Where more than one nominal voltage exists on the same premises, each ungrounded conductor shall be identified by a system. *Section 520-53(k)* has been written to cover single-pole separate connectors and provides explicit guidelines when using single-pole portable cable connectors. They must be listed as the locking type, and must meet the exact conditions that require listed sequential interlocking with a sequence of the equipment grounding, conductor connection first, the grounded circuit connection second, and the ungrounded conductor connections last. The disconnection requirements must be in the reverse order. A "CAUTION" notice must be provided adjacent to these connectors indicating the plug connection shall be in the following order: equipment grounding, conductor connectors, grounded circuit conductor connectors, if provided, and ungrounded conductor connectors, and disconnection in the reverse order. In *Section 520-53(o)*, new wording has been added that where the single conductor feeder cables are not installed in raceways and are used on multi-phase circuits, the grounded neutral conductor shall have an ampacity of 130% of the ungrounded circuit conductors feeding the portable switchboard. This requirement is needed because of the nonlinear harmonic producing solid-state equipment in which the current in the neutral sometimes exceeds that of the ungrounded conductor current. In *Section 520-53(p)*, the qualified personnel requirements for this section have been added with exceptions, requiring that the routing of portable supply conductors, the making and breaking of the connectors, and energization and de-energization of the supply conductors must be performed by qualified personnel. The portable switchboards must be marked to indicate this requirement in a permanent conspicuous manner, such as "ONLY QUALIFIED PERSONNEL PERMITTED TO PERFORM WORK ON THIS SYSTEM." There is an exception to this rule. *Section 520-61* covers arc lamp fixtures and now requires that they be listed. *Section 520-68* covering the conductors for portables, the conductor ampacity requirement has been extensively rewritten and an exception to cover the ampacitive conductors as provided in Section 400-5 that are not in direct contact with equipment containing heat producing elements are permitted to have their ampacity determined by Table 520-44, and maximum load current of these conductors shall not exceed the values in Table 520-44. The exception permits that where alternate current conductors are allowed in Exception 2 and 3 of Section 520-68(a), the ampacity shall be given in the appropriate table of this code for the type of conductors employed.

Article 530—Motion Picture and Television Studios and Similar Locations

Section 530-2, the definitions for portable equipment, plug in box and AC power distribution box (AC plugging box, scatter box). *Section 530-17*, portable arc lamp fixtures, (b) for portable noncarbon arc electric discharge lamp fixtures has been added to include enclosed arc lamp fixtures and associated ballasts are required to be listed, and the interconnecting cord sets and interconnected cords and cables must be of an extra-hard usage type and listed. *Section 530-18(b)*, electrical power distribution boxes used on sound stages and shooting locations must contain receptacles of a polarized grounding type. *Section 530-21* has been revised similarly to that of Section 520-53(k) for single-pole separable connectors, to require sequential connection and disconnection of the single conductor connectors to provide the safe connection and disconnection of these connectors.

Article 540—Motion Picture Projectors

No significant changes were made to this article.

Article 545—Manufactured Buildings

Section 545-7 now requires that service equipment shall be installed in accordance with Section 230-70(a). *Section 545-11* covering the bonding grounding requirements, requiring that prewired panels and building components shall provide for bonding or bonding and grounding of all exposed metals likely to become energized in accordance with Article 250, Parts E, F, and G.

Article 547—Agriculture Buildings

Section 547-1, the scope, has been revised as specified in (a) and (b). Excessive dust and water, (a), has been revised to clarify that where excessive dust and dust with water may accumulate, including areas of poultry, livestock, and fish confinement systems, and corrosive atmospheres. An area that is damp by reason of periodic washing or cleaning with water and cleansing agents and similar conditions may exist. *Section 547-4*, the wiring methods have been minorly altered to clarify that Article 347 and Article 351 are acceptable permitted wiring methods in this requirement. **Note**: Other wiring methods are permitted. All environmental elements must be considered when selecting the proper wiring methods and equipment for any location. *Section 547-5(a) and (b)* have been revised for excessive dust and dust with water and corrosive atmospheres to require that buildings described in the scope of this article must use dust-proof and weather-proof enclosures, and buildings described in the scope of this article with corrosive atmospheres must use water-proof and corrosive resistant enclosures. A fine print note has been added, stating cast aluminum and magnetic steels will corrode when in agriculture environments. *Section 547-8* covering the grounding and bonding and equipotential plane, (a) covering the grounding and bonding requirements to comply with Article 250, Exception 1 to this requirement, states permits the main bonding jumper is not required at the distribution panelboard in or on buildings housing livestock or poultry where all five of the following conditions are met. The condition in this exception for equipment grounding conductors has been modified, stating that where it is run with the supply conductors and is the same size as the largest supply conductor and of the same material or a different size adjusted with an equivalent size conductor, as per Table 250-95, if of different materials. The impact of this change is to permit a full-size conductor of the same or equivalent material as the supply conductors. *Section 547-8(d)* has been revised to cover both water pumps and metal well cases and requires the frame of a motor of any water pump to be grounded, as required in Section 430-142, and where a submersible pump is used in a well casing, the well casing must be bonded to the pump circuit equipment grounding conductor or to the equipment grounding bus of the panelboard supplying the submersible pump.

Article 550—Mobile Homes

In *Section 550-2*, covering the definitions, the laundry area has been defined as an area containing or designed to contain laundry tray, clothes washer, or a clothes dryer. Mobile home has been defined as a factory assembled structure or structures that are transportable in one or more sections built on a permanent chassis and designed to be used as a dwelling without a permanent foundation where connected to the required utilities, and includes the plumbing, heating, air conditioning, and electrical systems contained therein. Unless otherwise indicated, the term mobile home includes manufactured homes that are similar structure(s) designed to be used with a permanent foundation. A fine print note has been added below this definition to say the phrase "with a permanent foundation" indicates that the home is attached to a permanent foundation acceptable to the authority having jurisdiction such that moving the structure is not likely to occur. In *Section 550-5(a)*, a new Exception 2 has been added to this section, to say that manufactured homes constructed in accordance with Section 550-23(a), Exception 2. In *Section 550-7*, covering branch circuits, (b) has been added for small appliances, stating that small appliance branch circuits are to be installed in accordance with Section 210-52(b). In *Section 550-8(d)*, the last paragraph has been revised, clarifying that receptacles are not required in an area occupied by a toilet, shower, or tub or any combination thereof. *Section 550-8(e)* has been revised so the required outlets are to be provided in all rooms other than the bath, closet, and hall areas. *Section 550-8(f)*, which covers receptacle outlets not permitted, has been revised, clarifying that receptacles are not permitted in or within reach of 30 inches of a shower or tub space and that receptacles are not to be installed face up in any countertop. In *Section 550-15*, which covers outdoor outlets, (b) has been changed to clarify that outside heating equipment, as well as outside air conditioning equipment, or both, designed to energize this equipment when it is located outside shall have the branch circuit conductors terminating in a listed outlet box or disconnecting means located on the outside of the mobile home and be labeled with specific information for the user. That information can be found in Section 550-15(b). In *Section 550-23*, covering mobile home service equipment, (a) has been revised, adding the word "adjacent," clarifying that the mobile home service equipment is to be located adjacent to the mobile home. An Exception 2 has been added, which permits the service equipment to be installed on the manufactured home provided that it is installed by the manufacturer of the structure and it complies with Article 230 and a means is provided for the connection of the grounding electrode conductor to the service equipment routed on the outside of the structure. This new exception clarifies that this cannot be a field installed system. All field installed systems continue to be required to be mounted adjacent to the mobile or manufactured home but not on it. In *Section 550-23(g)*, marking requirements have been added, to state that where 125/250 volt receptacle is used in a mobile home, the service equipment must be marked "turn disconnecting switch or circuit breaker off before inserting or removing plug. Plug must be fully inserted or removed." The marking must be located on the service equipment adjacent to the receptacle outlet. In *Section 550-24*, covering feeder conductors, an exception has been added, stating that mobile home feeder located between service equipment and the mobile home disconnecting means as covered in Section 550-23(a) Exception 1 is permitted to omit the equipment grounding conductor where the grounded circuit conductor is grounded at the disconnecting means, as permitted in Section 550-24(a).

Article 551—Recreational Vehicles and Recreational Vehicle Parks

Under Part G covering recreational vehicle parts, a fine print note has been added to *Section 551-72*, warning that on 120/240 volt, three-wire, single-phase service and feeder, a neutral cannot be reduced in size if there are no 240 volt loads, because under the most severe conditions of unbalance the neutral will pull the same current as the ungrounded conductor supplying the load. *Section 551-73(a)* has been clarified, noting that loads for other amenities, such as but not limited to service buildings, recreational buildings and swimming pools, shall be sized separately and then added to the value calculated for the recreational vehicle sites where they are all supplied by one service. In *Section 551-73* has been further modified, listing the demand factors for feeders of service entrance conductors for park sites. In *Section 551-73(d)*, feeder circuit capacity, requirements have been modified, noting that the grounded conductor shall have the same ampacity as the ungrounded conductors. *Section 551-76* covering recreational vehicle site supply equipment, an exception has been added for pull-through sites, noting that it is permitted to locate the electric supply equipment at any point along the line from 16 feet forward of the rear of the stand to 30 feet forward from the rear of the stand.

Article 553—Floating Buildings

No major changes were made to this article.

Article 555—Marinas and Boat Yards

Section 555-6 covering the wire methods has been revised to clarify that the wiring method shall be a type identified for wet locations, extra-hard usage portable power cord listed for both wet and sunlight resistance are permitted where the flexibility is required. A new fine print note 2 has been added referencing NFPA 303-1990. In *Section 555-7*, covering the grounding requirements, (b) an exception has been added stating that the equipment grounding conductor of Type MI cable shall be permitted to be identified at the terminations. *Section 555-10* has been added, clarifying that the service equipment for floating docks and marinas must be located adjacent, not on or in, the floating structure.

Article 600—Electric Signs and Outlying Lighting

In *Section 600-2* covering the disconnects, (a) Exception 1 has been added, stating the disconnecting means is not required for an exit directional sign connected to a circuit within the scope of Article 600. Exception 2 has been changed to add outlying lighting signs or outlying lighting. In *Section 600-9*, the word "or sections" has been added to the title, and the words "or sections" has been added to the text, clarifying now that portable signs or sections, letters, fixtures, symbols, and other displays are covered. In *Section 600-9(b)* covering the cords, new wording has been added to cover junior hard service or hard service types of design in Table 400-4. *Section 600-11* covering outdoor portable signs has been rewritten to clarify that the wiring of an outdoor sign, portable or mobile, must be provided with factory installed ground-fault circuit interrupter protection for personnel, and the ground-fault circuit interrupter must be an integral part of the attachment plug or located in the power supply cord within 12 inches of the attachment plug. This clarifies that it is not sufficient to plug these outdoor portable signs into GFCI receptacles; they must contain their own GFCI protection integral to the sign itself. In *Section 600-32* covering transformers, (d) a new Exception 2 has been added, which requires that transformers for small portable signs, show windows, and similar locations listed for the purpose are permitted to be connected in series where they are equipped with secondarily permanently attached to the secondary winding within the transformer enclosure. *Section 600-34(b)* has been rewritten to clarify that all energized parts of tube terminals of conductors must be separated from grounded metal material and that soldering connection devices are not required for tube terminals to conductors where they are spliced or joined so they are mechanical and electrically secure.

Article 604—Manufactured Wiring Systems

In *Section 604-6* covering the construction, (a) cable or conduit types has been rewritten clarifying that cable shall be listed armor cable or metal clad cable containing 600 volt No. 10 or 12 copper insulated conductors with a bare or insulated copper equipment grounding conductor equivalent in size to the ungrounded conductors. *Section 604-6(a)(2)* also has been revised with this same wording.

Article 605—Office Furnishings

No major changes were made to this article.

Article 610—Cranes and Hoists

In *Section 610-11* covering the wiring methods, a new Exception 5 stating that where flexibility is required for power or control to moving parts, a cord suitable for the purpose shall be permitted, provided it provides suitable strain relief and protection from physical damage and if located in a Class 1, Division 2 hazardous location it must be approved for extra-hard usage.

Article 620—Elevators, Dumbwaiters, Escalators, Moving Walks, Wheelchair Lifts, and Stairway Chairlifts

Section 620-21 covering the wiring methods has been significantly revised, adding additional cable types, such as AC cable. The exceptions have been revised. In *Section 620-22*, the title has been revised and the car light source must now be a dedicated branch circuit that supplies the car lights and accessories on each elevator car. The air conditioning and heating source must be a dedicated branch circuit for each elevator car. *Section 620-34* covering the support requirements for cables for raceways in a hoistway within an escalator or moving walkway has been revised, noting that these wiring methods must be securely fastened to the guide rail, escalator, or moving walk track, or to the hoistway, wellway, or runway construction. *Section 620-37* has been rewritten and a new Exception 2, stating that feeders are permitted inside the hoistway for elevators with driving machine motors located in the hoistway or on the car or counterweight. *Section 620-51(a)* covering the type of disconnecting means and control has been revised to require that no provision can be made to open or close the disconnecting means from outside the hoistway, machine room, or machinery spaces and that fuses or circuit breakers provided for in the disconnecting means shall be selectively coordinated with any and all other supply side overcurrent devices. In *Section 620-51(b)* covering the location, (4) has been added stating that on wheelchair lifts and stairway chair lifts, the disconnecting means must be located within sight of the motor controller. The disconnecting means is permitted in the same enclosure with the motor controller. *Section 620-53* has been revised so elevators shall have a single means for disconnecting all ungrounded car light and accessories supply conductors for each unit. *Section 620-54* covering the heating and air conditioning disconnecting means now states that elevators shall have a single means for disconnecting all ungrounded car heating and air conditioning supply conductors and where the equipment, whether it is equipment for more than one car in the machine room, the disconnecting means must be numbered to correspond with the number of the elevator car and arranged to be locked in the open position and located in the machine room for that car. *Section 620-84* and *Section 620-85* have been added so that escalators, moving walks, wheelchair lifts, and stairway chair lifts must be grounded in compliance with Article 250 and that ground-fault circuit interrupter protection for personnel is provided for all 125-volt, single-phase, 15/20-amp receptacles installed in machine rooms, machinery spaces, pits and elevator car tops.

Article 630—Electric Welders

No significant changes were made to this article.

Article 640—Sound Recording and Similar Equipment

No significant changes were made to this article.

Article 645—Electronic Computer/Data Processing Equipment

In *Section 645-5(d)*, which covers under raised floors, wiring methods for branch supply conductors, rigid nonmetallic conduit have been added and liquidtight nonmetallic conduit. A new (5) has been added, stating that cables other than those covered in (2) must be listed as Type DP cable having an adequate fire resistant characteristics for use under raised floors of a computer room, effective July 1, 1994. Exceptions have been added to this section and two fine print notes. In *Section 645-11*, uninterruptible power supplies, an Exception 2 has been added stating that disconnecting means complying with Section 645-10 is not required for power sources capable of supplying 750 volt-amperes or less derived from UPS equipment or from battery circuits integral to the electronic equipment, provided the requirements of Section 645-11 are met. *Section 645-15* covering the grounding requirements has been modified so all exposed noncurrent-carrying metal parts of electronic computer and data processing equipment must be grounded in accordance with Article 250.

Article 650—Pipe Organs

No significant changes were made to this article.

Article 660—X-Ray Equipment

No significant changes were made to this article.

Article 665—Induction and Dielectric Heating Equipment

Section 665-23, covering warning labels and signs, has been revised to require that new wording "DANGER—HIGH VOLTAGE—KEEP OUT" must be attached to the equipment and it shall be plainly visible where unauthorized persons might come in contact with the energized parts, even when doors are open or when panels are removed from the compartments containing over 250 volts AC or DC.

Article 668—Electrolytic Cells
No significant changes were made to this article.

Article 669—Electroplating
No changes were made to this article.

Article 670—Industrial Machinery
Section 670-3 has been revised to require a permanent nameplate that lists the supply voltage, phase frequency, full-load current, and ampere rating of the short circuit and ground-fault protective device amp rating of the largest motor or load, short circuit interrupting capacity of the machine, and overcurrent protective device. *Section 670-4(b)* has been revised stating that the rating or setting the overcurrent device for the circuit supplying the machine may not be greater than the sum of the largest rating or setting of branch circuit overcurrent protective device with the machine plus 125% of the full-load current rating of all resistance heating loads plus the sum of the full-load currents of all other motors or apparatus that may be in operation at the same time. An exception has been added to cover instantaneous trip circuit breakers or motor short circuit protectors. In *Section 670-5*, the clearance requirements have been reduced to a minimum of 2½ feet. This section conflicts with Section 110-16 of the Code. An Exception has been added for where the enclosure requires a tool to open or where only diagnostic and troubleshooting testing is involved on live parts operating not over 150 volts line to line, the clearances are permitted to be less than 2½ feet.

Article 675—Electrically Driven or Controlled Irrigation Machines
No significant changes were made to this article.

Article 680—Swimming Pools, Fountains, and Similar Installations
Section 680-6 has had its title revised to cover ceiling fans. *Section 680-6(a)* has been revised, adding the requirements that the receptacles on the property shall be located 10 feet from the inside wall of the pool or fountain. The exception has been revised to add wall or pool or fountain. *Section 680-6(b)* covering light fixtures has been revised to add ceiling fans. In *Section 680-8*, covering the overhead conductor clearances, Exception 1 has been revised so supply lines or service drops are covered. In *Section 680-20(b)(1)*, the words "metal" and "threaded" were deleted now permitting wet-niche underwater fixtures to be equipped with conduit entries and no longer required to be metal and threaded. *Section 680-21(a)(1)* now permits junction boxes to be equipped with threaded hubs or bosses or nonmetallic hubs listed for the purpose. *Section 680-32*, an Exception has been added for light assemblies without a transformer with fixture lamps operating at not over 150 volts are permitted to be cord-and-plug connected, provided they meet all of the conditions required in that section. *Section 680-40(b)* has been revised to add ceiling fans. In *Section 680-40(d)(4)*, an Exception has been added covering the bonding requirements of small conductive surfaces not likely to become energized, such as air and water jets and drain fittings, where not connected to metallic piping, towel bars, mirror frames, and similar nonelectrical equipment is not required to be bonded. In *Section 680-42*, a new requirement stating that spas and hot tubs and associated electrical components are required to be protected by ground-fault circuit interrupters effective January 1, 1994.

Article 685—Integrated Electrical Systems
No changes were made to this article.

Article 690—Solar Photovoltaic Systems
In *Section 680-13*, an exception has been added that where circuit grounding connections are not designed to be automatically interrupted as part of a ground-fault protection system as required in Section 690-5, the switch or circuit breaker used as a disconnecting means shall not have a pole in the grounded conductor. The fine print note was added stating that it is intended that grounded conductor be permitted to have a bolted or terminal disconnecting means to allow maintenance or troubleshooting by qualified personnel. In *Section 690-31*, wiring methods permitted, (b) has been revised adding Type UF single conductor cable identified as sunlight resistant or Type USE cable to be used.

Article 700—Emergency Systems
Section 700-8(b) has been revised that where grounded circuit conductors connected to emergency sources connected to a grounding electrode conductor at a location remote from the emergency source there shall be a sign at the grounding location to identify all emergency and normal sources connected at that location. In *Section 700-9*, a new requirement for emergency wiring circuit(s) shall be designed and located to

minimize the hazards that might cause failure due to flooding, fire, icing, vandalism, and other adverse conditions. Section 700-26 covering the ground-fault protection equipment has been revised so that ground-fault indication of an emergency source shall be provided as per Section 700-7(d).

Article 701—Legally Required Standby Systems

Section 701-9(b) has been revised to require that where the grounded circuit conductor is connected at a remote location from the emergency source, that a sign be placed at that grounding location to identify the emergency and normal sources connected at that location.

Article 702—Optional Standby Systems

In *Section 702-8*, a similar change has been made to this article where the grounding electrode conductor is located remote from the emergency source, requiring the sign to indicate all connections, emergency as well as normal sources connected at that location.

Article 705—Interconnected Electrical Power Production Sources

This article, often referred to as "Co-generation," covers the requirements for power production sources located on premises or customer owned and is generally connected in parallel to the electrical utility or primary source of power. No significant changes were made to this article.

Article 710—Over 600 Volts, Nominal General

Minor changes have been made to the numbering of Section 710-2 where the definition of high voltage is now defined as more than 600 volts nominal. Renumbering is thus reflected so the former Section 710-2 is now 710-3 and Section 710-3 is now 710-4, and so forth. No significant changes were made to this article.

Article 720—Circuits and Equipment Operating at Less than 50 Volts

An Exception was added to the scope, stating that Articles 551, 650, 669, 690, 625, and 760 are not covered by this article. No significant changes were made to this article.

Article 725—Class 1, Class 2, and Class 3 Remote Control, Signaling, and Power-Limited Circuits

A new *Section 725-6* was added, clarifying that access to equipment shall not be denied by the accumulation of wires and cables that would prevent the removal of panels, including suspended ceiling panels. Similar changes were made in Article 300, Article 760, 770, and 810 of this Code. *Section 725-38* has been revised, stating that the insulation shall not be less than the requirements of *Section 725-50,* and the wiring methods must be installed in accordance with *Section 725-52. T*hese requirements were formerly contained in this section. *Section 725-50* has been significantly changed, covering both Class 2 and Class 3 cables. *Section 725-52* should be studied extensively for the many requirements found in this section.

Article 760—Fire Protective Signaling Systems

In *Section 760-17(c)(1)*, conductor materials, two Exceptions have been added, Exception 1 for stranded copper with a maximum of 7 strands for No. 16 and 18 are permitted, and Exception 2 where stranded copper with a maximum of 19 strands for No. 14 and larger is permitted. *Section 760-18(b)(1) and (2)* have been revised to note that the derating factors in Article 310, Note 8A of the Notes to the Ampacity Table 0–2000 apply to all conductors where the fire-protective signaling circuit conductors carry continuous loads in excess of the ampacity of each conductor, where the total number of conductors is more than three, and to the power supply conductors only where the fire-protective signaling conductors do not carry continuous loads in excess of 10% of the ampacity of each conductor and where the number of power supply conductors is more than three. *Section 760-22* has been added to cover the equipment, to require that the equipment shall be durably marked where plainly visible to indicate each circuit that is power limited, fire-protective signaling circuit.

Article 770—Optical Fiber Cables and Raceways

Section 770-1, the scope, has been revised to clarify that the article applies to the installation of optical fiber cables and raceways but does not cover the construction of optical fiber cables and raceways. *Section 770-5* has been added to cover optical fiber raceway systems. *Section 770-7* has been added to require that access be maintained to equipment and shall not be denied by the accumulation of wires and cables that prevent the removal of panels, including suspended ceiling panels. *Section 770-49* covering fire resistance of optical fiber cables, stating they must be installed as wiring within buildings and shall be listed as resistant to the spread of fire in accordance with Sections 770-50 and 770-51. *Section 770-50* now requires that optical fiber cables in buildings shall be listed as suitable for the purpose and the cables shall be marked in accordance with Table 770-50.

Article 780—Closed Loop and Programmed Power Distribution

This article was originally placed in the Code to cover smarthouse wired residences. No changes were made to this article. **Note:** Smarthouse is no longer a closed loop system and now must be wired in accordance with Chapters 1 through 4.

Article 800—Communications Circuits

In *Section 800-2*, definitions, wording has been added "See Article 100." For the purposes of this article, the following additional definitions apply to clarify that the definitions in Article 100 are applicable to this article, as well as the definitions found in this section: Point of entrance has been defined as the point at which the wire and cable emerge from an external wall from the concrete floor slab or from rigid metal conduit or intermediate metal conduit grounded to an electrode in accordance with Section 800-40(b). *Section 800-4,* an Exception has been added, stating that for other wires and cables, grounding conductors, and protectors, the listing requirement shall not apply to premise wiring components such as plugs, jacks, connectors, terminal block, cross-connect assemblies that were manufactured before October 1, 1990, and all other equipment manufactured before July 1, 1991. *Section 800-5* has been added to require access to electrical equipment and shall not be denied by the accumulation of wires and cables that prevents the removal of panels, including suspended ceiling panels. *Section 800-10,* covering overhead communication wires and cables, has been extensively revised. *Section 800-12,* covering circuits requiring primary protectors, has been revised. *Section 800-30,* covering protective devices, has been extensively revised and new fine print notes added. In *Section 800-40(b)*, a paragraph has been added stating that for the purposes of this section, mobile home service equipment or mobile home disconnecting means as described in Section 800-30(b) shall be considered accessible. *Section 800-49* has been revised clarifying that fire resistance of communication wires and cables installed within a building must be listed as resistant to the spread of fire in accordance with Section 800-50 and Section 800-51. *Section 800-50* has been revised to require that communication wires or cables installed within buildings must be listed as suitable for the purpose. Part C of this article covering electric light and power circuits has been extensively revised.

Article 810—Radio and Television Equipment

In *Section 810-2*, a new sentence has been added to require that where optical fiber cable is used, Article 770 shall apply, coaxial cable wiring for television receiving equipment must comply with Article 820. No significant changes were made to this article.

Article 820—Community Antenna Television and Radio Distribution Systems

In *Section 820-3*, definition has been added for the point of entrance and a statement "See Article 100," thereby requiring that the definitions in Article 100 do apply to Article 820. A new *Section 820-5* has been added to state that access to equipment shall not be denied by the accumulation of wires and cables that prevents the removal of panels, including suspended ceiling panels. *Section 820-6* has been added, covering the mechanical execution of work, requiring that all community antenna television and radio distribution systems be installed in a neat and professional manner. *Section 820-11* for conductors and cables and raceways entering buildings, Exceptions have been added that where service conductors or coaxial cables are installed in raceways that have metal cable armor, separation is not required. An Exception 2 has been added that where electrical light and power branch circuit or feeder conductors or Class 1 circuit conductor is installed in a metal raceway, metal sheath or metal clad Type UF, Type USE or coaxial cables have metal cable armor or are installed in raceways, are not required to be separated by at least 12 inches. In *Section 820-33*, a new paragraph has been added to clarify that grounding located at mobile home service equipment located inside and no more than 30 feet from exterior wall of the mobile home that is served or the mobile home disconnecting means grounded in accordance with Section 250-24 and located inside and not more than 30 feet from exterior wall of the mobile home shall be considered to meet the requirements of this section. *Section 820-40(5)*, physical protection requirements, are stating that where subject to physical damage, the grounding conductor shall be adequately protected. Where the grounding conductor is run in metal raceway, both ends of the raceway are required to be bonded to the grounding conductor or the same terminal or electrode in which the grounding conductor is connected. A new *Section 820-42* has been added for bonding and grounding at mobile homes. New *Section 820-49 and 820-50* have been added to cover the fire resistance requirements for CATV cable and the listing and marking and installation of coaxial cables.

Chapter 9—Tables and Examples

No significant changes were made to the tables; however, several changes were made to the examples and these examples are placed in Chapter 9 for your use when designing or calculating installations similar to those given in the examples and should be studied based on the application of the example.

Note: This list of changes does not cover all of the changes or the significant impact of these changes. For further information related to these changes, you should consult other books devoted specifically to this topic, such as *Illustrated Changes to the 1993 National Electrical Code®* by Ron O'Riley, published by Delmar Publishers Inc.

Appendix 4: Answer Key

SAMPLE EXAM QUESTION REVIEW

1. A. Ref.—Article 350
2. B. Ref.—Section 347-14
3. C. Ref.—Section 349-10(b)
4. C. Ref.—Section 370-28(a)(2)
5. B. Ref.—Section 300(6)(c)
6. B. Ref.—Section 364-5
7. C. Ref.—Chapter 9, Table 1
8. B. Ref.—Section 400-13 and 240-4, Exception 1
9. D. Ref.—Section 310-14
10. D. Ref.—Section 373-6(a)
11. D. Ref.—Section 336-15
12. A. Ref.—Note 3, 0 to 2000 Volts Ampacity Tables
13. D. Ref.—Section 300-5(d)
14. D. Ref.—Section 362-5
15. D. Ref.—Note 3, 0–2000 Volts Ampacity Tables
16. D. Ref.—Section 280-25
17. C. Ref.—Section 384-15
18. B. Ref.—Section 230-54 Exception
19. A. Ref.—Section 230-3(b)
20. B. Ref.—Section 230-9 Exception
21. C. Ref.—Section 230-24(a)
22. D. Ref.—Section 240-83(c)
23. D. Ref.—Section 110-16(e)
24. C. Ref.—Section 230-95
25. B. Ref.—Knowledge
26. D. Ref.—General Knowledge
27. B. Ref.—Section 430-38
28. D.
29. D. Ref.—Section 430-24 and Table 430-150, Table 310-16
30. A. Ref.—Section 430-22 and Table 430-150, Table 310-16
31. C. Ref.—Section 430-22, Table 430-150, Table 310-16
32. A. Ref.—Section 430-81(c)
33. B. Ref.—Section 380-8(b)
34. C. Ref.—Section 450-43(a)
35. A. Ref.—Section 300-7(a)
36. B. Ref.—Section 410-66(a)
37. C. Ref.—Section 410-76(c)
38. A. Ref.—Section 700-12(b)(2)
39. C. Ref.—Section 520-25(a) and (d)
40. D. Ref.—Section 501-5(a)
41. A. Ref.—Section 500-6
42. D. Ref.—Section 511-6(a)
43. B. Ref.—Section 517-18(b)
44. C. Ref.—Sections 675-15, 250-46
45. A. Ref.—Section 440-61, 440-3
46. B. Ref.—Section 250-94, Exception 1
47. D. Ref.—Section 250-81
48. C.
49. A. Ref.—Section 210-52, 220-4(b)
50. B. Ref.—Section 422-14
51. C. Ref.—Section 422-28e Exception
52. C. Ref.—Chapter 3. $P = EI$
53. A. Ref.—Chapter 3
54. A. Ref.—Section 250-95
55. B. Ref.—Chapter 9, Tables 1, 4, 8
56. B. Ref.—Section 430-24, Table 310-16
57. D. Ref.—Table 430-15
58. C. Ref.—Section 424-3b
59. B. Ref.—Chapter 3

Appendix 4   281

60. C. Ref.—Section 424-3b
61. D. Ref.—Section 90-2(b)
62. C.
63. C.
64. B. Ref.—Section 318-8C
65. B. Ref.—Section 373-2A
66. C. Ref.—Section 348-12
67. B. Ref.—Section 370-16(a) Table
68. D. Ref.—Table 430-91
69. D. Ref.—Section 410-73f
70. A. Ref.—Section 450-47
71. B. Ref.—Section 450-13, Exception 1
72. C. Ref.—Section 460-2
73. C. Ref.—Section 440-32, Table 310-16
74. C. Ref.—Table 430-152
75. D. Ref.—Section 450-21(a) and (b)

QUESTION REVIEW CHAPTER 2

1. No. However, there is an exception: Section 351-23(b) for electrical signs. Ref.—Section 351-23(b) Exception

2. Governmental bodies exercising legal jurisdiction over the electrical installation, insurance inspectors, or others with the authority and responsibility for governing the electrical installation. Ref.—Section 90-4

3. Ships, railway rolling stock, aircraft, or automotive vehicles. Ref.—Section 90-2(b). Four installations are not covered, for example, see Section 90-2(b)(1)-(5).

4. Fine print notes (FPN). Ref.—Section 90-5

5. Section 110-3(b)

6. When adopted by the regulatory authority over the intended use. Ref.—Pages 70-i, which states the Code is advisory as far as NFDA and ANSI are concerned but is offered for use in law for regulatory purposes.

7. No. It is the intent that all premise wiring or other wiring of the utility owned meter read equipment on the load side of service point of buildings, structures, or other premises not owned or leased by the utility are covered. It also is the intent that buildings used by the utility for purposes other than listed in Section 90-2(b)(5), such as office buildings, warehouses, machine shops, or recreational buildings, that are not an integral part of the generating plant, substation, or control center are also covered. Ref.—Section 90-2(b)(5), fine print note

8. There has been a change in this edition of the *NEC®* from the wording of the previous edition.

9. (a) 1993, (b) September, 1992.

10. 1881, The National Fire Engineers met in Richmond, Virginia, which resulted in the first document that led to the *National Electrical Code®*. Various meetings were held after that; then in 1897 some various allied interests of the church, electrical, and architectural concerns met, and in 1911 NFPA was set up as the sponsor.

11. Yes. Ref.—Article 305, Sections 305-8, 305-4(c)

12. Chapter 8. Ref.—Section 90-3

13. No, it is not a training manual for untrained persons. Ref.—Section 90-1(c)

14. No, only underground mines and rolling stock for surface mines. All surface mining other than the rolling stock are covered by the *NEC®*. Ref.—Section 90-2(b)(2)

15. Nowhere in the Code does it state that it is the minimum allowed. Only when adopted by a local jurisdiction would it be considered the minimum of that jurisdiction. When an installation is installed for the *National Electrical Code®*, it provides a safeguarding of persons and property, and the provisions are considered necessary for safety and compliance with the *NEC®* will result in an installation essentially free from hazards. However, it may not be efficient or convenient or adequate in some installations. Ref.—Section 90-1(a) and (b)

16. Definitions, Article 100, to continue for three hours or more

17. 1987. The *National Electrical Code®* is adopted every three years; there being a 1987, a 1990, and

a 1993. Therefore, not adopting the two editions, they would be using the 1987 *NEC*®.

18. Article 725 covers Class 2 wiring.

19. Article 517, Health Care Facilities, Section 517-3, Definitions

20. Section 250-81. The water pipe must be used if available, and supplemented with another grounding electrode of one of the types listed in Section 250-81 or 83.

21. Article 422 lists specific requirements for specific appliances. Ref.—Section 422-8(d)(1)a

22. Yes. Ref.—Article 770

23. Phase converters are covered in *NEC*® Article 455. Both static and rotary types are covered.

24. Article 426 covers fixed outdoor electrical deicing and snow melting equipment.

25. Article 680 for swimming pools, fountains, and similar installations. Fine print note to Section 680-1 clearly states that a fountain covers reflective pools, ornamental pools, and display pools. However, it does not cover a drinking fountain. Part E of that article must be applied.

QUESTION REVIEW CHAPTER 3

(Reduce fractions to their lowest terms.)

1. 1/2
2. 1/4
3. 1/10
4. 2/5
5. 3/4
6. 3/4
7. 1/2
8. 3/5
9. 3/5
10. 1/5
11. 1/4
12. 1/4

(Change mixed numbers to improper fractions.)

1. 3/2
2. 13/4
3. 43/8
4. 17/4
5. 57/8
6. 13/6
7. 23/4
8. 25/3
9. 46/7
10. 29/8

(Change improper fractions to mixed numbers.)

1. 4 1/2
2. 2 2/5
3. 12 4/5
4. 3 1/4
5. 9 2/3
6. 15 1/2
7. 1 2/3
8. 4 1/3

(Multiply whole numbers and fractions. Answer in proper reduced form.)

1. 4
2. 3/8
3. 1/40
4. 9/2 = 4 1/2
5. 21/4 = 5 1/2
6. 5/2 = 2 1/2
7. 9
8. 15/7 = 2 1/7

(Convert fractions to decimal equivalents.)

1. .25

2. .625

3. .75

4. .866

5. .375

(Resolve to whole numbers.)

1. 16

2. 81

3. 2500

4. 1

5. 1000

(Find the square root.)

1. 5

2. 9

3. 7

4. 56

5. 101

6. 29

7. 78

(Solve the following problems.)

1. .05.   Ref.—Ohm's Law

2. 100%.   Ref.—Chapter 3

3. 5 amperes.   Ref.—Chapter 3, Ohm's Law

4. 1.33 ohms.   Ref.—Chapter 3

5. 14.4 ohms.   Ref.—Chapter 3

6. 6 volts.   Ref.—Chapter 3

7. 33.34 ohms

8. 38.34 ohms

9. 23.81 ohms

10. 62.15 ohms

11. 64.3 ohms

12. 9.65 volts

13. 45.95 volts

14. 1.93 amps

QUESTION REVIEW CHAPTER 4

1. Yes.   Ref.—Section 90-2a

2. Chapter 8. Article 250 does not apply in Chapter 8 unless specifically referenced.

3. Article 410, Part L

4. False. Branch circuits are referenced in many articles in the *NEC®*, Index pages 70–87.

5. Article 516

6. Article 422 for appliances, specifically Section 422-10. Electrical heating is covered in Article 424 and heat pumps in Article 440.

7. Chapter 9, Tables 1 through 8 and all Notes

8. Lighting and power in mine shafts are not covered by the *NEC®*, Article 90, Section 90-2B

9. Article 710, Table 710, 4b, Over 600 Volt General Requirements

10. Article 225, which covers outside branch circuits and feeders, Section 225-8

11. Chapter 9, Table 10

12. Places of assembly, Article 518

13. *NEC®* Article 250, Section 250-6

14. Article 680, Section 680-22, Part B

15. Article 90, Section 90-3

16. These requirements appear in the building codes or the NFPA Life Safety Code; they are not found in the *NEC®*.

17. Article 760

18. Yes.   Ref.—Section 305-6(a)

19. Article 502—(a) Class 2, (b) Division 1. Article 500 for the definitions and Article 502 for the specific requirements.

20. It is a violation to put over 1,000 volt wiring in a residence. Neon signs requirements are found in Article 600. However, Article 410 requirements limit installation under 1,000 volts in dwellings. Ref.—Section 410-80(b). In this application, neon would not be covered in Article 600 since it is an ornamental (not informative) sign.

21. Article 500 defines this area. Article 501, Volatile Liquids, would be the article required in which to make the installation.

22. Yes, Article 210, Section 220-36

23. Article 300, Section 300-4d

24. Article 702

25. Article 705, which covers interconnected electrical power production sources, otherwise known as cogeneration

QUESTION REVIEW CHAPTER 5

1. 4 feet. Ref.—Section 110-16, Table 116A

2. Two. Ref.—Section 110-16(c)

3. Yes. Ref.—Section 110-16(e)

4. 6½ feet. Ref.—Section 110-16f

5. One. Where the required working space is doubled, only one door is required. Ref.—Section 110-16c, Exception 2.

6. No. The required disconnecting means and supplementary overcurrent device must have a working clearance in accordance with Section 110-16. Clear working clearance of 30 inches wide and 40 inches deep are required from the ground level up 6½ feet.

7. 30 inches. Ref.—Section 110-16(a)

8. Ungrounded. Ref.—Section 240-20

9. 60%. Ref.—Chapter 9 Table Notes

10. Clothes closet. Ref.—Section 240-24d

11. 1¼ inches. Ref.—Section 300-4(a)(1)

12. 600 volts. Ref.—Article 331-4(6)

13. Yes, with exceptions. Ref.—Section 230-43(14)

14. Yes. It shall be bonded to the electrode at each end and all intervening raceway boxes and enclosures between the service equipment and the grounding electrode conductor. Ref.—Section 250-71(a)(3)

15. Overcurrent device. Ref.—Section 250-95

16. Two. Ref.—Section 410-4(a) and (d)

17. No. Ref.—Article 680, Part G, except manufacturer's instructions may require Section 110-3(b).

18. Yes. Ref.—Section 210-52(a)

19. No. Ref.—Section 210-52(a) for site-built dwellings; Section 550-8(d) Exception 4 for mobile homes

20. Section 210-63. Ref.—Article 210

21. Article 373

22. Article 410, Section 410-4d

23. No. Section 310-4 states that parallel conductors shall not be permitted unless they are size 1/0 or larger.

24. 2/0. Ampacity Tables of 0–2000 Volts, Note 3

25. Chapter 9, Table 8

QUESTION REVIEW CHAPTER 6

1. 17.8 Amperes– $P = \dfrac{E \times 1.73}{I}$

2. 10, 80% loading. Ref.—Section 220-10(b)

3. 1120. 59 feet. Ref.—Chapter 3

4. 2. Ref.—Table 220-3(b), and Footnotes, also see Figure 6–4

5. 3—2 kitchen appliance circuits, 1 laundry circuit, and 1 heating circuit Ref.—Section 422-7, Section 220-4(b) and (c)

6. 2. Ref.—Sections 220-4(b) and 210-52(b)

7. Yes. Ref.—Sections 220-4(b) and 210-52(f)

8. 30 amperes. Section 422-28(e)

9. None. The *NEC*® does not require power to the structure. Ref.—Section 210-52

10. 2. Ref.—Section 210-52(e)

11. 1, if a GFCI circuit breaker is used to feed bathroom and outdoor receptacles; 2 if receptacles are located within six feet of the kitchen sink. Ref.—Section 210-8(a)

12. Yes. Ref.—Section 210-70(a)

13. No. Ref.—Section 210-63 and 70

14. Yes. Ref.—Section 210-63

15. 20 amperes. Ref.—Section 210-4(c)

16. No. Ref.—Section 422-7, Definition of Individual Circuit

17. No. Ref.—Section 410-8

18. Yes. Ref.—Section 210-52(b) Exception 5

19. No. Ref.—Definition of direct grade access, Section 210-8

20. No, 2 receptacles at direct grade access are required. Ref.—Section 210-52 and 210-8

21. Section 230-42(b), Table 220-11, Chapter 9, Example 1

    (a) Section 220-3(b) and Table 220-3(b); 28 × 40 = 1120 sq. ft. 3 VA × 1120 = 3360 volt-amperes

    (b) Section 220-4(b); 3000 volt-amperes

    (c) Section 220-3(c); 4500 volt-amperes

    (d) Section 220-4(c); 1500 volt-amperes

    (e) Section 220-18; 5000 volt-amperes

    (f) 0. Electricity not required; however, if supplied, the load would have to be included in calculation

    (g) Section 220-4(c); size not stated, therefore a 20-ampere branch circuit is recommended

    (h) 0. Section 220-3(b); Section 210-52(b) Example 2 could be general lighting or on appliance load

    (i) 0. None

    (j) 0. Gas range igniter can be fed from appliance circuit, Section 210-52(b) Example 5

    (k) 0.

22. 125%. Ref.—Section 430-22

23. 1 volt-ampere per sq. ft. or 100 × 200 = 20,000 VA Ref.—Table 220-3(b)

24. Two small appliance circuits minimum required Ref.—Section 220-4(b)

25. "Size"; Formulas in Appendix 2.

QUESTION REVIEW CHAPTER 7

1. 6. Ref.—Section 230-208

2. 94%. Ref.—Chapter 3

3. A disconnecting means is required at each building. Ref.—Section 225-8(a) and (b) Addition: However, overcurrent is not required at the second building if the feeder is adequately protected at the service. (Section 225-9 FPN)

4. Yes, you are limited to not more than six disconnects on this secondary side. Ref.—Section 450-3—conductors must be protected in accordance with Section 240-3.

5. No, however, Section 250-54(e) requires the conductors be separated where of a different potential.

6. No, Section 240-24(e) prohibits branch circuit overcurrent devices from being located within a residential bathroom. Ref.—Section 240-24(e)

7. No. However, SE cable is permitted when ranges and dryers where it originates in the service panel. SE cable can be used in the same manner at Type NM cable, however, it is not considered NM cable, and must comply with Articles 300 and 336. Ref.—Section 338-4(4)

8. Usually no. Yes, if the window is not designed to be opened, or where the conductors run above the top level of the window. Where windows are designed for opening and closing, then a minimum of 3 feet from the windows, doors, porches, fire escapes, or similar locations is required. Ref.—Section 230-9

9. (a) Yes. Ref.—Section 250-91(a), Exception 2

    (b) They shall be sized in accordance with Section 250-94 for the largest conductor entering each respective enclosure. Section 250-91, Exception 2

10. Requirement to open all of the conductors is a disconnect requirement and not an overcurrent requirement. Ref.—Section 240-20(b) gives the requirements for circuit breakers, Section 220-40 gives the requirements for fuses. Section 210-4 is the requirement for multi-wire branch circuits. Disconnecting, not overcurrent, is the purpose.

11. A separately derived system is defined in Article 100, and the grounding installation requirements are found in Section 250-26. Ref.—Article 100 and Section 250-26 12. Section 450-3—not more than six disconnects are permitted on the secondary side of the transformer.

13. No. Identification of the hi-leg is only required where the neutral is present. Ref.—Sections 215-8, 230-56, and 384-3(e)

14. No. The grounding electrode conductor must be continuous and the connection shall be made solidly and used for no other purpose. Ref.—Section 250-92(a) and (b)

15. Less than .25 inch. Ref.—Section 347-9 Exception

16. Section 90-5—mandatory rules of the Code are characterized by the word "shall." Explanatory material is in the form of fine print notes (FPN).

17. Grounded effectively is defined in Article 100 and appears in Section 250-81(b) in the form of a fine print note. It also may be found in Chapter 8. The authority having jurisdiction will make that determination.

18. Article 336, nonmetallic sheath cable. Although Romex is a trade name, the term can be found in the index of the *NEC*®.

19. Article 349 covers the requirements of flexible nonmetallic tubing. It is manufactured in sizes ½ inch and ¾ inch only. *Note:* This is a flexible metallic tubing that is liquidtight without a nonmetallic jacket. Rarely used in construction installation.

20. Article 380

21. Chapter 9, Table 1, Footnote 6 states that multiconductor cable of two or more conductors shall be treated as a single conductor cable, and cables that have an elliptical cross-section, the cross-section layering calculation shall be based on using the major diameter for the ellipse as the circle diameter.

22. Article 511 does not apply. Section 511-1 and 511-2 clearly exempt these parking structures. Therefore, Chapters 1 through 4 apply, and the wiring method would be taken from those four chapters.

23. Article 362, Part B

24. 180 volt amps each. Article 220, Section 220-4 covers the requirements

25. It is not permitted. Section 240-24(e) prohibits the installation of overcurrent devices in bathrooms of residences.

QUESTION REVIEW CHAPTER 8

1. Yes. Ref.—Section 240-20(b). However, handle ties are required.

2. No maximum length needed, provided they are outdoors. Ref.—Section 240-21(m)

3. No. Ref.—Section 240-60 Exception. Voltage to ground can exceed 300 volts on a wye connection and therefore would not be permitted.

4. It is not a limiting factor. However, the six disconnect rule applies. Ref.—Section 230-90(a) Exception 3.

5. 100%. Section 220-15

6. Yes. Ref.—Section 501-4, 351-9 Exception 2, 501-16(b), and must comply with Section 250-79

7. No. However, the overcurrent device would have to meet these requirements. Ref.—Sections 220-10(b), 240-3, 215-3, and 210-20

8. There is no limit. Ref.—Table 220-3(b) *Note:* Many local jurisdictions set limits on 15- and 20-ampere branch circuits in all installations. However, the *NEC*® does not specify a limit on dwelling branch circuits for general use electrical outlets.

9. No. Ref.—Sections 220-3(c)(5), 220-3(c)(6), also the Note at the bottom of Section 310-16 limits No. 12 conductors to a 20 ampere branch circuit. $16 \times 180\ VA = 2880 \div 120\ V = 24$ amperes.

10. No, you must use Section 310-15, the Note at the top of Appendix B. The Appendix is not a part of the *NEC*® requirements; it serves only for use as reference.

11. (a) No. Ref.—Section 220-21

    (b) Yes. Ref.—Section 220-21

12. No, only if installed in cable with 12 inches of cover. Otherwise, can be installed at the greater depth or in raceways. Ref.—Section 300-5 and Table 300-5

13. Yes. Ref.—Section 240-20(b)

14. No. Ref.—Section 215-2, 220-10(a) and (b)—Footnote: Branch circuits are required to be calculated at 125% for water heaters—Section 422-14(b)

15. No, No, No. Ref.—Section 240-21(a) through (n)

16. Yes. Ref.—Section 210-19

17. No. Ref.—Section 210-6—12 amperes or less must be 120 volts; over that can be 240 volts. 1400 VA ÷ 240 V = 5.833 amperes.

18. 180 VA. Ref.—Sections 220-4, 220-3(c)

19. Yes, it complies with the tap rules in Section 240-21(b) for the 10 foot tap rules.

20. Yes, it complies with the 25 foot tap rule. Ref.—Section 240-21(c).

21. No, it does not comply with any of the tap rules in Section 240-21.

22. True. Ref.—Section 364-11

23. B. Ref.—Section 240-21(b)

24. Yes. Ref.—Table 250-95 and Notes

25. A fuse must be installed. Ref.—Section 110-3(b). The manufacturer's instructions must be followed.

QUESTION REVIEW CHAPTER 9

1. Soldered. Ref.—Section 250-113

2. 14. Ref.—Section 250-97

3. Not required to be accessible. Ref.—Section 250-112 Exception

4. White or natural gray. Ref.—Section 200-6

5. 4/0. Ref.—Section 250-23(b).
   3-500 kcmil = 1500 kcmil or 3 each 500 kcmil = 1500 kcmil × 12½% = 187.5 kcmil
   1500 × 12.5% = 187.5 kcmil
   Chapter 9, Table 8—3/0 167.8 kcmil too small; therefore, use 4/0 211.6 kcmil

6. 4. Ref.—Table 250-95

7. Yes, bonding is not required. Ref.—Section 250-72(c)

8. They are required to be run in each nonmetallic raceway and are required to be sized in accordance with the sum of the paralleled phase conductors. Ref.—Sections 215-6 and 250-95 and Table 250-95.

9. Yes. Ref.—Section 250-43(k)

10. No, the language is specific. Ref.—Section 250-26(c)

11. Yes, it must be bonded. No, it is not sufficient to bond at one end. It must be bonded at both ends and at all enclosures in which it passes through. Ref.—Section 250-71(a)(3)

12. No, a driven rod is not acceptable. An equipment grounding conductor must be carried with the circuit. Ref.—Section 551-70(h)

13. No, the requirements for separately derived transformers are found in Section 250-26(c). These requirements do not amend the requirements of Section 250-81(a), but they are the specific requirements. Therefore, no supplementary grounding electrode is required.

14. No. 8 AWG copper or equivalent. Ref.—Section 551-56(c)

15. Yes. Ref.—Section 250-80(b). *Note:* The interior gas piping system is required to be grounded for safety. It can be grounded with the circuit conductor equipment ground most likely in which it would come in contact. For example, a central heating electric furnace may have No. 12 conductors supplying the fan motor. The No. 12 equipment grounding conductor run with those circuit conductors could be the conductor used to bond the gas piping system in this case. **Warning:** The underground gas piping system may not be grounded. Electrolysis may occur, creating dangerous underground gas leaks.

16. Yes, a No. 1 copper or a No. 2/0 aluminum or copper-clad aluminum wire is required in each conduit. Ref.—Table 250-95 and Section 250-95, which states that where run in parallel in multiple raceways or cables as permitted, the equipment grounding conductors also shall be run in parallel, each equipment grounding conductor shall be sized on the basis of the ampere rating of the overcurrent device protecting the circuit conductor in the raceway in accordance with Table 250-95.

17. No. Ref.—Section 110-14(b)

18. No. 8. Ref.—Table 250-95. This is a minimum size conductor and may have to be larger if the fault current available exceeds that for which the No. 8 conductor will carry.

19. Yes. The flexible metal conduit cannot exceed six feet for the combined two pieces of flexible metal conduit. Ref.—Section 250-91(b)

20. The maximum rating of the grounding impedance, no smaller than No. 8 copper or No. 6 aluminum. Ref.—Sections 250-23(b), 250-27(b)

21. By a ground fault circuit interrupter. Ref.—Part G, Article 680

22. Yes, all backfed circuit breakers must have a retainer kit. Ref.—Section 384-16(f)

23. Maybe. Ref.—Section 250-81 does not permit the water line as being used as a grounding electrode conductor. All connections must be made within the first five feet. However, the other types can be connected at any point and therefore the answer is maybe.

24. Section 250-32 Exception for Services and Section 250-33 Exception 4 for Feeders or Other Conductors permit these metal 90s where maintained at a burial depth in accordance with Table 300-5 need not be grounded. Otherwise, they must be bonded and at most practical places where they emerge from the ground, such as turning up into a concrete slab or turning up under a panelboard, or turning to go up the side of a pole or other structure. Bonding is required and provides a greater degree of safety. Ref.—Section 250-33 Exception, Section 250-32 Exception

25. Main grounding electrode conductor would be based on the four No. 750 kcmil service conductors, and they must be sized at size 2/0 copper or 4/0 aluminum. Switch A, the tap conductor from the grounding electrode conductor to Switch A, would be sized only on the conductor feeding that switch, No. 3, and the size of the conductor, based on Table 250-94, would be No. 8 copper or No. 6 aluminum. Switch B would be based on the same table, 250-94, and would be No. 4 copper or No. 2 aluminum. Switch C would be No. 8 copper or No. 6 aluminum, and Switch D would be the 2/0 conductor for the main going in. However, if you chose to tap that conductor to Switch D and run the main grounding electrode to one of the others, then based on the four No. 1/0s, that conductor would have to be a No. 6 copper or No. 4 aluminum. Ref.—Table 250-94

QUESTION REVIEW CHAPTER 10

1. Within 12 inches of the outlet or box, and every 4½ feet. Ref.—Section 338-4 and Section 336-15

2. Maximum number of conductors shall be permitted using the provisions of Section 370-16(b). Conduit bodies shall be supported in a rigid and secure manner. Ref.—Section 370-16(c)

3. Yes. Ref.—Section 351-24

4. Yes. Ref.—Section 345-12 Exception

5. Yes. Ref.—Section 331-11 Exception—Metal studs are the required support when they do not exceed 3 feet apart.

6. Type MC cable must be supported at intervals not exceeding 6 feet, except when installed in dwellings, where it shall be secured within 12 inches from every outlet box, junction box, cabinet or fitting, or where fished. Ref.—Section 334-10(a) Exceptions 1 and 2

7. NM cable shall be supported at intervals not exceeding 4½ feet and within 12 inches of every cabinet, box, or fitting. Two conductor cables shall not be stapled on edge, run through holes in wood studs or rafters shall be considered support. Ref.—Section 336-15 Exceptions 1 and 2

8. No. Ref.—Section 362-16

9. Above ground conductors must be installed in a rigid metal conduit, intermediate metal conduit, rigid nonmetallic conduit in cable tray or busway or cable bus or in other identified raceways, or as open runs of metal clad cable suitable for the use. Open runs of Type MB cable, bare conductors or bare bus bars also are permitted as they apply. Ref.—Section 710-4

10. Yes, provided that carpet squares not exceeding 3 × 3 are used in a wiring system in accordance with Article 328 Type FCC cable and associated accessories are used. Ref.—Article 328

11. Metallic wireway shall be established in accordance with Section 362-5 and shall not exceed 30

current-carrying conductors at any cross-sectional area. The sum of cross-sectional areas of all conductors at any cross-section of the wireway shall not exceed 20%. The derating factors in the 0–2000 Volt Notes, Note 8, do not apply. For nonmetallic wireway, Section 362-19 applies. The cross-sectional area contain conductors at any cross-section of the nonmetallic wireway shall not exceed 20% of the interior cross-sectional area. The derating factors in the 0–2000 Volt Tables, Note 8, do apply to the current-carrying conductors at the 20% specified. *Note:* This is a critical difference. Caution should be used when reading questions as they apply to a specific wiring method, metallic or nonmetallic, since the conditions and rules are very different.

12. 1¼ inches from the nearest edge of the stud. Ref.—Section 300-4(d)

13. No, the Electrical Metallic Tubing must be securely fastened in place at least every 10 feet and within 3 feet of each outlet box, junction box, device box, cabinet, conduit body, or other tubing termination. Support within 3 feet of each coupling is not required. Ref.—Section 348-12

14. There is no difference in the permitted uses or installation requirements in the two products. Article 345 and Article 346 are essentially the same. Ref.—Article 345 and Article 346

15. 1/0. Ref.—Section 300-4

16. Yes, in any building exceeding 3 floors, nonmetallic tubing must be concealed within a wall, floor, or ceiling where the material has at least a 15-minute finished rating as identified in the listings of the fire-rated assemblies. Ref.—Section 331-3(2)

17. No, they must be removed. They are not permitted for a period to exceed 90 days. Ref.—Section 305-3(b)

18. No, lengths not exceeding 4 feet can be used to connect physically adjustable equipment and devices that are permitted in the duct or plenum chamber. Ref.—Section 300-22(b)

19. .25 inches. Ref.—Section 347-9 Exception

20. Yes, even though Section 300-3 requires that all conductors of the same circuit, including the neutral and equipment grounding conductors, must be run within the same raceway, cable trench, cable, or cord; Exception 2 to that section permits that with column type panelboards; the auxiliary gutter, and pull box may contain the neutral terminations. Ref.—Section 300-3(b) Exception 2

21. No, each cable shall be secured to the panelboard individually. Ref.—Section 373-5(c)

22. No, a branch circuit or a lighting panelboard is one that 10% of its overcurrent devices are rated 30 amperes or less for which neutral connections are provided. A lighting and branch circuit panelboard, as required by the Code, shall have not more than two circuit breakers or two sets of fuses for disconnecting. Ref.—Section 384-14 and Section 384-16—A disconnect is required ahead of the lighting and branch circuit panelboard although the load is relatively small in this cabin.

23. It is prohibited by the Code. The reference you are looking for is Section 380-8(b). The switches must be arranged so the voltage does not exceed 300 volts, or permanently installed barriers between the adjacent switches must be installed. You are correct; it is a dangerous situation.

24. Yes, knob-and-tube wiring is still an acceptable wiring method and is covered in Article 324. Extensions are permitted in hollow spaces, walls, and ceilings. However, they are not permitted in the hollow spaces of walls, ceilings, and attics where the spaces are insulated by loose, rolled, or formed in place insulated material that covers the conductors. You can make the installation, but you will have to route the conductors along a running board above the insulation. Ref.—Article 324, Section 324-4

25. 1/0 or larger. Ref.—Section 318-3(b)(1)

QUESTION REVIEW CHAPTER 11

1. Section 430-6a. Ref.—Section 430-6a

2. 80 ampere. Ref.—Section 240-6, Tables 430-150 and 430-152

3. 175%. Ref.—Table 430-150 and 430-152

4. No. 8 THW aluminum. Ref.—Table 430-148, Section 220-10(a) and (b), Sections 430-24 and 430-25

5. Same. Ref.—Section 240-21

290 Master Electrician's Review

6. 20. Ref.—Section 240-3, 240-6 and Table 310-16
7. 40. Ref.—Section 240-3, 240-6 and Table 310-16
8. 100. Ref.—Section 240-3, 240-6 and Table 310-16
9. 200. Ref.—Section 240-3, 240-6 and Table 310-16
10. 125. Ref.—Section 450-3(b)
11. 175%. Ref.—Table 430-152
12. $V_D = 22 \times 2 \times 150 \times 1.24/1000 = 8.184$ volts
13. 3.5%
14. No. 8
15. 125%. Ref.—Article 430
16. 6. Ref.—Table 430-91
17. Indoors. Ref.—Section 410-73(e) and (f)
18. Fire protection or equipment cooling. Ref.—Section 450-47
19. Accessible. Ref.—Section 450-13, Exception 1
20. 3. Ref.—Section 460-2
21. 8. Ref.—Section 440-32 and Table 310-16
22. 200%. Ref.—Table 430-152
23. 112.5 kVA. Ref.—Section 450-21(b)
24. 600 watts. Ref.—Ohm's Law (*Clue:* resistance does not change.)
25. 87%. Ref.—Chapter 3

QUESTION REVIEW CHAPTER 12

1. Nonconductive, conductive, composite. Ref.—Section 770-4
2. False. Ref.—Section 511-6
3. Yes, where the raceway emerges from the ground, at both the office and at the dispenser. Ref.—Section 514-6a and b
4. False. Ref.—Section 700-1
5. Class 1, Division 2. Ref.—Section 511-3(a)
6. Parallel interconnection. Ref.—Section 700-6

7. ⅝. Ref.—Section 501-5(c)(3)
8. 12 inches. Ref.—Sections 710-4, 710-33, and Table 710-33
9. Two duplex or four single receptacles. Ref.—Section 517-18(b)
10. 6. Ref.—Section 220-10(a) and (b), Table 430-148 and 310-16
11. 3.05 meters (10 feet). Ref.—Section 600-37(c)
12. Length and design. Ref.—Section 600-33(a)
13. True. Ref.—Section 240-54(b)
14. False
15. False
16. True
17. False
18. False
19. True
20. True
21. (a) Electrician or electrical contractor
    (b) Field marking is required to state "Caution—Series Rated System, ___ Amperes Available. Identified Replacement Components Required." Ref.—Section 110-22.
    (c) Manufacturer
22. 6 amperes. Ref.—$I = \dfrac{HP \times 746}{E \times PF \times eff\%}$

    Note: When power factor is not noted it must be calculated at 100%.
23. 60. Ref.—Tables 430-150 and 152
24. 27. Ref.—Ohm's Law, Chapter 3, Table 220-20
25. 18. Ref.—Ohm's Law, Chapter 3

QUESTION REVIEW CHAPTER 13

1. Yes. Ref.—Sections 810-20(c), 810-21
2. No. Chapter 8 stands alone. There are no burial depth requirements for satellite dishes installed anywhere there is no safety hazard, and therefore the only consideration is for the customer's protection of his or her own equipment, *NEC*® Section 800-10. *Note:* Also look at Section 90-3.

Chapter 8 covers communication systems and is independent of the other chapters, except where they are specifically referenced. Chapter 8 stands alone, and other chapters cannot be applied unless specifically referenced.

3. Yes. Ref.—Section 800-33

4. 4. Ref.—Section 800-12b

5. It is an intrinsically safe circuit covered in Article 504 of the *National Electrical Code®*.

6. Yes, motor control centers are covered in Part H of Article 430. *Note:* The requirements in 384 for accessibility are also applicable as referenced in Section 430-1, Exception 1.

7. Definition of Article 100 defines a separately derived system. Section 250-26 covers the grounding requirements for this separately derived transformer.

8. Section 410-56c covers these isolated ground receptacles intended for the reduction of electrical noise. See also Section 250-75 Exception.

9. Article 530 covers film vaults. *Note:* Scope of Article 530-1 covers these areas in a portion of a building.

10. When a term is used only once in the *NEC®*, the definition appears where it is being used. The definition of a bathroom area appears where used in Section 210-8.

11. Yes. Ref.—Article 650

12. Yes, Article 110, Section 110-16 for installations of less than 600 volts, Section 110-32 for installations over 600 volts.

13. Yes. Section 100-12c clearly states that paint, plaster, and other abrasives may damage the bus bars, wiring terminals, insulators, and other surfaces in panelboards. Therefore, the internal parts of this equipment must be covered and protected.

14. Inspector was correct. Section 550-23 requires the service equipment to be located inside and not more than 30 feet from the exterior wall of the mobile home. An exception permits the installation of the service equipment on the mobile home when installed by the manufacturer. Electrical contractors, however, cannot make this installation.

15. Optional standby system is covered in Article 702 and is defined in Section 702-2.

16. Color coding is not specifically required; however, Section 210-4d requires the identification of ungrounded conductors on multi-wire branch circuits under certain conditions. This identification can be by color coding, marking tape tagging, or other equally effective means.

17. No. The footnotes to Table 310-16 are mandatory language, and the note limits 14 conductors to a 15-ampere overcurrent device.

18. Section 225-7b covers outdoor branch circuits for lighting equipment and there is no maximum number.

19. Yes. Article 547, Agriculture Buildings, covers this type installation. See also Section 250-24(a) Exception 2 and 250-24(c) Exception 2

20. Yes. Article 328 covers this type installation. However, the carpet must be laid in squares not to exceed 3 × 3.

21. Article 280, Surge Arrestors, covers that type of installation.

22. Article 305 permits a class less than would be required for permanent installation. However, there are time constraints. *Note:* 305-4a requires the service on that temporary pole to comply with Article 230.

23. 86%. Ref.—Chapter 3, Duff and Herman, *Alternating Current Fundamentals, 4E* by Delmar Publishers Inc.

24. 25 HP. Ref.—Loper, *Direct Current Fundamentals, 4E* by Delmar Publishers Inc.

25. 0.105. Ref.—Loper, *Direct Current Fundamentals, 4E* by Delmar Publishers Inc.

PRACTICE EXAM 1

1. Yes. Ref.—Section 511-10

2. No. Ref.—Section 680-71, however, Section 410-4(a) and (d) limits the use of pendant type fixtures over this area.

3. Yes. Ref.—Section 210-7(d) and Exceptions

4. No. Ref.—Section 210-8(a)(5)—only those serving the counter tops. *Note:* only very old ranges have receptacles mounted on them.

5. No. Ref.—Section 370-3, Exceptions 1 and 2

6. Yes. Ref.—Section 410-16(h)

7. 15-minute finish rating. Ref.—Section 331-3(3)

8. No. Ref.—Section 300-4(b)

9. Where run through the framing members ENT shall be considered supported. Ref.—Section 331-11, Exception 1

10. Yes. Ref.—Section 300-4b

11. Yes. Ref.—Sections 339-3(a)(4) and 336-26

12. No. Ref.—Section 517-10, Exceptions 1 through 3

13. Yes. Ref.—Section 210-52a

14. No. Ref.—Section 300-13b

15. Yes. Ref.—Sections 210-52(c) and 210-8(a)(5) *Note:* The GFCI requirement would not apply if the receptacle was located behind the refrigerator.

16. No. Ref.—Section 210-52(a) and (e)

17. No. The requirement is now 6½ feet. Ref.—Section 380-8(a)

18. See referenced exceptions and general rule. Ref.—Section 410-31, Exceptions 1–3

19. General and critical care patient bed location. Ref.—Section 517-18(b)

20. No. Ref.—Section 210-52(a)—shall be in addition to those required if over 5½ feet above the floor

21. Self-closing. Ref.—Section 410-57(b) Exception

22. No. Ref.—Section 680-25(b)

23. Yes, provided it is installed and supported in accordance with Article 336. Ref.—Section 336-3, 4, 10, and 15

24. No, they may be rubber or thermoplastic types. Ref.—Section 225-4

25. Yes, by meeting one of the exceptions. Ref.—Section 410-31

PRACTICE EXAM 2

1. No. Ref.—Only as environmental conditions require—Sections 547-1 and 547-7a, b, c

2. No. Ref.—Sections 333-10, 300-4

3. Extra hard usage portable power cables listed for wet locations and sunlight resistance. Ref.—Section 553-7(a) and (b)

4. No. Ref.—Section 680-25(c)—NM cable has a covered or bare equipment ground conductor, not insulated.

5. No. Ref.—Section 680-21

6. No, water heaters are not listed with an attachment cord. Ref.—Section 400-7 and 8(1)

7. No, generally. Ref.—Section 373-8

8. Voltage with the highest locked rotor KVA per horsepower. Ref.—Section 430-7(b)(3)

9. No. Ref.—Section 680-25(d)

10. Yes. Ref.—Section and Tables 210-21(b)(2) and (b)(3)

11. Yes. Ref.—Section 210-24

12. Not enclosed, buried, or in raceway. Ref.—Table 310-17

13. No. Ref.—Section 347-2 (FPN) is not mandatory, only explanatory

14. No, GFCI protection is still required. Ref.—Section 305-6(a)

15. Yes. Ref.—Section 210-8(a)(5)

16. No. Ref.—Section 210-8(a)(5) (FPN)

17. No maximum. Ref.—Table 220-3b and footnotes

18. 6. Ref.—Section 517-18b

19. Yes. Ref.—Sections 430-81(c), 430-14, 430-42(c)

20. Yes. Ref.—Section 230-70(a), Exception 1

21. Yes, one or more supplied by a 20-ampere current. Ref.—Sections 210-52(f), 220-4(c)

22. Yes. Ref.—Sections 410-17, 18 and 21

23. Yes, flexible metal conduit must be bonded to grounded conductor. Ref.—Sections 250-43(9) and (13), 250-32, and 250-61(a)

24. No, only interior metal water piping systems. Ref.—Section 250-80(a)

25. No, generally. Ref.—Sections 370-22, 370-16(a)

PRACTICE EXAM 3

1. No. Ref.—Section 370-27(b)

2. No. Ref.—Section 370-16(a) and (b)

3. No. Ref.—Section 210-52(e) only single and two-family dwellings

4. No, polarized or grounding type. Ref.—Section 410-42(a)

5. Yes. Ref.—Section 210-52(a)

    Note: Due to the arrangement of furniture, the receptacle located behind the door may be the only receptacle accessible for frequent use such as vaccuuming.

6. No. Ref.—Section 305-4(g)

7. 6, where integrated clamps, etc. are not used. Ref.—Table 370-16(a)

8. Table 370-16(b)—370-16(a)(2)

9. Yes. Ref.—Section 510-4(a)

10. No, however, it is a raceway. Ref.—Article 348

11. Yes, up to 5 feet. Ref.—Section 348-12 Exception

12. Yes. Ref.—Section 410-16(c)

13. No. Ref.—Definitions, Article 100, and Section 410-4

14. No, must be spaced 1½ inches from the surface. Ref.—Section 410-76(b)

15. Yes, with exceptions. Ref.—Section 410-65(c)

16. No, only where the grounded (neutral) conductor is present. Ref.—Sections 215-8, 230-56, and 384-3(e)

17. Yes. Ref.—Section 210-8(a)(5)

18. Generally no. Ref.—Section 373-8 and Exceptions

19. Yes. Ref.—Section 250-43(i), Part E Article 410

20. No. Ref.—Section 700-12(f)

21. No. Ref.—Section 210-8(a)(2) Exceptions 1 and 2

22. Yes "all." Ref.—Section 210-8(a)(1)

23. No; however, it cannot be counted as one of the countertop receptacles as required by Section 210-52(c). Ref.—Section 210-52(b) Exception 3

24. No. Ref.—Sections 370-1 and 410-64

25. With 12 inches of service head, goose neck, or connector and at intervals not exceeding 30 inches. Ref.—Sections 338-2, 230-51(a)

PRACTICE EXAM 4

1. 24 inches. Ref.—Table 300-5

2. Yes. Ref.—Table 210-21(b)(2)

3. No, you are not required to locate required receptacles behind furniture. However, the number required by Section 210-52(a) must be installed convenient to furniture arrangement. Ref.—Section 210-60 and Exception

4. Yes, but only if the box is listed for fan support. Ref.—Section 370-27(c) and Exception, and Section 422-18

5. See referenced Exceptions General Rule Section 410-31 prohibits. Ref.—Section 410-31 Exceptions 1, 2, and 3

6. Yes, generally. Ref.—Section 410-65(c) Exceptions 1 and 2

7. Yes. Ref.—Section 210-6(c)

8. Yes, it is required. Ref.—Section 700-12(f)

9. 150 kVA or less. Ref.—Diagram 517-41(3)

10. Yes. Ref.—Section 230-95

11. Yes, there is no exception for 15-volt lighting fixtures. Ref.—Section 680-41(b)(i) and Exceptions

12. Yes. Ref.—Section 348-1

13. No, it is not permitted. Ref.—Section 250-54

14. No. Ref.—Sections 110-16(a) and 384-4

15. Yes. Yes. Ref.—Sections 210-8(a), 220-4(b)
16. See Section 547-8(a) Exception 1—250-24(a) Exception 1 and 547-8(a)
17. No. Ref.—Part E, Article 230, specifically Section 230-70. *Note:* 1993 *NEC®* has added Section 225-8. If service is properly established at the pole the answer to this question is Yes.
18. 24 inches. Ref.—Table 300-5
19. No, unless the cord and plug are part of listed appliance. Ref.—Sections 110-3(b), 400-8(1)
20. (a) No minimal burial depth requirements. Ref.—Article 810

    (b) As per Section 810-21 and 810-15
21. Yes, provided they control the power conductors in the conduit. Ref.—Section 300-11(b) Exception 2
22. Yes. Underground water piping system must be at least 10 feet and be supplemented by an electrode type. Ref.—Section 250-81(a)
23. Yes. Ref.—Section 210-70
24. Entire fixture. Ref.—Section 410-4
25. No. Ref.—Sections 225-6 and 339-3(b)(9)

PRACTICE EXAM 5

1. Yes, limited for corrosion and voltage isolation. Ref.—Section 318-3(e)
2. Ref.—Section 410-8(d)

    (a) 12 inch incandescent, 6 inch fluorescent

    (b) 12 inch incandescent, 6 inch fluorescent

    (c) 6 inch incandescent, 6 inch fluorescent
3. No, generally four or less No. 14 AWG fixture wires are not required to be counted. Ref.—Section 370-16(a)
4. Yes, generally. Ref.—Section 110-16(c)
5. No. 4 copper. Ref.—Section 250-94 Exception 1
6. Yes. Ref.—Section 700-12(f), Exception
7. No, must be factory installed. Ref.—Section 550-23 Exception 2
8. Yes. Ref.—Section 680-25(d)
9. Yes, when smaller than $5/8$ inch must be listed. Ref.—Section 250-83(c)(2)
10. Both are acceptable. Ref.—Section 230-43
11. Exothermic welds. Ref.—Section 250-115
12. No. Ref.—Section 250-91(b) Exceptions 1 and 2
13. No, with exceptions. Ref.—Section 518-4
14. No. Ref.—Section 430-21
15. Yes. Ref.—Section 351-23(a) and (b)
16. Yes, but bare conductors are permitted. Ref.—Sections 230-41 and Exception
17. No. Ref.—Section 300-3(c) and 725-52(a)(2)
18. No, only circuits over 250 volts to ground. Ref.—Section 250-76
19. Yes. 1993 *NEC®* New Exceptions would permit this practice in some instances without raceway protection. Ref.—Section 300-22(c)
20. Yes. Ref.—Section 300-13(b)
21. 120 amperes. Ref.—Section 430-72(b) Exception 1—Table 430-72(b) Column B
22. Yes. Ref.—Sections 410-56(i), 250-74
23. Code uses nominal voltages. Ref.—See Definition Article 100 or Sections 220-2–220-2
24. No. 12 TW conductor in cable or raceway. Ref.—Table 310-16 = 25 amperes

    Four conductors in a raceway Note 8, 0–2000 Volt Ampacity Tables—four to six conductors in a raceway or cable shall be derated to 80% = 25 × 80% = 20 amperes the allowable ampacity.

    Ambient temperature correction factors shown at the bottom of Table 310-16 requires an additional deduction of .82 – 20 × .82 = 16.4 amperes. Section 210-21—outlet devices shall have an ampere rating not less than the load to be served. Table 210-21(b)(2) permits a 20-ampere branch circuit for a maximum of 16 amperes. 16.4 exceeds this requirement. The conductor will carry the maximum allowable load. *Note:* See obelisk note at the bottom of Table 310-16 that limits the overcurrent protection on a No. 12 conductor to 20 amperes.
25. 36 × 36. Ref.—Section 328-10

PRACTICE EXAM 6

1. Yes. Ref.—Sections 250-76, 250-72(d)

2. Six. Ref.—Section 230-71(a), Section 230-40 Exception 1

3. (a) Yes. Ref.—Section 600-11

   (b) No. Ref.—Section 600-11—There is no exception

4. Yes. Ref.—Section 545-2, 210-70 (a dwelling), Section 210-70(c) (other types of buildings)

5. Yes, if they meet all of the conditions. Ref.—Section 605-8

6. Yes. Ref.—Section 510-5(a), Article 500—Gasoline is a Group D, Class I hazard. Diesel is not. However, the sealing requirements are to prevent the passage of gases, vapors, and flames.

7. Yes. Ref.—Section 210-52(a)

8. (a) No. 2 copper or 1/0 aluminum. Ref.—Sections 250-23 Base on 300 Kcmil Service Drop Table 250-94

   (b) Bonded to the panelboard supplying each apartment size to Table 250-95 base and an overcurrent device 100 amperes No. 8 copper or No. 6 aluminum. Ref.—Section 250-80(a) Exception

   (c) No. Ref.—Section 250-81(a)—a supplementary grounding electrode only required where the underground water pipe is available. This installation is suggested by a nonmetallic water system.

9. Yes. Ref.—Section 210-23(a)

10. No, generally. Yes, for interior portion of a one-family dwelling. Ref.—Section 680-25(c) Exception 3

11. No, where are equipment grounding conductor and not over four fixture wires smaller than No. 14 enter the box from the fixture. Ref.—Section 370-16(a) Exception

12. Receptacle is not required specifically. Section 210-52(a) would require a receptacle near the wet bar. If located within 6 feet, a GFCI is required. Ref.—Sections 210-52(a), 210-8(a)(5)

13. Read heading for Table 220-19. Note 3 permits add nameplates of the two ranges. Use Column C. Ref.—Table 220-19 and Note 3

14. 4.5 KW. Ref.—Section 220-3(a)-(d). Section 220-18 is used for feeder load calculations.

15. 35 pounds. Ref.—Section 422-18

16. Table 350-3 for ⅜ inch. For ½ inch through 4 inch use Table 1, Chapter 9. Ref.—Section 350-7

17. No. Ref.—Sections 370-23(b), 370-23(a)–(g). *Note:* 300-11(a) the general method requirements give limited permission for branch circuits. However, junction boxes are covered by Article 370.

18. No, explosion proof not required or acceptable. Must be approved for Class 2 locations. Ref.—Section 502-1

19. No. THHN is not acceptable in wet locations. Ref.—Table 310-13 and Section 310-8

20. "Authority Having Jurisdiction." Ref.—Section 90-4, also see Definition Article 100 "Approved"

21. Yes. Ref.—Sections 210-52(f) and 220-4(c)

22. Yes. Ref.—Section 210-63

23. Column A is the basic rule to be applied when an exception does not apply. Ref.—Section 430-72(b)

24. 83%. Ref.—*Alternating Current Fundamentals*, Delmar Publishers Inc., and Chapter 3 of this book.

25. Qualified yes "Metal Multioutlet Assembly." Ref.—Section 353-3

PRACTICE EXAM 7

1. Lighting circuit serving the area. Ref.—Section 700-12(f)

2. Yes, if they are "Classified" for purpose and meet all conditions of Section 318-7(b). Ref.—Section 318-7(a) and (b)

3. No. Ref.—Section 352-22

4. If the required work space is unobstructed or doubles. Ref.—Section 110–110-16(c), Exceptions 1 and 2

296 Master Electrician's Review

5. No, the conductors in MC cable are wrapped in a nylon wrapping. AC cable conductors are individually wrapped in kraft paper. Ref.—Section 335-9 Armored Cable, 334-12 Type MC Cable

6. Generally, no, if exposed to physical damage. Ref.—Section 331-4(3) and (7). *Note:* If these conditions do not exist, then ENT could be installed.

7. 18 inches. Ref.—Table 300-5

8. Large capacity multibuilding industrial installations under single management. Ref.—Section 225-8(b) Exception

9. No for Class 2 Div. 1, Yes for Class 2 Div. 2. Ref.—Section 502-4(a) and (b)

10. No. Ref.—Section 410-31 Exception 1, 2, and 3

11. Grounded (neutral) conductor must be insulated generally. Ref.—Sections 310-12; 250-23 and 24, 250-60 and 61. The rules for an insulated neutral or grounded conductor for a feeder, although not clearly written, the service neutral is often bare and Section 250-60 permits dryers and ranges originating in the service panel to be Type SE, which has a bare grounded conductor.

12. No. Classifying structures are the responsibility of the building inspector or the fire marshall generally. They rely on the adopted building code or the NFPA 101 Life Safety Code. Ref.—Article 518

13. No. Ref.—Section 220-22, also see Note 10 to the 0–2000 Volt Ampacity Tables

14. None. Ref.—Table 310-16 Ambient Temperature Correction factors at bottom of table.

15. Table 370-16(b), 370-26(a)(2)

    Example: An outlet box contains a 120-volt 20-amp single receptacle and a 120-volt 30-amp receptacle. The 20-ampere receptacle is fed with a 12-2 W/GRO NM cable that extends from the outlet on to additional outlets located elsewhere. The 30-ampere receptacle is fed by an individual branch circuit with 10-2 W/GRO NM cable. What size box is needed?

    Answer: 370-16(a)

12-2 W/GRO Feed	2-12 @ 2.25 cu in	4.5 cu in
12-2 extending on	2-12 @ 2.25 cu in	4.5 cu in
10-2 W/GRO feed	2-10 @ 2.5 cu in	5 cu in
Receptacle	2-10	5 cu in
GRO	-10	2.5
Clamp	-10	2.5
		23.5

    A box with 23.5 cubic inches is required.

16. 10 feet. Ref.—Section 230-24(b)

17. No, it is considered by the *NEC*® as "Other Space Used for Environmental Air." Ref.—Section 300-22(c) and Exceptions, Definition of "Plenum" Article 100

18. No, this area must be unobstructed. Ref.—Section 410-8(a)

19. Equipment grounding conductor required. Ref.—Section 250-57(a) and (b), Section 250-58(a)

20. Yes. Ref.—Section 410-4(a)

21. By interpolation. Ref.—Section 430-6(a)

22. 53%. Ref.—Chapter 9, Table 1 and Note 6

23. Yes. Ref.—Note 3, 0–2000 Volt Ampacity Tables

24. No, motors are computed in accordance with Article 430. Ref.—Sections 220-14, 680-3

25. Yes, if supplied with electricity. Ref.—Sections 210-8(a)(2), 210-52(g)

PRACTICE EXAM 8

1. Yes, generally. Ref.—Section 410-73(e)

2. As per Section 547-8 in compliance with Article 250. Ref.—Section 250-24 Exception 2 and Section 547-8 Exceptions 1 and 2

3. Yes, if known, and they occupy a dedicated space with the individual branch circuit. Ref.—Table 220-19 "Heading." If they are plugging into one of the "small appliance" outlets, then they would not have to be added to the load calculations.

4. Yes, for controlled starting. Ref.—Section 430-82(b)

5. No. Ref.—Section 553-2 and 553-4

6. No. Ref.—Article 600, 210-6(a)

Appendix 4   297

7. Yes. Ref.—Section 210-21

8. Yes, limited application. Ref.—Article 517-10, 13, 30(6)(b), Exceptions 3 and 4

9. No, generally. Ref.—Section 240-3 and 310-15

10. Yes. Ref.—Sections 240-83(d), 380-11

11. There shall be two deductions for the largest conductor connected to each device, and one deduction of the largest conductor entering the box for each of the following: the clamps, the hickeys, all of equipment grounding conductors, etc. Ref.—Section 370-16(a)(2)

12. No. Ref.—Section 250-72

13. No, the cable can be fished; however, Article 370 does not permit the fished NM cable without a box clamp. Ref.—Section 336-15 Exception 1, Section 370-17(c) Exception, Section 300(4)(d) Exception

14. 150 amp will meet all conditions required for a general use switch. Ref.—Section 440-12, Section 430-109 Exception 3(c)

15. No. Ref.—Section 350-3 Exceptions 2 and 5, and Section 250-91b(5) Exception 1, Section 410-77c

16. I P/E $I = 10,000/240$ $I = 41.667$

    $R = E/I$ $R = 240/41.667$ $R = 5.76$

    $P = IE$ $P = 208 \times 36.1$ $P = 7.511$ KW

    $I = E/R$ $I = 208/5.76$ $I = 36.1$ amperes

    36.1 amperes—Chapter 3, Ohm's Law

    $I = P/E$ $I = 10,000/240$ $I = 41.667$

    $R = E/I$ $R = 240/41.667$ $R = 5.76$

    $I = E/R$ $I = 208/5.76$ $I = 36.1$ amperes

17. Yes. Ref.—Section 680-20(a)(1), 680-5(a), (b), and (c)

18. Chapter 9, Table 1 unless durably and legibly marked by manufacturer in cubic inches. Ref.—Section 370-16c

19. By the markings on the fixture by the listing agency. Ref.—Section 410-65c, Exception 182

20. Conductors only. Ref.—Section 240-21

21. Table 110-16a, Condition 3. Ref.—Section 110-16a

22. Yes, if cord and plug connected but not required or advisable. Generally, these appliances are wired to the requirements in Article 422. Ref.—Section 210-52(b) Exceptions 1–5 and (2)

23. 10 feet. Ref.—Section 680-8(1)

24. Yes, no depth requirement but is required to be protected from severe physical damage. Ref.—Section 250-92a

25. Water meter and all insulating joints. Ref.—Section 250-81(a)

PRACTICE EXAM 9

1. Yes, but not advisable. One should isolate any possibility of imposing faults on those circuits. Ref.—Sections 680-5(b), 680-20(a)

2. (a) Yes. Ref.—Sections 250-5(b), 250-23(a) and (b)

    (b) In each of the paralleled raceways. Ref.—Sections 300-3b, 250-23(a) and (b)

    (c) No. 2 copper in each conduit. Ref.—Sections 250-23, 250-94

3. 21 inches × 24 inches × 6 inches. Ref.—Section 370-28(a)(2)

4. Yes, by Exception. Ref.—Section 373-8 Exception

5. Yes. Ref.—Section 348-12 Exception

6. Either system installation is okay where a three pole transfer switch is used. See Section 250-5(d)(FPN). Grounding electrode is not required for this installation. Ref.—Section 250-5d, 250-26

7. No. Ref.—Section 300-5(i), Exception 2, although grouping of the conductors is preferred.

8. This installation is a violation; you cannot mix these conductors, unless the control is a Class 1 or power circuit. Ref.—Section 300-3(g)(1), 725-52(a)(2)

9. Permit, however it should not be located directly over the switchboard for the best coverage. Ref.—Section 384-4(FPN-2)

10. Yes. Ref.—Section 250-81. Steel can be bonded to the water pipe within 5 feet of where the water pipe enters the building. However, the 1993 Code prohibits the water pipe from being used as a grounding conductor.

11. 20-ampere breaker. Ref.—Section 310-16 obelisk note, Sections 210-22, 220-3(a)

12. No, if over 6 feet-6 inches; however, the receptacles required by Section 210-52(e) still must be installed at direct grade access and be GFCI protected. Ref.—Section 219-8a(3)

13. Yes, if listed for 2. Ref.—Sections 110-3b, 373-5(a), (b), and (c)

14. No. Ref.—Section 230-42(a), (b), and (c)

15. Yes. Ref.—Section 430-109, Exception 5

16. Article 310 Note to Ampacity Tables 0–2000 Volts; Ref.—Same

17. No, neutral is not used at the switch. Ref.—Section 200-7 Exception 2, Section 300-3(b)

18. Yes, but where they enter the building they must be enclosed in a raceway. Ref.—Section 339-3(a)(2), (5)

19. Yes, area may be declassified by AHJ where there are four air changes per hour. Ref.—Sections 511-7(a), 511-3(a)

20. No. Neutral must be capable of carrying the maximum unbalanced load. Ref.—Section 210-4(b) and (c); Section 250-60(a); Section 210-21(b)(20), and Table 210-21(b)(2)

21. Yes. Ref.—Section 350-3, Exception 2

22. Maybe AHJ is the building inspector. Ref.—Local adopted building code—Section 518-2(FPN)

23. Yes, but it must be connected to the same circuit as supplies the lighting for the area to be served. Ref.—Section 700-12

24. No. This wet location is defined in Section 517-3. Ref.—Section 210-8(a) and (b) Dwelling locations only

25. Yes; No. Ref.—Section 422-7 and 21(b) individual branch circuit is defined in Article 100.

PRACTICE EXAM 10

1. Yes, but storage of combustibles should be considered. Ref.—Part B, Article 410

2. Hot tubs and spas are defined in Article 680. Ref.—Section 680-4

3. No. Ref.—Section 511-4, 501-5(a)(4) Exception

4. 150, No. $P = EI = 480 \times 104 = 49,920$ single phase

    $49920/480 \times 173 = 60.12$ amperes

    $60.12 \times 250\% = 150$

    Ref.—Section 450-3(b)(2)

5. Each section 5 feet or fraction thereof shall be calculated at 180VA in other than dwellings. Ref.—Section 220-3c(6) Exception 1

6. Yes. Ref.—Part L, Article 430

7. Yes, possibly. Ref.—Section 430-72 and Exceptions

8. Yes, except in an industrial facility. Ref.—Section 364-11

9. No. Ref.—Section 339-3b(1), (9), 300-5(d)

10. Yes. Ref.—Section 250-5 and 23(a) and (b)

11. No. Ref.—Section 250-26 covers the requirements for a separately derived system. Nearest building still is preferred.

12. No, you cannot tap a tap. Although a tap is not defined, the rules cover the application sufficiently, so the definition is not necessary. Ref.—Section 240-3(d).

13. Yes. Ref.—Part C, Article 424, Section 24-20(b) Exception

14. No. 2. Ref.—Chapter 9, Table 8, Section 250-23(b), Table 250-94

15. Yes. However, consideration must be made for corrosive elements. Ref.—Section 300-6, 501-4a, 346—Non-Ferrous Rigid Metal Conduit.

16. No. Ref.—Section 700-12(p)

17. No. Ref.—Section 250-60

18. Yes, but there are installation requirements for disconnecting overcurrent. Ref.—Section 230-2 Exception 1, 430-31, NFPA 20

Appendix 4   299

19. Bond services together.   Ref.—Section 250-54

20. No.   Ref.—Section 680-71, 410-4(a)–(d)

21. No.   Ref.—*NEC*®. Most requirements for GFCI protection appear in Section 210-8. Other requirements appear in Articles 305, 511, 517, 550, 551, 553, 555, and 680.

22. No.   Ref.—Section 410-28(d), 410-31 Exceptions 1–3

23. Yes.   Ref.—Section 331-3(5), 331-3(1)(a). The 15-minute finish rating is a wall finish rating established by Underwriters Laboratories and can be found in the "UL Fire Resistant Directory."

24. No.   Ref.—Article 370, Section 370-16(a) and (b), Table 370-6(b). $4 \times 2\frac{1}{8} = 30.3$ cubic inches. 8 conductor = 3 cubic inches. Therefore, ten No. 8 conductors would be allowed without considering devices or other appertances.

25. They must be fastened.   Ref.—Section 410-16(c)

PRACTICE EXAM 11

1. Yes.   Ref.—Section 339-3(a)(4)

2. No.   Ref.—Section 680-20(a)(1)

3. Yes.   Ref.—Section 700-12(f) Exception

4. No.   Ref.—Section 300-22(c), Exception 5. Perpendicular, yes; parallel, no.

5. No, they must be equipped with a factory installed GFCI.   Ref.—Section 600-11

6. In each disconnect means as per Section 250-91(a) Exception 2.   Ref.—Section 250-91(a) Exception 2

7. No.   Ref.—Section 250-91(a) Exception 2

8. As a wireway in accordance with Article 362. Ref.—Sections 362-5, 362-7

9. Yes.   Ref.—Sections 250-32, 250-75, 250-79, 250-114

10. No. It can be insulated or covered.   Ref.—Section 680-25(c) Exception 3

11. Yes. This installation is not in violation of the code. Ref.—Sections 240-3, 310-16, 220-10, 384-13.

    Section 240-3(b)—The 400-ampere breaker properly protects 500 kcmil conductors that permit the next higher overcurrent device.

    Section 220-10—Conductors 500 kcmil are large enough to carry the 180-ampere load.

    Section 384-13—It is a power panel, not a lighting and appliance branch circuit panelboard.

12. Yes, unless not required by NFPA 99.   Ref.—Section 517-25 (FPN)

13. Generally, no—Table 220-11, and 220-13. Ref.—Section 220-3(a), (b), and (c), 220-10(a), (b)

14. Yes, No.   Ref.—Section 517-3 definition of patient care areas in hospitals.   Ref.—Section 517-13(a)

15. Yes, there are no conditions.   Ref.—Sections 600-5, 250-43(g)

16. No.   Ref.—Section 250-26(c)

17. Yes.   Ref.—Section 310-4

18. They must be derated to 80%.   Ref.—Notes to Ampacity Tables of 0–2000 Volts.

19. Yes.   Ref.—Section 210-6(c)

20. Yes.   Ref.—Definition "Accessible Readity," Definition of "Direct Grade Access"—210-8(a)(3)

21. Yes. No, Prohibition   Ref.—Sections 210-52(d), 210-8(a)

22. Yes, if they do not exceed allowable load of circuit. They are not continuous loads.   Ref.—Sections 210-19(c), 210-22(a), 210-23

23. No.   Ref.—Section 210-52(g) only applies to dwellings (See Definition Article 100)

24. Yes, Yes.   Ref.—Section 210-70(a) Exception 2

25. Overcurrent device, Yes.   Ref.—Section 220-10(a) and (b)

PRACTICE EXAM 12

1. Only branch circuits.   Ref.—Section 220-15

2. No. Table 220-19 Heading Column A applies to all notes as applicable, Columns B and C apply to Note 3.   Ref.—Same

3. Yes.   Ref.—Sections 225-8(b), 230-71 and 72

300 Master Electrician's Review

4. Location is not specified; where bonded to different electrodes they must then be bonded together. Ref.—Section 250-54.

5. Yes, but may be wired as nonhazardous. Ref.—Section 511-31(D) and 514-1.

6. ½ inch. Ref.—Section 410-66(a)

7. System ground. Ref.—Section 810-21(b)

8. No, Section 511-6 applies. There is no permission to use Chapters 1–4 wiring methods. Ref.—Section 511-3(c).

9. Yes. Ref.—Section 250-76 Exception

10. No. Ref.—Section 680-25(c)

11. No, only as taps (from raceway system to fixture). Ref.—Section 350-3 Exception 2

12. No. Ref.—Section 680-20(c), 680-7

13. Yes. Ref.—Section 680-40 Exception 2 and 680-41 Exception

14. Section 210-70(a) specifically states "at outdoor entrances or exits." Ref.—Section 210-70(a)

15. 65%—Column C, Note 3—add 3 + 6 = 9 × 65% = 5.85 KW

16. Yes. Ref.—Section 210-52(f). At least one but only for the laundry area.

17. Yes, all. Ref.—Section 210-8a(1)

18. No, except when using Class 1 or power circuits for control. Ref.—Section 300-3(c)(1), 725-52(a)(2)

19. Yes. Ref.—Section 800-33

20. Bonding jumper shall be the same size as the grounding electrode conductor. Ref.—Section 250-79(d)

21. Yes, for wet bar sink; no, for laundry sink. Ref.—Section 210-8(a)(5)

22. 25 ft. tap = No. 1 THW 20 ft. tap = No. 10 THW. Ref.—Section 240-21, Table 310-16 and obelisk note at bottom of Table

23. No. Ref.—Section 210-70a. At least one switch

24. 75°C, because there are no circuit breakers rated 90°C. Ref.—Sections 110-14c, 310-15c

25. Those areas determined by governing body of facility. Ref.—Section 517-3 Definitions, Patient Care Areas of a Hospital

PRACTICE EXAMINATION

1. C. Ref.—Article 100, Definition
2. A. Ref.—Section 310-11(c)
3. E. Ref.—Section 725-12
4. D. Ref.—Section 373-8
5. B. Ref.—Sections 240-51(a) and 240-53(a)
6. D. Ref.—Section 760-51(a)
7. C. Ref.—Chapter 9, Note 3 to Tables
8. D. Ref.—Section 240-80
9. C. Ref.—Section 720-5
10. D. Ref.—Table 402-5
11. A. Ref.—Chapter 9, Table 8, Col 6
12. B. Ref.—Section 760-52(a)(2)
13. B. Ref.—Section 110-1
14. D. Ref.—Section 346-12 Exception No. 1 and Table 346-12
15. D. Ref.—Article 100, See Definition Switch
16. D. Ref.—Section 725-18
17. B. Ref.—Section 250-118
18. C. Ref.—Article 100, Definition
19. D. Ref.—Section 110-17(a)
20. B. Ref.—Section 110-6
21. B. Ref.—Section 90-3
22. B. Ref.—Article 100, Definition
23. B. Ref.—Section 725-38(a)(2)
24. B. Ref.—Article 100, Definition
25. A. Ref.—Section 210-5(a)
26. B. Ref.—Section 110-19
27. A. Ref.—Section 424-39
28. C. Ref.—Section 660-5
29. A. Ref.—Section 225-14(c)

30. D. Ref.—Section 110-16(f)
31. A. Ref.—Section 220-3(c)(5) Exception 1
32. A. Ref.—Section 810-11 Exception
33. B. Ref.—Section 230-79(c), also see 230-42
34. A. Ref.—Section 430-89(b)
35. D. Ref.—Section 225-4
36. D. Ref.—Section 710-53
37. D. Ref.—Section 230-84(b) Exception
38. A. Ref.—Section 210-23(a)
39. D. Ref.—Section 374-6, copper (assumed) is 1000A per square inch. $4 \times \frac{1}{2} = 2$ square inches, $1000 \times 2 = 2000$ amps max.
40. D. Ref.—Section 305-7
41. D. Ref.—Section 430-32(b)(1)
42. D. Ref.—Section 550-11(a)
43. C. Ref.—Section 600-10(b)
44. D. Ref.—Section 820-40(a)(1,3,5)
45. E. Ref.—Section 110-32
46. C. Ref.—Section 517-15(a)
47. D. Ref.—Section 320-8
48. A. Ref.—Article 100, Definition
49. A. Ref.—Section 225-6(a)(1)
50. C. Ref.—Section 600-32(b)

FINAL EXAMINATION

1. B. Ref.—Section 800-12b
2. B. Ref.—General Knowledge
3. D. Ref.—General Knowledge, IEEE Dictionary
4. A. Ref.—Chapter 3
5. A. Ref.—Chapter 3
6. C. Ref.—Chapter 3
7. C. Ref.—Chapter 3
   $30P = I (E \times 1.73)$ $P = 83KW$
8. A. Ref.—General Knowledge
9. C. Ref.—Chapter 3
   $.5/1$ ohm $= \frac{1}{5}$
   $2.5 = 1$
10. A. Ref.—Chapter 3
11. A. Ref.—Chapter 3
12. A. Ref.—Section 210-22b
13. C. Ref.—Chapter 3
    $16 \times 4 = 64 \times 40 = 10 \times 4 = 40$
14. D. Ref.—Section 250-79(d)
    Note: $12\frac{1}{2}$ minimum or not smaller than required in Table 250-94, individual bonding jumpers could be run from each conduit sized 1/0
15. B. Ref.—Chapter 9, Table 8
    Ohm's Law
    Chapter 3
    240 volts $\times 3\% = 7.2$ volts drop permitted
    $R = Ed/I$ $R = 7.2/58$ $R = .124$ ohms for 500 ft.
    No. 3 $= .254$ ohms for 1000 feet
16. A. Ref.—Chapter 9, Tables 1, 5, and 4
    $2/0 = .2265$ sq inches $\times 2 = .4530$
    No. 1 $= .1590$
    $.4530 + .1590 = .6120$ Table 4 $= 1\frac{1}{2}$ inches
17. B. Ref.—Chapter 9, Tables 1, 4, and 5
    2—.1590
    1—.1893
    1—.2265
    .7338 sq inches   Table 4, $1\frac{1}{2} = .82$
18. D. Ref.—Section 220-30, Chapter 9 Example
19. B. Ref.—Table 250-94
20. C. Ref.—Section 230-95(a)
21. B with Exception.   Ref.—Section 230-9
22. B. Ref.—Duff, *Alternating Current Fundamentals, 4th Ed.*, © Delmar Publishers Inc.
23. B. Ref.—Note 8, 0–2000 Ampacity Tables
24. A. Ref.—Table 310-16
25. A. Ref.—Table 310-16
26. D. Ref.—Note 8, Ampacity Tables 0–2000 Volts
27. B. Ref.—Chapter 9, Tables 1, 5, and 4

Table 5 sq. inches:

3 × .0845 = .2535

4 × .2367 = .9468

1.2003   Table 4 = 2 inches

28. B. Ref.—Chapter 9, Tables 1 and 3B

29. C, Authority Having Jurisdiction. Ref.—Section 90-4

30. C. Ref.—Section 230-70(a)

31. C. Ref.—Appendix 1

32. C. Ref.—Appendix 1

33. D. Ref.—Basic Math, Chapter 3

34. B. Ref.—Section 110-9

35. B. Ref.—Section 430-52

36. D. Ref.—Article 430 Sections 430-37, 38 and 39, Table 430-37.

37. A. Ref.—Section 430-126

38. A. Ref.—Article 430, Part I

39. D. Ref.—Section 430-89

40. D. Ref.—Section 250-60

41. B. Ref.—Section 250-42(a)

42. C. Ref.—Section 250-83(c)

43. C. Ref.—Section 250-42(f) Exception 2

44. D. Ref.—Article 430

45. D. Ref.—Section 225-18

46. C. Ref.—Section 230-95(c)

47. A. Ref.—Section 430-24

Largest 5.6 × 1.25 = 7

4.5 × 1.00 = 4.5

4.5 × 1.00 = 4.5

16 amperes—minimum ampacity of conductors

48. C. Ref.—Section 690-4(a), Article 705, 230-82 Exception 6

49. C. Ref.—Section 250-53(b)

50. D. Ref.—Electrician's Handbook

51. B. Ref.—Section 380-8(a)

52. A. Ref.—Section 430-36

53. D.

54. C.

55. A.

56. B.

57. C.

58. A.

59. A.

60. D.

61. D.

62. B.

63. C.

64. A.

65. B.

66. D.

67. A.

68. B.

69. B. Ref.—*Direct Current Fundamentals,* Delmar Publishers Inc.

70. B. Ref.—*Direct Current Fundamentals,* Delmar Publishers Inc.

71. A. Ref.—Basic Math, Power Formula, Ohm's Law

Clue: Resistance is constant

I = P/E   2400/240 = 10

R = E/I   240/10 = 24

I = E/R   120/24 = 5

P = EI   5 × 120 = 600

72. B. Ref.—Basic Math, Ohm's Law

73. C. Ref.—Basic Math

74. C. Ref.—Tables 430-150 and 430-152

75. C. Ref.—Ohm's Law, 100% load

Index

A
APL (Applied Research Laboratories), 21
Applied Research Laboratories (APL), 21

B
Block & Associates, 1
Bonding, requirements, 102
Boxes, pull/junction, wiring, 121
Branch circuits, 64
    general purpose, 65–66
    ground-fault protection, 67–71
    outside, 71–73
    question review, 73–79
    wiring, 114

C
Cable trays, 116
Calculators, math and, 28
Circuit breakers, adjustable trip, 92
Circuits
    branch, 64
        general purpose, 65–66
        ground-fault protection, 67–71
        outside, 71–73
        question review, 73–79
        wiring, 114
Communication
    installations, 143
    question review, 143–49
Conductors
    derating factors and, 55–56
    grounding, 101–2
    insulation, equipment termination ratings and, 54
Conduits
    metallic, flexible, 120
    PVC rigid, 120

D
Decimals, 31
Derating factors, conductors and, 55–56

E
Electrical question review, 150–244
Electrical Testing Laboratories (ETL), 21
Electrical Testing Service, 1
Equipment
    special, occupancies and, 137–42
    terminations, 54–55
        conductor insulation and, 54

ETL (Electrical Testing Laboratories), 21
Examinations
    national testing organizations and, 1
        question bank development, 1–2
    practice
        final exam, 227–44
        no. 1, 151–56
        no. 2, 157–62
        no. 3, 162–67
        no. 4, 167–72
        no. 5, 172–77
        no. 6, 177–83
        no. 7, 183–88
        no. 8, 188–93
        no. 9, 193–99
        no. 10, 200–205
        no. 11, 205–13
        no. 12, 213–18
        no. 13, 218–26
    preparing for, 2–3
    sample, 4–19

F
Feeders, 64
    outside, 71–73
    question review, 73–79
Final examination, 227–44
Fractions
    improper
        changing to mixed numbers, 29
        reducing mixed numbers to, 29
    mathematics, 28–29
    multiplication of, 31
Fuseholders, 92–93
Fuses, 92–93

G
Ground-fault protection, branch circuits, 67–71
Grounding, 101–4
    question review, 104–10

H
High rated conductors, derating factors and, 55–56

I
Improper fractions
    changing to mixed numbers, 29
    reducing mixed numbers to, 29

J
Junction boxes, 121

L
Local codes, 21

M
Mathematics
  calculators and, 28
  decimals and, 31
  fractions, 28–29
    multiplication of, 31
  Ohm's law and, 32–34
  power factors, 34
  powers and, 31–32
  question reviews, 35–41
  square roots, 31
Metallic conduits, flexible, 120
Mixed numbers
  changing improper fractions to, 29
  reducing to fractions, 29
Multiplication, fractions, 31

N
National Assessment Institute, 1
National Electrical Code® (NEC®)
  general requirements, 42–43
  introduction to, 42
  National Fire Protection Association, 20–21
  question review, 21–27, 44–50
  wiring and protection, 44
National Fire Protection Association
  National Electrical Code® (NEC®), 20–21
  standards, 20
National testing organizations
  examinations and, 1
    question bank development, 1–2
NEC® (National Electrical Code®), 20–21
  general requirements, 42–43
  introduction to, 42
  question reviews, 44–50
  wiring and protection, 43–44
Numbers, mixed, changing improper fractions to, 29

O
Occupancies
  special equipment and, 137
    question review, 137–42
Ohm's law, 32–34
OSHA, 21
Outside feeders, 71–73
Overcurrent protection, 92–93
  question review, 94–100

P
Power factors, 34
Powers, 31–32
Pull boxes, 121
PVC rigid conduits, 120

Q
Question reviews
  branch circuits, 73–79
  communication installation, 143–49
  electrical, 150–244
  feeders, 73–79
  grounding, 104–10
  mathematics, 35–41
  National Electrical Code® (NEC®), 21–27
  NEC® (National Electrical Code®), 44–50
  occupancies, special equipment and, 137–42
  overcurrent protection, 94–100
  sample examination, 4–19
  services, 86–91
  utilization equipment, 131–36
  wiring
    methods of, 122–28
    requirements, 44–50
  see also Examinations, practice

S
Services
  over 600 volts, 86
  question review, 86–91
  600 volts or less, 80–86
Short circuit current, 93
Square roots, 31

T
Temperature
  ratings, importance of, 52–53
  termination, wire and, 52
Trays, cable, 116

U
UL (Underwriters Laboratories), 21
Underwriters Laboratories (UL), 21
Utilization equipment
  question review, 131–36
  use of, 129–31

V
Voltage drop, calculations/formula, 32

W
Wiring
  boxes, pull/junction, 121–22
  branch circuits, 114

cable trays and, 116
conduits and, 120
methods of, 111–22
   question review, 122–28
requirements, 51–57

conductors/derating factors, 55–56
equipment terminations, 54–55
importance of ratings, 52–53
question review, 58–63
temperature termination, 52